The Drama of Conservation

Carolyn M. King · D. John Gaukrodger
Neville A. Ritchie
Editors

The Drama of Conservation

The History of Pureora Forest, New Zealand

Department of Conservation
Te Papa Atawhai

Springer

Editors
Carolyn M. King
University of Waikato
Hamilton
New Zealand

D. John Gaukrodger
Thames
New Zealand

Neville A. Ritchie
Department of Conservation
Te Papa Atawhai, Hamilton Office
Hamilton
New Zealand

The original version of the book frontmatter was revised: For detailed information please see Erratum. The Erratum is available at 10.1007/978-3-319-18410-4_16

ISBN 978-3-319-18409-8 ISBN 978-3-319-18410-4 (eBook)
DOI 10.1007/978-3-319-18410-4

Co-published with the New Zealand Department of Conservation
NZ ISBN: 978-0-478-15044-5 (soft cover, New Zealand only)

Library of Congress Control Number: 2015941145

Springer Cham Heidelberg New York Dordrecht London

Printed on acid-free paper

Springer International Publishing AG Switzerland is part of Springer Science+Business Media (www.springer.com)

Preface

The Aims and Scope of this Book

The central North Island of New Zealand is one of the most significant sites of natural and cultural history interest in New Zealand. It has a rich Maori history, followed by a period of well-documented European exploitation of the seemingly inexhaustible resources of high-quality native timber in the mountains west of Lake Taupo. Over time, a vigorous forestry industry established many sawmills handling huge volumes of logs, accompanied by some important early experiments in forestry science, silviculture and selection logging.

Forestry has long been a vital source of rural employment and timber products supporting the development history of New Zealand. At the same time, forestry has contributed directly or indirectly to extensive biodiversity loss, in complex and inter-related ways. Conservation of biodiversity in the last remaining tall forests is critical for ensuring fully representative environmental preservation.

In the 1970s, Pureora Forest, the largest remaining area of native forest in the west Taupo region, was the location of a pivotal moment in New Zealand conservation history, focussed on an iconic endemic bird at severe risk of extinction. Since the last of the timber mills was closed in the 1970s and early 1980s, Pureora Forest Park (PFP) has become a nationally important site for experiments in conservation research and management, practical ecological restoration and outdoor recreation.

Development Versus Conservation

Conflicts arising between people and wildlife came into sharp focus in the timber industry at Pureora in 1978, when the rumbling of protests against continued logging of prime conservation habitat suddenly became a roar that the Government could no longer ignore. A series of well-publicised and dramatic tree-sitting demonstrations forced the Minister of Forests to break two binding legal contracts in order to declare a moratorium on logging.

The competition for press attention and public sympathy between aggrieved loggers and environmentalists over the next few years was intense, but PFP was gazetted in 1978, and the moratorium was made permanent in 1981. In 1987, the previously powerful New Zealand Forest Service was itself disbanded, and control of the park passed to the newly created Department of Conservation.

The Pureora story is nationally significant because it illustrates in one compact area a fundamental conflict of interests that applies in many other contexts. This book discusses the events that opened up important practical challenges and moral dilemmas associated with the massive historical deforestation of the central North Island forests; the 1978 wildlife conservation interventions in forests available for logging; the human consequences of radical decisions on forest conservation; and the dramatic changes in attitudes to conflicts over forest use, leading to the establishment and management of the Forest Park.

We begin with the geological history of the area, especially the impact of recent volcanism, and the species of fauna and flora protected within the park. We look at the questions of fundamental ecology crucial to developing effective management solutions: the controversies surrounding the establishment of a Recreational Hunting Area on Conservation land, and of pig-hunting with dogs in areas that could still be supporting kiwi; at the increasingly difficult challenges posed by the spread of introduced predators (especially rats and stoats): their impacts on native species, and how to control them.

We discuss how a massive effort to understand the behavioural ecology of the introduced Australian brushtail possum has informed the national campaign to minimise the spread of bovine tuberculosis and protect the forests from browsing damage; and how the attitudes and actions of foresters, hunters, urban environmentalists and amenity managers differ with respect to the arguments for and against the use of 1080 poison to achieve effective landscape-scale control of possums and rats.

All the factual information we present has been taken from the sources cited in the references, preferably the published literature. We have had access to important unpublished official archives, including the hitherto unknown significant details of the massive compensation packages paid to the timber companies after the moratorium forced the cancellation of their logging contracts. Where necessary, we have also quoted from newspapers because, although we are well aware that non-refereed reports are not always reliable, they are often the only sources of contemporary information about events that were not recorded anywhere else.

The text is designed to be easy to dip into by the many, many people—the timber workers and their families, foresters, scientists, hunters, birdwatchers, photographers, mountain bikers and trampers—who have loved, worked in or visited, and/or fought for Pureora over the years. We provide a fascinating background story for educational groups, especially those based at the Pureora Forest Lodge and the Tihoi Venture School, and users of the new Timber Trail for cyclists and walkers.

We quote a tiny selection of the hundreds of people-centred stories and historic illustrations known to us. Many more are available from knowledgeable local historians Ken Anderson, Ron Cooke, Audrey Walker and their many colleagues. Citations to these invaluable sources often appear here and will lead interested readers to masses of further information held in public libraries.

We aim to inform the academic community by providing all the necessary references, but have not attempted to provide a comprehensive review of biodiversity conservation in native forest generally. Rather, we aim to demonstrate how ecological study, often long-term, combined with a respect for people and for natural history, plus a flexible, interdisciplinary approach to current scientific priorities, can be welded into a consistently effective strategy to answer the pressing forest-ecology questions of our time.

Place Names

This book is about the eastern half of the King Country centred on Pureora Mountain. Pureora Forest village lies in the northern half of PFP, between the northern Hauhungaroa Range and the southern Rangitoto Range (Fig. 0.1).

Place names used both by Maori and by Pakeha (Europeans) usually reflect the perspectives of their authors. Originally, only the southern half of the Hauhungaroa Range was given that name, and the northern part was called Hurakia (and is so-called on many old maps and records). The two ranges meet at the central watershed, Weraroa. In Pakeha times, the entire range has taken the name Hauhungaroa, and the name Hurakia has been reserved for some of the westerly ridges making up the former Hurakia State Forest.

When the North Island Main Trunk Railway was being constructed, Europeans referred to the Pureora area as the Rangitoto-Tuhua country, since those were the ranges most visible from the railway route. Later, Auckland timber interests called it the Mangapehi bush, because that was the origin of the timber being cut by the mills at Mangapehi and sent to Auckland on the railway, whereas southern timber interests called it the Taumarunui bush or the Tuhua country, which was closer to Wellington. Looked at from the Lake Taupo side, it became the west Taupo ranges, and the Forest Research Institute at Rotorua thought of it as the west Taupo forest. All these names refer to the subject of the book, home of one of the largest blocks of continuous North Island forest remaining after 1900.

To everyone who knows it, Pureora is a very special place, and its story is well worth the telling.

Glossary and Abbreviations

For a note on our use of the Maori language, a General and Maori glossary, and lists of abbreviations and of scientific names of animals and plants mentioned in the text, please refer to the Appendices.

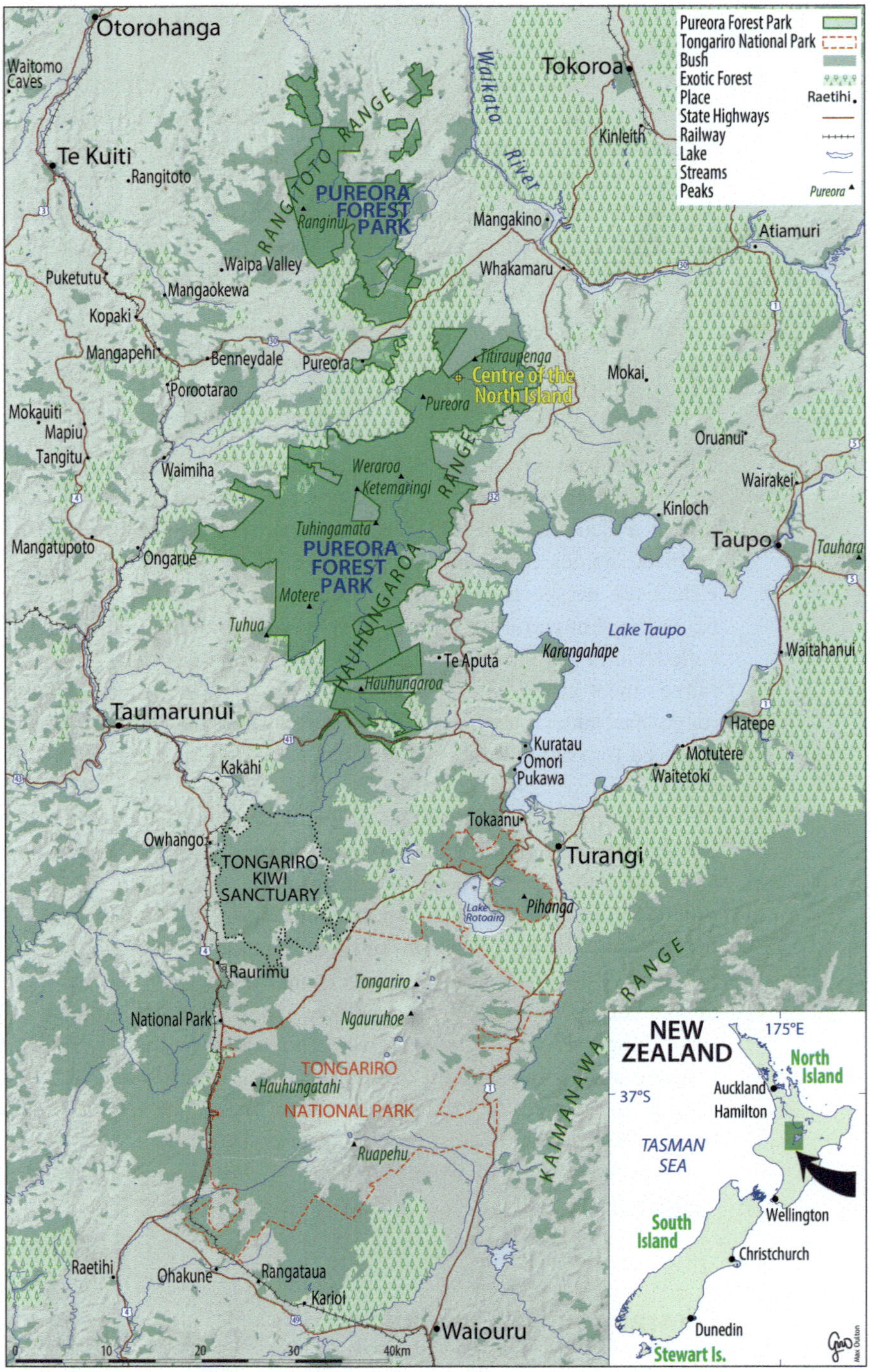

Fig. 0.1 Location of Pureora and the surrounding country, central North Island, New Zealand. The boundaries of forest areas (native and exotic), Pureora Forest Park, Tongariro Forest Kiwi Sanctuary and Tongariro National Park, roads (with their State Highway numbers) and the North Island Main Trunk Railway are shown, plus prominent ranges, peaks and rivers. Settlements mentioned in the text are marked here or on later, larger scale maps. All details shown as of 2014. *Max Oulton*

Acknowledgements

The authors and editors gratefully acknowledge the many people and organisations who have helped them in the following ways.

Principal funding:

Department of Conservation, Hamilton
NZ Lottery Grants Board, Environment and Heritage Fund
Warwick Silvester, Tane's Tree Trust
University of Waikato, Dean of the Faculty of Science and Engineering
University of Waikato, Earth Sciences Programme

Original archive and taped interviews:

The late Owen Wilkes, Department of Conservation

Contributors:

Tony Beveridge, former NZ Forest Service
Rod Hay, Department of Conservation
John Innes, Landcare Research
John Leathwick, Conservation Planning & Management Consultant
David J. Lowe, University of Waikato
Graham Nugent, Landcare Research
Tom Roa, University of Waikato
Mark Smale, Landcare Research

Cartography:

Max Oulton, University of Waikato

Artists and cartoonists:

Priscilla Barrett, Bob Brockie, Mark Garlick, Eric Heath, Pauline Morse, Burton Silver

Librarians and IT:

Ellen Clarkson, Dave Clements, Neil Fitzgerald, John Laurie, Bruce Patty, Cheryl Ward
Page references to 19th Century books are given to the online digitised versions, accessible from the University of Auckland Libraries and Learning Services and The Early New Zealand Books Collection, in the care of John Laurie

Loans of University theses:

Universities of Auckland and Canterbury
Victoria University of Wellington

Other information:

Bob Brockie, Marie Brown, Stacey Foster, Stephen King, Barry Murphy, Adrian Pittari, Guy Salmon, Kevin Tregoweth, Kerry-Jane Wilson
Barryville Historical Society (Graeme and Jenifer Reinhardt), Department of Conservation (Jo Burnard, Dave Smith, Ray Scrimgeour, Tertia Thurley, Owen Wilkes), Fletcher Trust Archive (Dorothy Neilson), Hamilton City Library (Perry Rice and Mark Caunter), SCION (Tracey Fenton and Tiena Jordan), Putaruru Timber Museum (Graham Hall); elders of Tūhourangi Ngāti Wāhiao of Te Pakira Marae, Whakarewarewa, Rotorua

Reviewers:

Peter Archer, Ken Ayers, Bruce Clarkson, Rob Guest, Ray Harlow, Stephen King*, John Leathwick, Andy Leigh, Chris Lusk, John Mason, Ray Scrimgeour, Dave Smith, Clare Veltman, Corrine Watts

*Not related to any other contributors with the same surname

Every effort has been made to contact copyright holders of material cited, but a few could not be found. Readers who can offer information leading to confirmed corrections or inclusion of missing acknowledgements are invited to contact us.

Lottery Grants Board
Te Puna Tahua
LOTTO FUNDS FOR YOUR COMMUNITY

Tāne's Tree Trust
Native Trees for the Future

THE UNIVERSITY OF
WAIKATO
Te Whare Wānanga o Waikato

Contents

Contributors

A.E. Beveridge Auckland, New Zealand

D.J. Gaukrodger Thames, New Zealand

J.R. Hay Department of Conservation, Te Papa Atawhai, Christchurch, New Zealand

J.G. Innes Landcare Research NZ Ltd., Hamilton, New Zealand

C.M. King School of Science, University of Waikato, Hamilton, New Zealand

J.R. Leathwick Hamilton, New Zealand

D.J. Lowe School of Science, University of Waikato, Hamilton, New Zealand

G. Nugent Landcare Research NZ Ltd., Lincoln, New Zealand

N.A. Ritchie Department of Conservation, Te Papa Atawhai, Hamilton Office, Hamilton, New Zealand

T. Roa School of Maori and Pacific Development, University of Waikato, Hamilton, New Zealand

M.C. Smale Landcare Research NZ Ltd., Hamilton, New Zealand

A Dramatic Landscape

1

D.J. Lowe and C.M. King

Abstract

This chapter introduces the story of Pureora Forest Park (PFP), in the central North Island, New Zealand, by describing the extremely violent Taupo eruption of c. AD 232 and its consequences for the surrounding forests and mountains. It gives a broad-scale local geological history, detailing the origins of some important local sedimentary rocks and landforms with a bearing on the story, including limestone caverns and coal deposits. It describes the location of the future PFP on the western edge of the Taupo Volcanic Zone, and how the history of volcanic activity, together with erosion, have determined much of the character of its landscape, the radial drainage pattern and deep entrenchment of its rivers, the distribution of its vegetation, and its long isolation from human access and permanent settlement.

Keywords

Lake Taupo · Pureora · Titiraupenga · Taupo eruption · Pyroclastic flow · Taupo ignimbrite · Plinian eruptions · Geology of west Taupo · Whakamaru ignimbrite · Break-out floods · Carbonised forest · Pumice soils · Pureora climate

Introduction

Lake Taupo (623 km^2 surface area, 357 m above sea level, and up to 185 m deep) is a justly famous mecca for tourists, boat-owners, and trout anglers in the centre of New Zealand's North Island. It is the largest body of freshwater in Australasia, and is cradled by green mountains both east and west, and the permanent snows of Tongariro National Park to the south. One of the earliest views of it, painted by Ferdinand von Hochstetter in 1859 [11], illustrates well the appeal of this majestic landscape, still as striking as it was in Hochstetter's day (Fig. 1.1).

The scene looks peaceful, but is deceptive. Taupo is also the most frequently active and productive rhyolite volcano on Earth [44, 50].

D.J. Lowe (✉) · C.M. King
School of Science, University of Waikato, Private Bag 3105, Hamilton 3240, New Zealand
e-mail: d.lowe@waikato.ac.nz

C.M. King
e-mail: cmking@waikato.ac.nz

C.M. King et al. (eds.), *The Drama of Conservation*,
DOI 10.1007/978-3-319-18410-4_1

Fig. 1.1 Lake Taupo viewed across to the southern shore in 1859. Prominent peaks: Ruapehu, Tongariro, Ngauruhoe (*left*); Pihanga (*centre*); and Karangahape (*far right*). *From Hochstetter* [11: 364], *courtesy of University of Auckland Libraries, Early New Zealand Books Collection*

The story of this beautiful but potentially dangerous place provides a dramatic introduction to the story of Pureora Forest.

The Taupo Eruption

About AD 232, long before human explorers discovered New Zealand, the Taupo volcano produced the most violent volcanic eruption known in the world in the last 5000 years [12, 46]. Towards the climax of the eruption, a towering eruption column, described by volcanologists until recently as 'ultraplinian' [41] but now simply 'plinian' [14], threw around *23 cubic kilometres* of loose volcanic material high into the atmosphere and stratosphere, about 35–40 km above ground [14] (about twice as high as the eruption column of Mt St Helens in 1980). Ash was blown eastwards across the North Island, well beyond Hawke's Bay and over the Pacific Ocean.

The column then collapsed catastrophically back to earth, generating deadly pyroclastic flows during the eruption's final climactic and most destructive phase (Fig. 1.2). Their burning, broiling fronts radiated outwards from the vent area at speeds exceeding 150 m/s (540 km/h) for about 80 km (±10 km) in all directions, and the flows stopped only when the material ran out. The flows covered about 20,000 km^2 of surrounding country *in less than seven minutes* [45].

All the surrounding mountains—the entire Hauhungaroa and Kaimanawa Ranges, the volcanic peaks of Pureora, Titiraupenga, and Tongariro, and much of the adjacent area except southwest Ruapehu—were blanketed in multiple suffocating layers representing the products of successive phases of the eruption (Figs. 1.3 and 1.4). The total output from the volcano through all stages of the

Fig. 1.2 Artist's impression of the catastrophic final phase of the Taupo eruption of c. AD 232. The towering eruption column in the background (at Lake Taupo) is collapsing and generating the roiling 'clouds' of very hot gas, ash, pumice, crystals, and rock fragments racing radially outward at high speed across the land as a ground-hugging pyroclastic flow, engulfing forests and animals in its path. The resulting non-welded ignimbrite deposit covers an area surrounding Lake Taupo to a radius of about 80 km [45]. *Painting by Mark Garlick specially commissioned for this book by David J. Lowe and Adrian Pittari, School of Science, University of Waikato*

eruption was about 105 cubic kilometres of loose volcanic material. For some more detailed technical explanations, see Box 1.1.

After the c. AD 232 eruption, Lake Taupo refilled over a period of about 15–40 years to a level some 30–40 m higher than today's (about where Taupo town now stands, at 400 m above sea level), held back by a temporary dam. When it collapsed, about twenty cubic kilometres of water was suddenly released down the Waikato River, an equivalent volume to that of the Mississippi River in flood [22].

Break-out flood deposits from this event can be traced along the Waikato River valley for up to 220 km downstream from Lake Taupo, burying huge areas that were otherwise little affected by the eruption. Parts of the modern city of Hamilton are built on the break-out flood deposits alongside the present Waikato River. Other North Island rivers, including the Whanganui, Rangitaiki, Mohaka, and Ngaruroro, were choked with fall deposits and Taupo ignimbrite-derived materials [24, 25].

All the original forest, about one cubic kilometre of timber, standing within about 80 km of the vent (the zone of Taupo ignimbrite emplacement), was almost instantly engulfed and incinerated [15], and the carbonized remains buried under a temporarily sterilised duvet (Fig. 1.5). Only a few remnants of pre-eruption forest survive, fortunately protected by rocky outcrops, such as the patch of silver beech that still grows in a gorge in the headwaters of the Mangatu Stream, in the Waihaha Forest.

From about 80 km to about 170 km east from the eruption centre, ash-fall deposits were thinner and cold, and generally caused progressively less devastating damage. Fires started during the

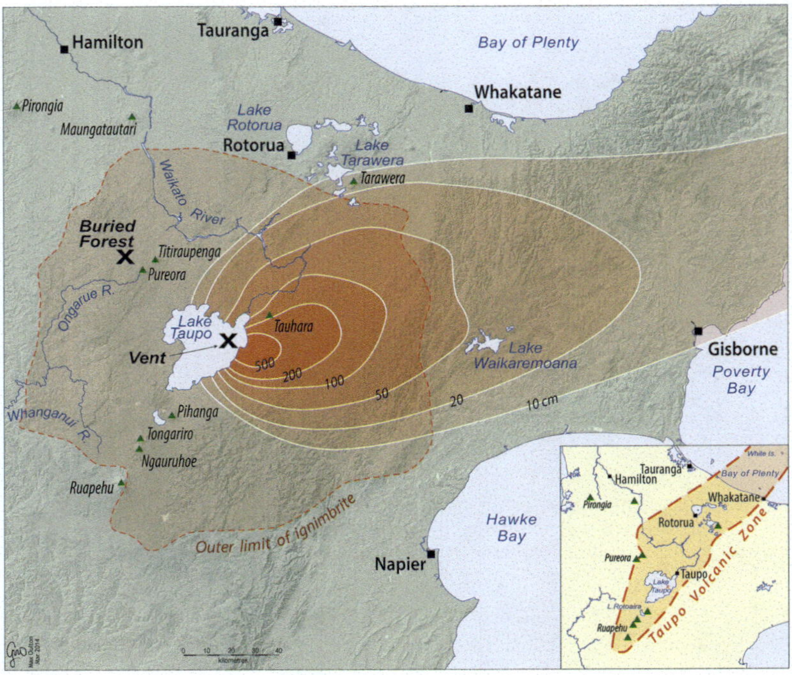

Fig. 1.3 The consequences of the powerful Taupo eruption of c. AD 232, showing the position of the vent, now submerged under Lake Taupo; the area covered by ashfall to a depth of 10 cm or more, with thickness contours (isopachs) in cm; the final extent of deposits from the extremely energetic pyroclastic flow, which spread a layer of loose ignimbrite across 20,000 km^2, covering all neighbouring peaks except Ruapehu in the southwest; and the position of the Buried Forest at Pureora. *Redrawn by Max Oulton from Wilson and Leonard* [44: 171]. NB: by convention, the bay itself is Hawke Bay, whereas the adjacent land is the Hawke's or Hawkes Bay region. Inset: Map of the Taupo Volcanic Zone (*TVZ*), extending from south of Ruapehu to White Island (Whakaari). The TVZ includes many volcanoes and hot springs and several geysers. *Redrawn by Max Oulton from Wilson and Leonard* [44: 168]

eruption burned for decades [43]. The collective effects of the eruption rendered uninhabitable an area now occupied by >200,000 people.

Dramatic as it was, the Taupo eruption was only the latest act in a very long play. Prolonged volcanic activity over thousands of years has left its mark on the modern composition and distribution of the soils and on the forests they support. Those earlier events help to explain much about the contemporary environment in this area.

Geological History of the West Taupo Area

The basement layers of sediments that now form greywacke rocks (mostly hard sandstone) under the Hauhungaroas were laid down on the sea bed in the Late Jurassic, about 155 million years ago (Mya), and represent complex deep marine deposits added onto the eastern Gondwana margin [19]. In places, such as at the base of the Waihuka Falls, fossickers can still find rare fossil shells as evidence of that distant time.

The sediments were uplifted tectonically after deposition, and additionally modified to form part of Zealandia, a large chunk of new land on the edge of the ancient continent, thickly covered with huge forests and peat swamps. Then, starting about 80 Mya, further tectonic movements split Zealandia off into a substantial continental fragment that slowly drifted eastward into permanent isolation in the southwest Pacific [36].

Widespread faulting across this ancient landscape in the mid-Tertiary period (c. 40–30 Mya) formed basins which became filled with

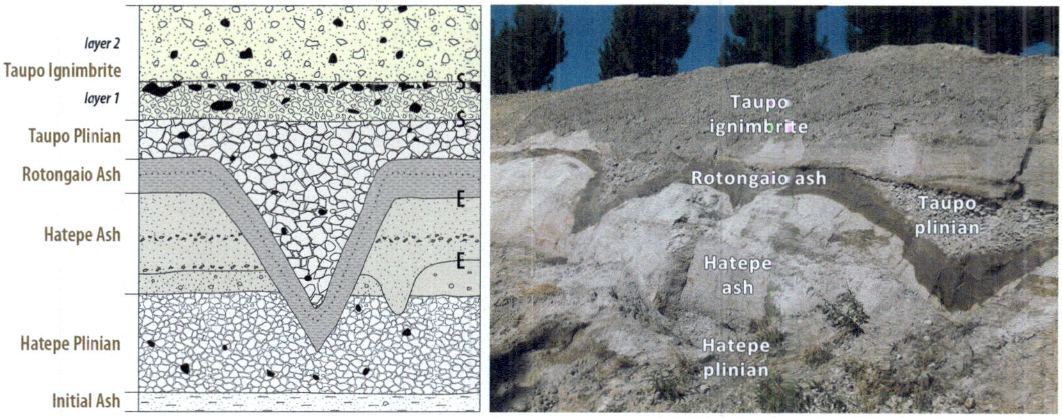

Fig. 1.4 Deposits from six of the seven distinct phases of the Taupo eruption are shown in this section, exposed in a forest road cutting. The first five units are all fall deposits, starting with Initial Ash (not deposited at this site) followed by Hatepe Plinian, Hatepe Ash, Rotongaio Ash, and Taupo Plinian. Taupo Ignimbrite was deposited at the climax of the eruption from a hot, fast-moving, ground-hugging pyroclastic flow. The V-shaped incisions into the Hatepe Ash mark a break of up to c. 3 weeks when torrential floods from intense rain carved rills and gullies into the soft surface (denoted by 'E'), into which the 'muddy' Rotongaio Ash was deposited. The ferocious emplacement of Taupo ignimbrite stripped the top of the Taupo plinian deposits, resulting in a 'planed' appearance (marked by 'S'). For further details see Box 1.1. *Diagram redrawn by Max Culton after Houghton & Wilson* [13: 59]*; photo by David J. Lowe*

Fig. 1.5 Two charred logs lying within pumiceous Taupo ignimbrite at a site near Wairakei. Such carbonized logs have been measured at up to 1 m in diameter and 5 m in length. The trees were broken off by the pyroclastic flow or preceding blast, and then incorporated into the flow. Their pattern of orientation radially around the eruption vent provided a means of deducing its position [7]. The logs were probably charred in situ at temperatures between at least 270° and 400 °C [15]. Lens cap 5 cm in diameter. *David J. Lowe*

sediments shed from adjacent higher ground. River flood plains and peaty swamps developed, thickly covered with forest and swamp vegetation. When the land subsided, the old land surface was flooded by the sea, and then buried and compressed by accumulating marine sediments that eventually formed mudstones, sandstones, and limestones [5, 16].

These sedimentary deposits and rocks play a less dramatic role in this story than that of the volcanoes, but still an important one. The limestones were corroded into the cave systems of the Waitomo district, which preserve a veritable zoo representing the prehistoric fauna that once roamed the central North Island forests, including many extinct birds and invertebrates [51].

The remains of the ancient swamps eventually became coal, and are preserved as the Waikato Coal Measures [34]. In their time they were an essential resource for steam-driven engines and mills, and were mined during the 20th Century at Mangapehi and Benneydale (Chap. 5).

About 25 Mya, the Australian-Pacific convergent plate margin began to develop, and new tectonic movements began to rearrange the geography of Zealandia. The greywacke basement rocks were uplifted to form the backbones of both main islands, now exposed in the Pureora area as alternating bands of sandstones and mudstones, split and bounded by northeast-southwest trending faults. The Rangitoto Range to the north of Pureora village, and part of the Hauhungaroa Range, which runs more or less continuously for some 70 km north to south along the west side of Lake Taupo (Fig. 0.1 in Preface), are both examples of uplifted, fault-bound blocks of greywacke rocks [5, 19]. Large areas of these ranges have been overlaid by the more recent products of cataclysmic volcanic eruptions (Box 1.1).

Box 1.1 Volcanic events and deposits in the central North Island

Volcanic activity in the central North Island volcanic region (part of the Volcanic Plateau) has had a significant influence on the history and geography of PFP, and so some additional explanation of the more violent events and ensuing deposits may be helpful.

Terminology

Pureora, Titiraupenga, Ruapehu, and Tongariro are all *cone volcanoes*, which erupt generally *andesitic* to *dacitic* lavas and pyroclastic (fragmental) material derived from moderately viscous magma (molten rock) of intermediate silica content (52–69 wt% SiO_2). Numerous small effusive and explosive eruptions build a cone made up of both lava flows and pyroclastic deposits around a vent area.

Taupo is a *rhyolite caldera volcano* producing highly viscous and gas-rich *rhyolitic* magma rich in silica (>69 wt% SiO_2). Rhyolitic eruptions are relatively infrequent but potentially very violent and explosive. Some (e.g., following the Oruanui and Taupo eruptions) were sufficiently voluminous to empty the underlying magma chamber and cause collapse of the ground surface, forming a caldera. Steep-sided rhyolite *lava domes* are commonly emplaced by effusive eruptions of de-gassed lavas following the explosive phase of rhyolite eruptions.

Tephra is a collective term (derived from a Greek word meaning 'ashes') for explosively erupted, loose, fragmental volcanic material that includes particles (pyroclasts) of different sizes, ranging from *ash* (<2 mm in diameter) and *lapilli* (2–64 mm), to *blocks* (angular) and *bombs* (rounded) (>64 mm).

Pyroclastic flows are also known as lateral pyroclastic density currents (pyroclastic means literally 'fiery fragmental'), and are gravity-controlled, laterally-moving mixtures of pyroclasts and gas with high particle concentrations, generating deposits called ignimbrites.

Ignimbrites are typically rhyolitic in composition and contain glass shards, pumice pieces, crystals (mineral grains), and rock fragments (lithics) all buoyed and carried along by very hot gases during their violent emplacement from swift, ground-hugging density currents. The term ignimbrite literally means 'fiery storm-clouds', an apt term that reflects its mode of origin. Pyroclastic flows infill valleys forming extensive, thick sheets of fragmental rhyolitic material, and mount ridges forming thin veneer deposits. After deposition, thick, very hot (600–700 °C) deposits (such as those infilling valleys) can weld or sinter together to varying degrees of hardness. At one end of the hardness spectrum, some ignimbrites are rock solid (referred to as densely-welded ignimbrite), and at the other end, some can be excavated with a spoon (referred to as weakly welded). Others may be entirely non-welded: for example,

Taupo ignimbrite and Oruanui ignimbrite are both non-welded ignimbrites [35].

Explosive eruptions and products: a summary

More than a dozen ignimbrites were erupted from Mangakino caldera, referred to collectively as Pakaumanu Group [3, 19]. Mangakino caldera has since been completely infilled by rhyolite lava domes. The ignimbrite cap on flat-topped Hikurangi is the Mangakino-derived Ongatiti ignimbrite, aged c. 1.23 Mya [5].

Later, the large Whakamaru caldera, which lies north of Lake Taupo, erupted pyroclastic materials in an episode of intense volcanism, generating Whakamaru Group ignimbrites around 350,000 years ago. The Whakamaru 1 and 2 ignimbrites comprise about 1500 and 500 km^3 of eruptive material (volumes are given as 'dense rock equivalent', DRE), respectively [50]. Both therefore qualify as products of so-called 'super-eruption' volcanic events [49]. The Whakamaru ignimbrites are very extensive in the Pureora area (including much of the Forest Park), both west and east of Lake Taupo, and further north in the Tokoroa area [19].

The next major event was the Oruanui eruption, another 'super-eruption' episode at Taupo volcano, which generated about 530 km^3 of material (DRE) or 1170 km^3 as loose material [48]. Also known as the Kawakawa eruption, the event is dated to c. 25,400 years ago [39]. The complex eruption produced a number of pyroclastic flow deposits, everywhere soft and non-welded (i.e. they comprise pumice clasts in a coarse ash matrix), known as Oruanui ignimbrite. In addition, some fall deposits were blown at least 1500 km beyond the New Zealand archipelago. The Oruanui ignimbrite is found both to the northeast and southeast of the Pureora area, and it is extensive around the flanks of Taupo volcano [19, 48]. The Oruanui eruption was centred on the northern part of modern-day Lake Taupo (including Western Bay), and formed a large, 35-km-wide elliptical caldera forming the modern shape of the northern lake. The Waikato River, which for tens of millennia had flowed north across the Hauraki Plains into the Firth of Thames and beyond, was diverted westward into the Hamilton Basin around 22,000 years ago, partly because volcaniclastic debris from the Oruanui eruption washed down and eventually blocked the river at Piarere, near the junction of SH 1 and SH 29 [23, 27, 33].

Since the Oruanui/Kawakawa eruption, Taupo volcano has erupted a further 28 times, most recently as the so-called Taupo eruption [47, 50]. Rhyolitic ash materials from many of these 28 eruptions, and also equivalent distal deposits from rhyolitic eruptions in the Okataina Volcanic Centre near Rotorua, would have blanketed or 'dusted' the Pureora area when wind directions were suitable. At the same time the andesitic volcanoes of Tongariro Volcanic Centre to the southeast of Pureora (Ruapehu, Ngauruhoe, Tongariro) have also been active, erupting very frequently since the earliest events dating back to c. 275,000 years ago [17]. Multiple, thin, andesitic ash beds of varying thickness, often only a few millimetres or so, were frequently deposited on the Hauhungaroa and southern Rangitoto Ranges.

Several prominent andesitic ash-bed deposits of Holocene age (the Holocene comprises the past 11,700 years) derived from volcanoes in the Tongariro Volcanic Centre include the Mangamate tephra (erupted c. 11,200 years ago), Papakai tephra (c. 5000 years ago), and Mangatawai tephra (c. 3000 years ago). Occasionally, ash from Mt Taranaki (Egmont volcano), which began activity about 130,000 years ago [38], was blown as far east as Pureora and beyond, one recent thin bed being the Burrell ash (c. AD 1655) [28]. The most recent eruptions of the area (since the Taupo eruption), and their possible effects on Maori society, are summarised by Lowe et al. [21].

Taupo eruption, its products and impacts

The complex Taupo eruption sequence of c. AD 232 has been divided into seven phases. The products of each are described as subunits Y1–Y7 of unit Y by Wilson [46, 47]. The eruption was centred on vents near the now-submerged Horomatangi Reefs in the northeastern part of Lake Taupo (Fig. 1.3). Four eruptions of varying character and eruption style, generating Initial ash (subunit Y1), Hatepe plinian (Y2), Hatepe ash (Y3), and Rotongaio ash (Y4) took place, the wide variations relating to differences in discharge rate and the degree of interaction between the magma and water in proto-Lake Taupo and from rainstorms [45, 46]. The fifth event in the sequence, Taupo plinian (Y5), generated a towering column of ash about 35–40 km high [14] comprising about 23 km^3 of loose material. Ash from this plinian column was blown eastwards across the country well beyond Hawke's Bay and into the Pacific Ocean (Fig. 1.3). During this phase, minor intraplinian pyroclastic flows were generated near the vent area (Y6). The column then collapsed catastrophically to generate powerful ground-hugging pyroclastic flows (density currents) of hot gas, ash, pumice, crystals, and rock fragments that raced radially outward at speeds of more than 150 m/s (200–300 m/s near the vent) to form the Taupo ignimbrite (Y7), containing at least 30 km^3 of loose material [45]. See Fig. 1.4.

The soft Taupo ignimbrite is non-welded, and was emplaced entirely within about 400 seconds [45]. Its temperature was about 380–500 °C at more than c. 40 km from vents; within c. 30–40 km of the vents it was about 150–300 °C [15, 26]. The extreme violence of the emplacement of Taupo ignimbrite caused the deposit to be spread thinly over the landscape to an average thickness of only c. 1.5 m, and the high energy release ($\geq$150 ± 50 megatons of TNT equivalent: [20]), enabled the Taupo ignimbrite to rush over hills and mountains up to c. 1500 m above the vent—no other pyroclastic flow is known to have climbed higher [6]—and the only mountain within its range that it did not ascend entirely was Ruapehu itself. That is why the beech trees on the southern and western slopes of Ruapehu were the only ones within 80 km of Taupo to escape complete destruction, and hence survived to spread and re-clothe the central volcanic mountains with beech forests similar to those that grew there before the eruption.

It is likely that this climactic ignimbrite-emplacement phase generated an atmospheric shock wave, producing a volcano-meteorological tsunami that reached coastal areas worldwide [20]. The total eruptive bulk (loose) volume for the Taupo eruptives has been estimated at c. 105 km^3 (equivalent to c. 35 km^3 DRE) [50].

Following the Taupo eruption, Lake Taupo refilled and reached a higher level than today's, as is evident from the semi-continuous, wave-cut bench and highstand shoreline deposits [22, 24]. Dramatic, sudden failure of a pumiceous pyroclastic dam led to the release of a peak discharge of 20,000–40,000 m^3/s. Tonnes of loose pumice and other materials were washed down the Waikato River as a break-out flood event [24], choking the river bed and depositing sediment many metres thick along the margins [37].

Pre-existing sediments were cannibalised in part and transported as well as the mainly pumiceous materials. Temporary dams formed on the Ongarue River, followed by flash floods as the dams collapsed [40].

Riparian terraces formed from the Taupo eruption-derived pumiceous alluvium (known geologically as Taupo Pumice Alluvium) are now common throughout many central North Island waterways [25, 30].

The violent emplacement of the Taupo ignimbrite devastated the forests, and charred logs as large as 1 m in diameter may be found in situ close to the vents [7]. Hudspith et al. [15] estimated that about 1 km^3 of forest timber was almost instantly incinerated.

The degree and nature of vegetation disturbance arising from the Taupo fallout deposits (rather than ignimbrite emplacement) varied according to the thickness of ashfall, local topographical features, and probably the vigour of the forest. Immediately after the eruption, stands of bracken and other seral taxa flourished. Revegetation was complete within about 200 years of the eruption, even at sites overwhelmed by the Taupo ignimbrite [43].

Explanation of Fig. 1.4

Products of most of the main phases of the eruption are shown here. Hatepe plinian, Hatepe ash, Rotongaio ash, and Taupo plinian are all fall deposits, comprising pyroclastic materials blasted into the air that were then blown by the wind to fall over the land surface like a blanketing snowfall. The characteristics of each unit relate in part to the rate of magma eruption and the amount of water entering the vent, so that the ensuing deposits range from coarse pumice clasts (open symbols), small rock fragments (closed symbols), and coarse to fine ash (fine stipple and dashes). The final deposit, Taupo ignimbrite, is the material that was spread across the landscape by the pyroclastic flow generated by the collapse of the towering plinian eruption column [14, 45].

The Hatepe ash phase of the eruption was abruptly terminated (possibly when lake water flooded deep into the vent), and running water from torrential rain carved rills and gullies into the deposits (marked by 'E', indicating erosion by flowing water). After a break of less than about three weeks [47], fine, dark-grey 'muddy' Rotongaio ash, and then pumices of the Taupo plinian phase, were deposited over the gullied landscape, forming the distinctive V-shaped pattern evident in the sequence. The emplacement of the Taupo ignimbrite generated another erosion surface (marked as 'S'), by shearing beneath the moving pyroclastic flow, resulting in a 'planed' or 'scalped' appearance to the top of the Rotongaio ash and Taupo plinian deposits [13]. The section shown is on the western side of High Level Road in the Kaingaroa Forest about 1 km north of the Napier-Taupo highway.

From about 1.6 million to about 900,000 years ago, successive explosive eruptions generated voluminous pyroclastic flows from the Mangakino volcano, and from the Whakamaru volcano around 350,000 years ago (Box 1.1). Very hot, fast-moving pumice-rich material, buoyed by superheated gases, raced across the landscape. Where the deposits were hottest and thickest, the pyroclastic material hardened into sheets of welded ignimbrite. They have since eroded to form distinctive landscapes with prominent cliffs, often with vertical joints, as seen for example along the Mangakahu Valley (Fig. 1.6). A cap of densely welded ignimbrite remains on the nearby flat-topped Hikurangi, 10 km northeast of Taumarunui, giving an effect described by Hochstetter (Chap. 6) as "the top appearing as if cut smooth with a knife" [11: 355] (Fig. 1.7).

About 25,400 years ago, the Taupo volcano produced the Oruanui/Kawakawa 'super-eruption' event [49]. It was an earlier and much larger event than the better-known Taupo eruption, centred on the wide northern part of modern-day Lake Taupo (including Western Bay). It generated about *1170 cubic kilometres* of

Fig. 1.6 A welded (hard) ignimbrite of the Whakamaru Group, erupted about 350,000 years ago as a pyroclastic flow from Whakamaru caldera volcano, with well developed vertical and subhorizontal cooling joints, exposed in the Mangakahu Valley east of Ongarue. *Photo from* [5: 40] *with permission: GNS Science image 140102 (c) Steve Edbrooke, GNS Science*

Fig. 1.7 Hikurangi, an eroded remnant of a former landscape, is capped by a now-welded ignimbrite erupted as a pyroclastic flow from the Mangakino caldera 1.23 Mya [5]. *C.M. King*

loose volcanic material, distributed mainly as pyroclastic flow and fall deposits. Some of the fall-out was blown at least 1500 km beyond the New Zealand islands [48]. The magma chamber producing all this material emptied, leaving the overlying section of the Earth's crust unsupported. It collapsed into a massive near-circular caldera 35 km across (a caldera is a large-scale ground collapse feature resulting from instability caused by the eruption of magma) [35].

Much of the accumulated volcaniclastic debris from this eruption washed down the ancestral Waikato River and blocked its then-normal flow north into the Firth of Thames [23]. Around 22,000 years ago, the river diverted sharply westwards, to reach the Tasman Sea via its present mouth at Port Waikato. From its junction with State Highway (SH) 1, SH 29 runs toward Tauranga through the old, choked-off valley of the ancestral Waikato River. Modern vehicles drive across its old bed, flanked on both sides by the tops of welded ignimbrite rock walls whose feet, which once stood at the river side, are now buried deep below the surface.

Of the succeeding 28 eruptions, all but three within the last 12,000 years, the Taupo eruption is by far the most famous. It formed a second, smaller caldera within the older Oruanui/Kawakawa caldera [47]. The northern part of the lake now occupies both of them.

Pureora and Titiraupenga

Pureora (1165 m) and Titiraupenga (1042 m) are the two principal volcanic peaks defining the skyline of the Hauhungaroa Range along the west shore of Lake Taupo, visible from both east and west (Fig. 1.8). They are andesitic volcanoes [8], aged about 1.6 and 1.9 Mya, respectively [9].

Titiraupenga is a dominant landmark, distinguished by a great spire of naked rock at its summit (a plug marking the position of the main volcanic vent). Just a few kilometres to the southwest is the gently sloping cone of Pureora. The characters of these mountains and their surrounding landscape have been largely determined by the prolonged and recent volcanic activity of the area, not just from the Taupo caldera, and from on-going erosion as well.

These peaks are encircled by skirts of welded rhyolitic ignimbrite derived from both the Mangakino caldera (dating from c. 1.6 Mya to c. 0.9 Mya) and the Whakamaru caldera (dating to c. 350,000 years ago) [19]. These volcanic materials, explosively poured and flung over the much more

Fig. 1.8 Pureora Forest, including Titiraupenga (*left*) and Pureora (*right*) mountains on the skyline, and the two steam-driven sawmills at the western end of the village, April 1951 (see Fig. 8.5). View ESE across the Ruapehu District. *Whites Aviation photograph WA-27381-F. Alexander Turnbull Library, Wellington, New Zealand*

Fig. 1.9 A typical landscape in Pureora Forest Park. Thick forest in the Waipapa Ecological Area surrounds a waterfall cascading over the edge of an ancient welded ignimbrite sheet. Another sheet of ignimbrite lies above the river level. *Crown Copyright, Department of Conservation Te Papa Atawhai (1995). Photographer: John Mason. DOC image library 10067771*

ancient (c. 150 Mya) greywacke basement, have been carved by river erosion and other geomorphic processes, including headward erosion and mass movement, into steep gullies radiating in all directions from the peaks (Fig. 1.9). All along the roads around the peaks and Lake Taupo, sheer cliffs, crags of solid volcanic rock, and exposed layers of ash and both welded and non-welded ignimbrite, bear witness to the violent past of this area.

The young soils developed from this surface material, interacting with the cool, wet climate of the central high country (Box 1.2), eventually helped to influence the patterns of the vegetation that repeatedly re-clothed the devastated land. Young soils on loose materials are also very vulnerable to erosion, which easily leads to damage requiring expensive repairs, as on the track to the summit of Pureora (Fig. 1.10).

Fig. 1.10 Track repairs on Pureora summit. The pumice deposits covering the peak up to about 1 m deep are soft Taupo ignimbrite erupted c. AD 232. The very high energy release of the eruption enabled the ground-hugging, ignimbrite-generating pyroclastic flow to ascend > 1000 m to overtop Pureora (see Box 1.1). The weakly weathered loose pumice, once exposed and lacking vegetation cover, is very vulnerable to erosion by heavy rain and human traffic. *Crown Copyright, Department of Conservation Te Papa Atawhai. Photographer: John Mason*

Box 1.2 Climate and soils of the Pureora area

Climate

Over many years, the meteorological station at PFP Headquarters (NZMS station C85551, altitude 549 m) returned data contributing to the calculations of climate normal for 1941–1970 [10, 29]. Climate change since then gives these data considerable historical value.

The central position of Pureora, well above sea level and far inland, gives it what meteorologists call a cool-temperate and damp climate. Pureora residents and visitors more often called it cold and wet, and NZFS managers were often faced with problems of maintaining relocatable buildings placed in a more severe climate than they were designed for.

The average annual temperature during 1941–70 was 10.3 °C, ranging from 16 °C in summer to 6 °C in winter. Daily extremes ranged from +30 °C to −9 °C. The lowest temperature ever recorded was −17 °C ("and I was there then", says ornithologist Rod Hay, personal communication, 1986).

The slightly continental character of the climate is illustrated by the fact that the July mean temperature is about 2° cooler than would be expected from the elevation alone. Ground frosts were recorded on an average of 87 days per year, at any season, and snow on two days a year.

The annual rainfall in Pureora village in 1941–70 averaged 1830 mm, falling on 180 rain days (probably up to 2500 mm and on more days on the higher land). Wind flows were generally light, although occasional, very destructive, storms have been known. One storm, some time before 1915, uprooted countless hundreds of trees (Chap. 7). Another in 1958, well remembered by forest scientist Tony Beveridge when he was doing field work there, stripped heavy seed crops from rimu crowns, with serious consequences for seed-eating birds.

Soils

Most of the soils in the Pureora area are developed mainly in pumice derived from the Taupo eruption event, notably the loose deposit of Taupo ignimbrite. Sequences of older tephra beds and buried soils are easily recognisable in cuttings and exposures, and river valleys are floored with soils derived from primary or reworked coarse pumice and ash deposits. Thin dustings of andesitic ash deposits overlie Taupo ignimbrite in places.

Over time the landscape has softened as the deep layers of tephra, eroded from higher ground and filling in the lower areas, gradually developed into infertile but usually free-draining soils. The tephra-derived soils generally become thicker eastwards, and are often composed of a thin layer of material derived from the Taupo eruption overlying sequences of older ash beds and associated, now-buried soils called paleosols.

These light, friable surface soils are very loose and easily eroded, and so as the landscape recovered from the eruptions, the regenerating forests played a crucial role in protecting the soils and reducing run-off into the rivers by helping to stabilise the landscape. Since then, any stripping of vegetation and living litter and humified cover on the soil surface has always carried the risk of accelerated soil erosion, which would not only clog valley drainages but also cause excessive sedimentation in the cave systems of the Waitomo area [18] as well as degrading the soils themselves.

Fortunately, hill slopes in PFP are generally less than 30° except in the steep-sided gullies, and the porous pumice soils seldom suffer from excessive run-off so long as they retain some vegetation [32].

Botanists are of course much more concerned about *what* vegetation remains, but for soil protection, anything with strong roots and deep litter will normally do the job. The hydrologist Colin O'Loughlin has concluded that forest management activities have had little effect on the important water resources of the region [31].

Rivers and Valleys

The geographic centre of the North Island lies on the Hauhungaroa Range, not far from Pureora village (Fig. 14.1). So the mountains of the park are as far as it is possible to be from any coast, which is why they provide the upper watersheds of some of New Zealand's most important rivers draining away in all directions (Fig. 0.1 in Preface). They have tended to develop a radial pattern of deeply incised gorges where they cross young pumice surfaces, and some have formed spectacular waterfalls where they spill over the edges of welded sheets of ignimbrite (Fig. 1.9).

North of Pureora village and SH 30, along the two parallel greywacke ranges of Rangitoto and Pukeokahu, the main peaks are Rangitoto (862 m) and Ranginui (983 m), drained to the east by the Waipapa and other Waikato tributaries, and to the west by the headwaters of the Waipa River. Both the Waipa and Waikato then flow north, and join at Ngaruawahia.

All the ranges and both the older cones of Pureora and Titiraupenga were mantled with pumice, including the Taupo ignimbrite, from the Taupo eruption (Fig. 1.10). To the south of Pureora and Titiraupenga lies the main block of the Hauhungaroas (about 70 km north to south, and about 25 km east to west at their widest), sprawling square across the middle of the North Island between the east-west ribbons of SHs 30 and 41.

The ridge and its highest point, Weraroa (1091 m), is the watershed between streams running down the steeper eastern face of the range, and those draining the broad rolling ridges of its western face. The eastern streams flow into Lake Taupo, and from there their waters join the Waikato River. The western streams have cut deep incisions on their way to join the Ongarue, Waimiha, and Okauaka Rivers, which are in turn important tributaries of the upper Whanganui River, which runs south through a steep-walled canyon.

To the southwest, semi-detached from the Hauhungaroas and forming a prominent corner-peg to one of the more jagged stretches of the park boundary, is Tuhua, a visually impressive non-volcanic mountain. It was an important landmark for travellers, both for Maori hunters and for the nineteenth-century Pakeha explorers following Maori trails (Chaps. 3 and 4).

The scenic qualities of this area deeply impressed early European visitors, none more so than the surveyor Laurence Cussen.

> Probably in no part of New Zealand can be found landscape so varied and picturesque as may be seen in favourable weather from some of the lofty peaks in this part of the district. Viewed under the conditions in which I first saw it, it would be difficult to conceive of a landscape of greater variety and grandeur. It was at sunrise on a clear frosty morning toward the end of May, from the summit of Pureora,

> 3,800 feet above the sea, overlooking all the surrounding country. The high mountainous district to the south was covered with snow; Taupo Lake seemed to be spread out at our feet, its 425 square miles of clear mirror-like surface reflecting the shadows of the eastern hills and promontories of the lake, cast across it by the rising sun…. To the north and east wound the valley of the Waikato River; along its course columns of steam arose from the hundreds of hot springs, fumaroles and puhias [sic] of the Taupo volcanic zone…. To the southwest, at a distance of 98 miles, the snow-clad sugarloaf peak of Taranaki (Mount Egmont) reared its head high above the surrounding landscape [4: 319–320].

Some decades before Cussen, Ernst Dieffenbach waxed lyrical about the scenic qualities of the Taupo area and its future potential as a tourist centre (Chap. 4). Dieffenbach was also an observant scientist, and was the first to record what the pumice-rich soils and layers of tephra he found meant about recent volcanic history in the area.

How Geography Helps Determine History

The crests of the west Taupo ranges reach only just over 1000 m above sea level, about 500 m above the surrounding valley bottoms and plains, and are not romantically rugged by New Zealand standards. But they are large enough to be intimidating, and the deeply dissected gorges surrounding them offer a considerable barrier to most forms of human transport. In early times, navigable rivers were few, so nearly all communication routes have always gone round the west Taupo ranges rather than over them. Travelling across them with a large, heavy load was virtually impossible.

The combination of Pureora's central location and its geography explain why it has always been a difficult place to access. That, plus the aftermath of the 1860–1864 Waikato War and the closure of the King Country (Chap. 4), deterred European exploration and exploitation for decades.

The north-south link provided by the North Island Main Trunk Railway line from Auckland to Wellington past the western flank of Ruapehu was long delayed by the need to bridge dozens of deep east-west gorges. It was not fully completed until 1908, after construction of some spectacular viaducts and the remarkable Raurimu Spiral. SH 4 from Te Kuiti to Ohakune via Mapiu, still a bush track in 1911, was not completed until 1928 [2]. South of the Hauhungaroas, SH 41 started as a bulldozed track in 1942, and was still unsealed in 1960 [1: 51]; the only east-west road through the future park across the ranges, SH 30, reached Benneydale from Te Kuiti in 1939, but was not connected through to Mangakino and Taupo until 1955 [42]. SH 32, running north-south along the western bays of Lake Taupo, became a through road in the late 1960s (Fig. 0.1 in Preface).

Taken these factors help explain why so many of the critical events described in this book are surprisingly recent.

References

1. Anderson, K. (2000). Servicing caterpillar. Taumarunui: CAS Publications.
2. Brake, T., & Cooke, R. (1986). Bush track to highway. Taumarunui: C & S Publications.
3. Briggs, R. M., Gifford, M. G., Moyle, A. R., Taylor, S. R., Normaff, M. D., Houghton, B. F., & Wilson, C. J. N. (1993). Geochemical zoning and eruptive mixing in ignimbrites from Mangakino volcano, Taupo Volcanic Zone, New Zealand. Journal of Volcanology and Geothermal Research, 56, 175–203.
4. Cussen, L. (1888). Notes on the physiography and geology of the King Country. Transactions of the New Zealand Institute, 20, 316–332.
5. Edbrooke, S. W. (2005). Geology of the Waikato area: Geological map 4, Scale 1:250,000. Lower Hutt: Institute of Geological and Nuclear Sciences.
6. Francis, P. (1993). Volcanoes: A planetary perspective. Oxford: Clarendon Press.
7. Froggatt, P. C., Wilson, C. J. N., & Walker, G. P. L. (1981). Orientation of logs in the Taupo Ignimbrite as an indicator of flow direction and vent position. Geology, 9, 109–111
8. Froude, D. O., & Cole, J. W. (1985). Petrography, mineralogy and chemistry of Titiraupenga volcano, North Island, New Zealand. New Zealand Journal of Geology and Geophysics, 28, 487–496.
9. Graham, I. J., Cole, J. W., Briggs, R. M., Gamble, J. A., & Smith, I. E. M. (1995). Petrology and petrogenesis of volcanic rocks from the Taupo Volcanic Zone: A review. Journal of Volcanology and Geothermal Research, 68, 59–87.
10. Hessell, J. W. (1986). The climate of Pureora Forest. In B. Veale & J. G. Innes (Ed.), Ecological research in

the central North Island Volcanic Plateau region (p. 27). Rotorua: Forest Research Institute.
11. von Hochstetter, F. (1867). New Zealand, its physical geography, geology and natural history. Stuttgart: J. G. Cotta.
12. Hogg, A. G., Lowe, D. J., Palmer, J., Boswijk, G., & Ramsey, C. B. (2012). Revised calendar date for the Taupo eruption derived by ^{14}C wiggle-matching using a New Zealand kauri ^{14}C calibration data set. The Holocene, 22, 439–449.
13. Houghton, B. F., & Wilson, C. J. N. (1986). Explosive rhyolitic volcanism: The case studies of Mayor Island and Taupo volcanoes. New Zealand Geological Survey Record, 12, 33–100.
14. Houghton, B. F., Carey, R. J., & Rosenberg, M. D. (2015). The Taupo eruption: "Ill wind" blows the ultraplinian type event down to Plinian. Geology, 42, 459–461.
15. Hudspith, V. A., Scott, A. C., Wilson, C. J. N., & Collinson, M. E. (2010). Charring of woods by volcanic processes: An example from the Taupo ignimbrite, New Zealand. Palaeogeography, Palaeoclimatology, Palaeoecology, 291, 40–51.
16. King, P. (2015). From source to sink. In I. J. Graham (Ed.), A continent on the move: New Zealand geoscience in the 21st Century, 2nd edition (p. 132–135). Geoscience Society of New Zealand, Miscellaneous Publication 141.
17. Lee, J. M., Bland, K. J., Townsend, D. B., & Kamp, P. J. J. (2011). Geology of the Hawke's Bay area: Geological map 8, Scale 1:250,000. Lower Hutt: Institute of Geological and Nuclear Sciences.
18. Leigh, A. H., & Clegg, S. E. (1989). Maniapoto District wild animal management plan, 1989 to 1999. Hamilton: Department of Conservation.
19. Leonard, G. S., Begg, J. G., & Wilson, C. N. J. (2010). Geology of the Rotorua area: Geological map 5, Scale 1:250,000. Lower Hutt: Institute of Geological and Nuclear Sciences.
20. Lowe, D. J., & de Lange, W. P. (2000). Volcano-meteorological tsunamis, the c. AD 200 Taupo eruption (New Zealand) and the possibility of a global tsunami. The Holocene, 10, 401–407.
21. Lowe, D. J., Newnham, R. M., & McCraw, J. D. (2002). Volcanism and early Maori Society in New Zealand. In R. Torrence & J. Grattan (Ed.), Natural disasters and cultural change (p. 126–161). London: Routledge.
22. Manville, V., White, J. D. L., Houghton, B. F., & Wilson, C. J. N. (1999). Paleohydrology and sedimentology of a post-1.8 ka breakout flood from intracaldera Lake Taupo, North Island, New Zealand. Geological Society of America Bulletin, 111, 1435–1447.
23. Manville, V., & Wilson, C. J. N. (2004). The 26.5 ka Oruanui eruption, New Zealand: A review of the roles of volcanism and climate in the post-eruptive sedimentary response. New Zealand Journal of Geology and Geophysics, 47, 525–547.
24. Manville, V., Hodgson, K. A., & Nairn, I. A. (2007). A review of break-out floods from volcanogenic lakes in New Zealand. New Zealand Journal of Geology and Geophysics, 50, 131–150.
25. Manville, V., Swegschneider, B., Newton, E., White, J. D. L., Houghton, B. F., & Wilson, C. J. N. (2009). Environmental impact of the 1.8 ka Taupo eruption, New Zealand: Landscape responses to a large-scale explosive rhyolite eruption. Sedimentary Geology, 220, 318–336.
26. McClelland, E., Wilson, C. J. N., & Bardot, L. (2004). Paleotemperature determinations for the 1.8-ka Taupo ignimbrite, New Zealand, and implications for the emplacement history of a high-velocity pyroclastic flow. Bulletin of Volcanology, 66, 492–513.
27. McCraw, J. D. (2011). The wandering river: Landforms and geological history of the Hamilton basin. Geoscience Society of New Zealand Guidebook, 16, 1–88.
28. Moebis, A., Cronin, S. J., Neall, V. E., & Smith, I. E. M. (2011). Unravelling a complex volcanic history from fine-grained, intricate Holocene ash sequences at the Tongariro Volcanic Centre, New Zealand. Quaternary International, 246, 352–363.
29. New Zealand Forest Service (1980). Pureora State Forest Park management plan. Auckland: New Zealand Forest Service, Auckland Conservancy.
30. Nicholls, J. (1986). A descriptive overview of the central North Island volcanic upland. In B. Veale & J. G. Innes (Ed.), Ecological research in the central North Island Volcanic Plateau region (p. 2–17). Rotorua: Forest Research Institute.
31. O'Loughlin, C. (1978). Hydrology of the Hauhungaroa Range state forests in relation to forest management practices. Management proposals for State Forests of the Rangitoto and Hauhungaroa Ranges, Central North Island, Taupo, 28–30 March 1978. Wellington: New Zealand Forest Service.
32. Selby, M. J. (1973). The erodibility of pumice soils of the North Island, New Zealand. Journal of Hydrology (NZ), 12, 32–56.
33. Selby, M. J., & Lowe, D. J. (1992). The middle Waikato Basin and hills. In J. M. Soons & M. J. Selby (Ed.), Landforms of New Zealand, 2nd edition (p. 233–255). Auckland: Longman Paul.
34. Sherwood, A., & Edbrooke, S. (2015). Rocks that burn. In I. J. Graham (Ed.), A continent on the move: New Zealand geoscience in the 21st century, 2nd edition (p. 326–329). Geoscience Society of New Zealand, Miscellaneous Publication 141.
35. Smith, R., Lowe, D. J., & Wright, I. (2012). Volcanoes. Te Ara—The Encyclopedia of New Zealand, http://www.TeAra.govt.nz/en/volcanoes, updated 9-Nov-12.
36. Sutherland, R., & King, P. (2015). Leaving Gondwana behind. In I. J. Graham (Ed.), A continent on the move: New Zealand geoscience in the 21st century, 2nd edition (p. 130–131). Geoscience Society of New Zealand, Miscellaneous Publication 141.

37. Thornton, J. (1985). Field guide to New Zealand geology. Auckland: Reed Methuen.
38. Townsend, D., Vonk, A., & Kamp, P. J. J. (2008). Geology of the Taranaki area: Geological map 7, Scale 1:250,000. Lower Hutt: Institute of Geological and Nuclear Sciences.
39. Vandergoes, M. J., Hogg, A. G., Lowe, D. J., Newnham, R. M., Denton, G. H., Southon, J., Barrell, D. J. A., Wilson, C. J. N., McGlone, M. S., Allan, A. S. R., Almond, P. C., Petchey, F., Dalbell, K., Dieffenbacher-Krall, A. C., & Blaauw, M. (2013). A revised age for the Kawakawa/Oruanui tephra, a key marker for the Last Glacial Maximum in New Zealand. Quaternary Science Reviews, 74, 170–194.
40. Walker, A., & Cooke, R. (2003). Waimiha: People of character. Waimiha: Waimiha Reunion 2001 Committee.
41. Walker, G. (1980). The Taupo pumice: Product of the most powerful known (ultraplinian) eruption? Journal of Volcanology and Geothermal Research, 8, 69–94.
42. Wilkes, O., Gaukrodger, J., & Ritchie, N. (2011). Chronology of Pureora. Hamilton: Department of Conservation.
43. Wilmshurst, J. M., & McGlone, M. S. (1996). Forest disturbance in the central North Island, New Zealand, following the 1850 BP Taupo eruption. The Holocene, 6, 399–411.
44. Wilson, C., & Leonard, G. (2015). Slumbering giants. In I. J. Graham (Ed.), A continent on the move: New Zealand geoscience in the 21st century, 2nd edition (p. 168–171). Geoscience Society of New Zealand, Miscellaneous Publication 141.
45. Wilson, C. J. N. (1985). The Taupo eruption, New Zealand II. The Taupo ignimbrite. Philosophical Transactions of the Royal Society, London A, 314, 229–310.
46. Wilson, C. J. N., & Walker, G. P. L. (1985). The Taupo eruption, New Zealand. I. General aspects. Philosophical Transactions of the Royal Society, London A, 314, 199–228.
47. Wilson, C. J. N. (1993). Stratigraphy, chronology, styles and dynamics of Late Quaternary eruptions from Taupo volcano, New Zealand. Philosophical Transactions of the Royal Society of London A, 343, 205–306.
48. Wilson, C. J. N. (2001). The 26.5 ka Oruanui eruption, New Zealand: An introduction and overview. Journal of Volcanology and Geothermal Research, 112, 133–174.
49. Wilson, C. J. N. (2008). Supereruptions and supervolcanoes: Processes and products. Elements, 4, 29–34.
50. Wilson, C. J. N., Gravley, D. M. Leonard, G. S., & Rowland, J. V. (2009). Volcanism in the Central Taupo Volcanic Zone, New Zealand: Tempo, styles and controls. In T. Thordarson, S. Self, G. Larsen, S. K. Rowland & A. Hoskuldsson (Ed.), Studies in volcanology: The legacy of George Walker (p. 225–247). London: Geological Society, London, Special Publications of IAVCEI 2.
51. Worthy, T. H., & Holdaway, R. N. (2002). The lost world of the moa. Bloomington: Indiana University Press.

Forests and Native Wildlife

2

C.M. King, J.R. Hay, M.C. Smale, J.R. Leathwick and A.E. Beveridge

Abstract

This chapter describes the composition of the modern forests; the discovery of the Pureora Buried Forest and the significant information about ancient environments and extinct invertebrates it preserved; and the past and present wildlife of the native forest community, especially the rare native species now protected within Pureora Forest Park.

Keywords

Pureora Buried Forest · Dating the Taupo eruption · McKelvey's volcanic succession hypothesis · North Island podocarp forest · Vegetation dynamics and patterns · Recolonization of volcanic surfaces · Mountain mires · New Zealand native wildlife · Extinct avifauna · Kiwi · Kokako · Tuatara · *Mystacina* · *Dactylanthus* · *Tymnopiptus*

Early European Descriptions

Eye-witness descriptions of any scene are always framed by the previous knowledge of the author and the interests of the intended audience. Maori developed an intimate understanding of the forest and its food resources, a product of many generations of settled dependency by the people of the land, the tangata whenua (Chap. 3).

Modern forest ecologists view and interpret the forest environment quite differently, even when they are referring to the same places as did Maori. Both views are or were quite different from the reactions of the early European explorers, whose accounts give us the nearest picture now available of what the land and the

C.M. King (✉)
School of Science, University of Waikato,
Private Bag 3105, Hamilton 3240, New Zealand
e-mail: cmking@waikato.ac.nz

J.R. Hay
Department of Conservation, Private Bag 4715,
Christchurch 8140, New Zealand
e-mail: rhay@doc.govt.nz

M.C. Smale
Landcare Research NZ Ltd, Private Bag 3127,
Hamilton 3240, New Zealand
e-mail: smalem@landcareresearch.co.nz

J.R. Leathwick
Conservation Planning & Management, 197
Silverdale Road, Hamilton 3216, New Zealand
e-mail: john.leathwick@gmail.com

A.E. Beveridge
19/10 Gerard Way, Auckland 1072, New Zealand

C.M. King et al. (eds.), *The Drama of Conservation*,
DOI 10.1007/978-3-319-18410-4_2

forests looked like after they had recovered from the Taupo eruption, and how people used them during the classic period of Maori culture.

The early European visitors who first ventured into the forest in the mid nineteenth century found an environment totally different from anything they had ever seen before. Naturally, they searched for comparisons with familiar scenes that would make sense to the readers of their detailed reports. Some of their wide-eyed descriptions are intensely moving, especially when we remember that they were describing these forests before the arrival of ship rats, possums and deer.

Not all available accounts refer specifically to forests closely similar to those found in Pureora, and their terminology has often been updated since, but the general picture of magnificent ranks of immense trees and ferns reappears in many different accounts.

> There is a solemn grandeur in these primeval forests, with their strange and luxuriant vegetation. Most of the trees are of the pine tribe, and grow to an enormous height. ....Beneath and among these and the other lords of the forest, are seen the less aspiring plants; the beautiful tree fern, reaching sometimes to the height of thirty feet..... the venerable ratu [rata] tree [is] often forty feet in circumference, and splendid with its dazzling scarlet blossoms; while graceful creepers, with their various coloured flowers, spread from tree to tree, and form an almost impenetrable barrier [53: 6–7].

> There is no country in the world so rich in ferns as New Zealand–the variety and elegance of their forms from the most minute species, to the giant tribe, is astonishing–some attain a height of forty feet, whilst others of exquisite beauty are extremely small. Two examples of the tree-ferns are figured in the accompanying scene [Plate VI: Fig. 2.1] –the *Cyathea medullaris*, and the *Cyathea dealbata*; the pulp of the former, at certain seasons of the year, is used as food by the natives, and when boiled, resembles apple sauce.
>
> During night, the forests frequently present a most beautiful appearance–the decaying and fallen trees, and the whole surface of the ground, covered with decomposed vegetable matter, sparkles with phosphorescence in every direction. So exuberant is vegetable life in these damp and gloomy forests, that it is difficult to find a single space, even on the trunks of the largest trees, not covered with plants; the warm and silent dells, eternally shaded from the sunbeams, by their lofty canopy of foliage, and fed with the ceaseless moisture that drops from every spray, are filled with palms, ferns, and countless parasites–all luxuriant to excess...... A vast portion of New Zealand, is covered with forest-clad mountains, yielding some of the finest timber, and the most ornamental and elegant woods in the world [1: 22–23].
>
> I cannot call to mind any tropical forests which excel those of New Zealand in beauty, for here there is magnificent timber, without the jungle of undergrowth which obstructs the view in more torrid climes. Brilliant parasites and creepers hang from the uppermost boughs of the loftiest trees, straight as cathedral bell-ropes, or, winding from stem to stem with fantastic curves, interlace distant trees, in the very extravagance of their luxuriant beauty. The lofty Totara, and the Rimu with its delicate and gently weeping foliage, and the shade-loving tree fern, the most graceful of all forest trees. Wild flowers are few and rare, but the ferns are more numerous and varied than in any other country [41: 117].

Fig. 2.1 Scene in a New Zealand forest, painted by Georg Angas in 1844. *From* Angas [1: 22] *Plate VI, courtesy University of Auckland Libraries, Early New Zealand Books Collection*

In parts of the landscape affected by Maori fires, thick forest gave way to open country.

> This large area [the Tihoi Plains] comprising nearly 1000 square miles, was the country described upon the maps as covered with dense bush; and where we had expected to travel through primeval forests we found magnificent open plains, clothed with a rich vegetation of native grasses, and composed of some of the best soil we had met with during our journey. As we rode over these plains, the scenery was magnificent, as much by reason of the vast scope of country that stretched before us as by the variety of mountain scenery that surrounded the plains in every direction [28: 319–320].

These descriptions are especially valuable because contemporary scientific descriptions can see only the natural vegetation cover that has survived to the present. Contemporary forests often look very different from the descriptions left by the early explorers 150 years ago. Fortunately, the Rangitoto and Hauhungaroa Ranges still support extensive and diverse tracts of native-dominant vegetation, including some of the few substantial remnants of North Island dense podocarp forests (native southern conifers) remaining anywhere.

The Forests Today

The forests in what is now Pureora Forest Park (PFP) were explored by the Polynesian ancestors of the Maori soon after their arrival in the central North Island some 700 years ago. From then until now, the forests have been appreciated from widely differing perspectives—for their provision of food and shelter, for their spiritual significance, for their timber, and more recently for their conservation value. They also contain some of New Zealand's most intensively studied forest ecosystems.

In 1963, Peter McKelvey [39] produced the earliest definitive account of the broad forest pattern, based on extensive plot data collected during the 1946-55 National Forest survey (Chap. 6). This work provided the basis for much of the subsequent utilisation of the native timber contained in these forests, first through clear felling, and then through early attempts at sustainable silviculture.

In the early 1980s, once logging had ceased, two ecological areas at Pureora (Waipapa and Pureora Mountain) were used by Forest Research Institute (FRI) plant ecologists as test sites within which to develop and demonstrate the use of 'modern' methods for the description of vegetation patterns. The first of these studies was carried out on Pureora Mountain, using semi-quantitative plots to describe both the forests [33] and the extensive wetland located near Bog Inn (Fig. 14.9) [7]. This work was later extended to cover the forests, shrublands and wetlands of the Waipapa Ecological Area [32]. These descriptions, along with the numerical analysis of forest/climate relationships by John Leathwick and Neil Mitchell [34], emphasise the complex interplay between climate, topography and disturbance that drives the broad vegetation patterns in PFP. Information from all of these sources has been used to compile the following account. For a general description of forest types, see Box 2.1.

Box 2.1 Revised forest classifications describing the principal forest types in Pureora Forest Park. From Nicholls [47]

The large tree species present in PFP are broadly divided into two major groups: conifers (gymnosperms) and broadleaved trees (angiosperms). The traditional forestry terms for these two groups are softwoods and hardwoods respectively, but this can be misleading because the actual hardness of the wood varies widely within and between these categories. A third group, the southern beeches, is almost entirely absent from PFP but dominant on most of Ruapehu, on the axial ranges of the North Island, and across much of the South Island.

The family Podocarpacae includes rimu, miro, matai, totara, Hall's totara, kahikatea, tanekaha, toatoa, mountain toatoa and bog pine. They are classed as southern conifers, but are quite unlike the conifers of the Northern Hemisphere; to avoid confusion with those, we retain the collective term "podocarps". Broadleaved trees belong to many different families, and those typical of PFP include tawa, rewarewa, hinau and pukatea, black maire, pokaka, broadleaf, tawheowheo and kamahi. Unlike the native broadleaved trees of the Northern Hemisphere, all these species are evergreen. Scientific names are listed as an Appendix.

Broad scale forest mapping (scale 1: 63,360) was produced by Peter McKelvey [39] to accompany his 1963 report. More detailed vegetation maps (1:10,000) were produced for the Pureora Mountain [33] and Waipapa Ecological Areas [32] after intensive field surveys by FRI staff working to develop methods for describing vegetation pattern in scientific reserves.

John Nicholls' scheme of forest types [47] is somewhat oversimplified, but was widely used for decades and is still mentioned in literature relating to Pureora, so a quick summary may be useful here.

Class L, Podocarp forest (Southern Conifers)

Class L is restricted to tall forest where podocarps are abundant, and to dense conifer pole stands that are successional to tall forest. The tallest of these magnificent trees grow up to 65 m, their branches draped with epiphytes and lianes. The drooping, shaggy bronze foliage of rimu is instantly recognisable; darker crowns of matai, miro and totara accompany them. Their flawless straight columnar trunks, 1–2 m across at waist height, soar up to the closed canopy above. The forest floor is covered with ferns, lichens and liverworts, and fallen logs are rapidly draped with mosses and saplings. Broadleaved trees are commonly present but confined to the understoreys. Appreciable areas of this class grow only in the central districts of the North Island.

Class M, Rimu-Matai-Broadleaved forest

In this class, the giant southern conifers, matai and rimu, are still impressive in number and size, but no longer so dominating. They appear as scattered emergents above a lower canopy of smaller broadleaved trees and ground plants. On the lower slopes, tawa dominates the many species of the subcanopy, and is replaced at higher elevations by kamahi. The class differs from other conifer-broadleaf classes in the presence of matai, totara, and kahikatea.

Class D, Rimu-Tawa forest

Podocarps tend to be very occasional in this class, although rimu, the commonest, is usually a large tree. The broadleaf northern rata is the only other large tree, no commoner than rimu generally and becoming sparse in some areas. Rata can (or could, before the arrival of the introduced Australian brushtail possums: Chap. 12) grow up to magnificent trees 30 m tall, but start as epiphytes perched on the branches of other established trees. The rata sends down roots on all sides, which eventually fuse and enclose the host tree; by the time it dies, the rata can stand alone. The beautiful scarlet flowers of the rata are a vital source of nectar for native honey-eating birds, and a favourite food of possums (Fig. 12.2). Other broadleaved trees are always abundant, with tawa predominating almost throughout.

Class F, Rimu-General Broadleaved forest

Forests of this class grow in a few, widely separate localities immediately above the altitudinal limits of taraire and tawa. Montane forests generally are different, not only because of their species composition, but also because they suffered the least damage from historic Maori fires [50]. Podocarps may be occasional to frequent; rimu is the commonest, but miro and Hall's totara are more numerous than is usual at lower altitudes. Broadleaved species are usually abundant, but in places many have died out in the last 10 to 20 years, especially rata and kamahi.

Class G, Lowland Steepland and Highland Podocarp-Broadleaved forest

This class contains many different mixtures of podocarps and broadleaved species, and any of these may be occasional to locally abundant. Forests of this type grow mainly where beeches are absent on the high country and above the altitudinal limit of rimu. There it is usually low forest, with stunted podocarps and malformed and shrub broadleaved species often prominent in the canopy. Also included are podocarp-broadleaved forests below the altitudinal limit of rimu on broken steep country or exposed ridges; that is to say, on sites where kauri or beeches are normally present. Where rimu grows it is usually small, and outnumbered by miro, Hall's totara, or tanekaha. Tawa usually persists on these sites, but several characteristic lowland broadleaved species do not.

The hill-country forests are generally the most predictable, with a strong and generally consistent relationship between forest composition and elevation. Most of the mature forests at low to middle elevations consist of scattered large podocarps, mostly rimu and matai, emergent over a canopy of broadleaved trees. Tawa, pukatea, kamahi and northern rata generally dominate the canopy at low elevations, along with occasional rewarewa, although rata has declined in abundance since the invasion of PFP by possums (Chap. 12). Tawa-dominant forests, often with pure stands of rewarewa along the margins, are most widespread in the north of PFP (from the northern parts of the Waipapa EA north, including in the Okahukura and Mangatutu catchments), and west of the Hauhungaroa Range in the southern part of PFP.

With progression to higher elevations, first pukatea and then tawa drop out, and dominance shifts to hinau and various low-stature tree species. Tawheowheo and kamahi gradually become dominant at high elevation, e.g., on upper slopes of Pureora and along the Hauhungaroa Range. Similarly, rimu, matai, kahikatea and miro are gradually replaced by Hall's totara and mountain toatoa. Tawheowheo forms locally pure stands in these upper elevation forests, particularly on sites with extreme soils, including around the margins of mires or on steep ignimbite escarpments, where it is often accompanied by scattered toatoa e.g., above the headwaters of the Ongarue River in the Pureora Mountain EA (Fig. 11.6). Along the tops (e.g., towards the summit of Pureora), the canopy gradually reduces in height and the scrubby vegetation is dominated by broadleaf, haumakaroa, stinkwood and mountain fivefinger.

There are several notable local exceptions to this general pattern. One high-elevation site just south of the hut at the southern end of the Hauhungaroa Range (Fig. 14.9) supports a distinctive community of silver pine growing in company with a range of mire plants. Isolated stands of silver beech survive in steep ignimbrite gorges on the southern and eastern flanks of the Hauhungaroa Range. Seeds from such remnants in the headwaters of the Mangatu Stream have dispersed by water transport to establish several riparian stands of young silver beech downstream, including where the Waihaha track crosses this stream about 2 km west of SH 32. An unusually dense stand of northern rata surviving on the western slopes of the Hauhungaroa Range, adjacent to the Piropiro Flats, has been protected as the Rata-nu-nui Ecological Area.

The forest pattern becomes much more complex on flat sites, reflecting the interactions between topography and the drainage of cold air and water. Measurements of frost intensity in the Waipapa EA found very steep temperature gradients over short distances, particularly where cold air drains downslope during frosts. Water drainage often follows similar patterns, so that some sites experience both water-logged soils and periodic extreme cold air. Such conditions favour podocarps rather than the broadleaved trees that dominate hill-country forests at the same elevation —the famed dense podocarp forests of Pureora.

In both the Waipapa and Waihaha Ecological Areas, cold air drainage is locally severe enough to produce inverted treelines. Then, broadleaved trees such as tawa, rewarewa and hinau are excluded from cold basin sites, even though they happily occupy the surrounding slopes. One of the most accessible examples can be observed along the Waipapa Stream downstream of the Pureora Forest Park Lodge, where forests growing on well-drained alluvial deposits of Taupo Pumice on the valley floor are dominated by the most cold-tolerant podocarps and broadleaved trees. Matai, kahikatea and black maire form a dense high canopy over an impenetrable understorey of frost-tolerant divaricate shrubs and lianes.

On the extensive plateau country of the Waipapa EA, well drained sites support abundant tall podocarps, including rimu, kahikatea, miro, totara (Fig. 2.2) and matai, emerging above diverse broadleaved canopies of tawa, hinau, kamahi and pokaka. The understoreys are diverse with abundant mahoe, fuchsia, pate, raukawa, wheki, mountain horopito, and climbers such as supplejack, lawyer and climbing ratas. On more poorly drained sites, rimu and kahikatea are generally dominant over a stunted canopy of tawheowheo and/or tawa, and locally abundant tanekaha (celery pine).

Disturbance by fire has, at times, profoundly affected the vegetation pattern of PFP, particularly around the eastern margins. Peter McKelvey [39] provided an excellent summary of its long-standing effects. In the North Block, fire history is clearly reflected in the dense stands of rewarewa along the park margins, e.g., along the southern part of Ranginui Road, and in the extensive area of secondary vegetation in the centre of the Waipapa EA. There, the complex patterns of vegetation succession after fire are intimately linked to topography and its controlling effect on frost intensity. Silver tussock and monoao dominate the coldest basin sites, while various mixtures of divaricate shrubs, kohuhu and lancewood dominate the intervening low ridges. On steeper sites, post-fire successions are often dominated initially by bracken, then by a gradual transition to broadleaved shrubs such as fivefinger, plus rangiora on the least frost-affected sites.

In the southern part of PFP, similarly complex patterns of forest succession after recurring fires are visible in the Waimanoa EA, with

Fig. 2.2 A fine specimen of a totara in Pureora, 1956, before it was felled. It stood 127 feet (38.7 m) tall, with 47 feet (13.3 m) to the first limb. *Photographer unknown, copyright assumed SCION 4536*

complications added by the complex topography. Some of the dissected rhyolitic landforms, including prominent knolls with skeletal and presumably drought-prone soils, support dense stands of tanekaha and/or toatoa. Further south, extensive secondary shrublands in the Waihaha EA can be readily observed along the track from

SH 32 to the Waihaha Hut. Here the vegetation pattern is broadly similar to that of the Waipapa, although on a grander and frostier scale. It includes dense stands of young conifers along the margin of the mature forest, with tanekaha particularly prominent. The mature forests on the basin floors closer to the Hauhungaroa Range support impressive stands of frost-tolerant podocarps, dominated by matai and totara on the flats, with few broadleaved trees.

Basin sites that have been infilled to varying degrees by water-transported Taupo pumice have developed into wetlands, which are numerous throughout the park from about the Waipapa EA south. The extensive mires in the Waipapa EA are dominated by square sedge and tangle fern, plus upright cutty grass, which has a very limited distribution elsewhere in the North Island.

The Mires

Mires are natural peat-forming wetlands, and seven of them lie within the Pureora Ecological District and PFP (Fig. 2.3). All the mires of PFP sit on a metre-thick layer of Taupo ignimbrite covering the blasted-flat remains of the former forest, some of them at unusually high elevations [6].

The most common plants in the mires are sedges and ferns, mainly square sedge, straw sedge and tangle fern. Two mires in the Waipapa Ecological Area have charred logs of bog pine on the peat surface, showing that a former shrub mire was burnt at some time in the recent past.

The Pureora mires are especially important places for palaeo-botanists, because they preserve the remains of long-vanished vegetation in the form of plant macrofossils and pollen deposited among dateable layers of debris and mud. Surveys of the living plants compared with analyses of what lies underneath them can offer vital information with which to reconstruct the history of the forest before the Taupo eruption, and the recovery and probable future trends in the vegetation of the park [6]. Hence, the historical records retrieved from the Pureora buried forest are of regional and international significance.

The Buried Forest

Krakatoa is justly famous as an example of how life returned to an island totally sterilised, at least temporarily, by the drastic consequences of an enormous volcanic eruption. Pureora is a mainland version of the same kind of cataclysmic

Fig. 2.3 A natural wetland, the Ongarue A montane mire, in November 2009. *Bruce Clarkson*

event, except that, unlike Krakatoa, the area devastated by the Taupo eruption (Chap. 1) retains clear evidence of what life was like there beforehand, and was not isolated from potential sources of recolonisation.

That fact makes Pureora one of the most exciting places in the world for ecologists and foresters to study the history of forests and their inhabitants—and the responses of natural habitats and biota to cataclysmic destruction. Ancient and modern plant assemblages can be compared, in order to illustrate some of the differences between the present and the pre-human, pre-rat past.

The extremely violent surge of the Taupo ignimbrite across the landscape in AD 232 ± 10 generated an air-blast shockwave that flattened the forest over the surrounding 20,000 km^2, and the pyroclastic flow that followed engulfed and incinerated most of the fallen trees (Box 1.1). At Pureora, some plant materials were turned to charcoal at temperatures of up to 364 °C; but most of the fallen vegetation was only slightly or not at all charred, perhaps protected by the wet leaf litter, or the cooling of the flow [24, 37]. Intact litter layers and prone trees all lay as they fell, facing away from the vent, with their bark intact and undamaged. All the remains have been beautifully preserved since AD 232 under the sterile blanket of Taupo ignimbrite, right down to leaves, seeds, fruits and invertebrates [4]. How did this happen?

Where the normal drainage channels were blocked by volcanic deposits, the water table rose to the surface. Layers of peat then gradually developed on the boggy ground above the ignimbrite. The whole assemblage from the top down became permanently waterlogged, and the buried trees and their associated fruits and leaves remained saturated, immune to decay in the anaerobic conditions, and so were preserved intact.

Several such post-eruption wetlands lie on the western flanks of the Hauhungaroas, northwest of Lake Taupo (Fig. 2.4). One of them covers about 37 ha of open ground, only 2.5 km from the Department of Conservation (DOC)'s Pureora Base. In 1983 a bulldozer driver was sent to dig drainage ditches through it. His digger repeatedly snagged on buried logs, which caused him a lot of irritation and prevented him from finishing his work on time. He complained to his boss, John Gaukrodger, who called on Rob Guest (NZFS District Forester based at Te Kuiti) to go with him to have a look. They recognised what they saw, and Guest called for advice from the experts at the Forest Research Institute in Rotorua.

Fig. 2.4 Mires in Pureora Forest Park have been well studied because they preserve valuable evidence of past vegetation patterns, often at unusually high elevations. *Redrawn by Max Oulton from* Clarkson et al. [6: 34)

They discovered that the logs causing all the trouble were the remains of the pre-eruption forest, flattened and buried under a metre-thick bed of pumice. Geologists identified the overlying material as comprising layers of soft Taupo ignimbrite [22, 24]. Subsequent excavations at Pureora, and at a second site at Benneydale about 20 km away, caused great excitement among scientists (Fig. 2.5) and comment from the

Fig. 2.5 A coordinated field trip to the Pureora Buried Forest site by geoscientists, botanists, ecologists and others on 14 February 1984. The presence of a TV crew recording the scene (*right*, with camera behind microphone) illustrates the high national interest stirred up by the discovery. *David J. Lowe*

Fig. 2.6 The discovery of the Buried Forest in 1983 caused much excitement, not least among bushmen who had recently been told there was no more standing timber available for logging. *Cartoon by Eric Heath, published in The Dominion, 1 February 1984*

interested public (Fig. 2.6). The site became an important destination for field trips for students and members of natural history clubs. To guard against too much trampling, all the ditches except one were re-flooded in 1985 to preserve the forest remains.

Detailed descriptions of preserved plant macrofossils (leaves, stems, fruits and seeds) have been used to interpret the pre-eruption vegetation at three sites along a 20 km transect, from Pureora westward to Benneydale. All the stands were on flat or undulating terrain and were dominated by podocarps, mostly rimu and tanekaha [9]. The mean width of the growth rings of 18 trees from the buried forest was 0.71 mm, suggesting that they grew only slowly. Broadleaved trees were generally more abundant at the Benneydale site, which is at lower elevation and has a milder climate. So the three samples of buried forest illustrate a gradual change in forest composition, from Pureora where horopito was prominent, to Benneydale, where rewarewa and northern rata were important canopy components.

Fig. 2.7 Samples of plant macrofossils. Miro and matai seeds (*above*), and rimu leaves (*below*) were preserved under Taupo ignimbrite at the Pureora Buried Forest. *David J. Lowe*

Dating the Taupo Eruption

The exact date of the eruption has been debated for years, but recent research has pinned the date down to AD 232 ± 10 (1718 ± 10 calendar years before present, 'present' taken as AD 1950). This date was derived by matching 25 high-precision ^{14}C dates from decadal samples of tanekaha logs from the Pureora buried forest against high-precision, first-millennium AD subfossil kauri calibration data [22].

Roman and Chinese literature recording red skies and poor summers in about AD 186, and a Greenland ice-core date of AD 181 ± 2 recording a layer of sulphate generated by an eruption, were for a long time assumed to have been correlated with this event. More recent research by Alan Hogg et al. [22] shows that the true date was in fact about 50 years later and, furthermore, that those historic Northern Hemisphere records have no connection to the Taupo eruption.

Taupo erupted in the late summer or early autumn of the year, because the buried forests at Pureora and Benneydale have preserved seasonal fruit and seeds (Fig. 2.7), and there is no latewood (formed late in the growing season) on the outermost rings of the flattened trees [8, 49]. Even more interesting, the assemblage of preserved insect remains lying under the pyroclastic flow deposits, excavated near what is now Link Road, suggested to Bev Clarkson et al. [8: 433] that the final phase of the lengthy eruption sequence (Fig. 1.4: the phase that produced the ignimbrite) was late in the day.

Some of the most interesting information derived from the excavation of the buried forest sites was the picture they painted, not only of the process of forest recovery, but also of differences in composition between the pre-Taupo and living forests. Were these differences related to geographical conditions, or was there a historical element as well? The answer has important implications for understanding the contemporary ecology of the forests.

Pollen diagrams from surrounding areas show that after the Taupo eruption, the earliest colonising vegetation dominated by bracken had stabilised the surface within five years [55]. It gave way to a shrubland with bog pine, then mountain toatoa, followed later by totara and matai; tall forest had returned within about 200 years.

At Pureora, the contemporary forests are dominated by matai and rimu, which contrasts strongly with the pre-eruption forests, which grew on old, infertile soils favouring the dominance of rimu and tanekaha, including several species that prefer a climate warmer than now [7]. Furthermore, tawa, kamahi and mahoe, which are all common dominants in the contemporary forests surrounding the site, were not recorded in any buried forest.

At Benneydale, the dynamic nature of these forests was demonstrated again by a pollen diagram constructed from samples representing the roughly 6000 years between the even older Mamaku tephra and the Taupo ignimbrite [10]. Pollen from the basal zone of the peat core

(c. 8000 years old) indicates forest quite different in composition from that immediately preceding the Taupo eruption. Hence, the forest at Beneydale around 8000 years ago had included a markedly different mix of species from that destroyed by the Taupo eruption in AD 232.

The Volcanic Succession Debate

One of the most contentious debates related to the forests of Pureora concerns the nature of dense podocarp forests and the place that they occupy in a linear succession initiated by the most recent Taupo eruption.

This debate was sparked by McKelvey's suggestion, in his seminal 1963 monograph, that the dense podocarps closest to Lake Taupo were the youngest forests, and represented a pioneering stage in the eastward colonisation of the sterile pumice surface by tree species spreading from the less damaged forests further west (Fig. 2.8). In addition, he argued that, with time, further succession would lead to a gradual replacement of the podocarps in these stands by less commercially valuable broadleaved tree species. In making this suggestion he was influenced by then-prevailing theoretical ideas of linear succession [48], in which an orderly and predictable change in the dominant species leads inevitably towards a 'climax' forest dominated

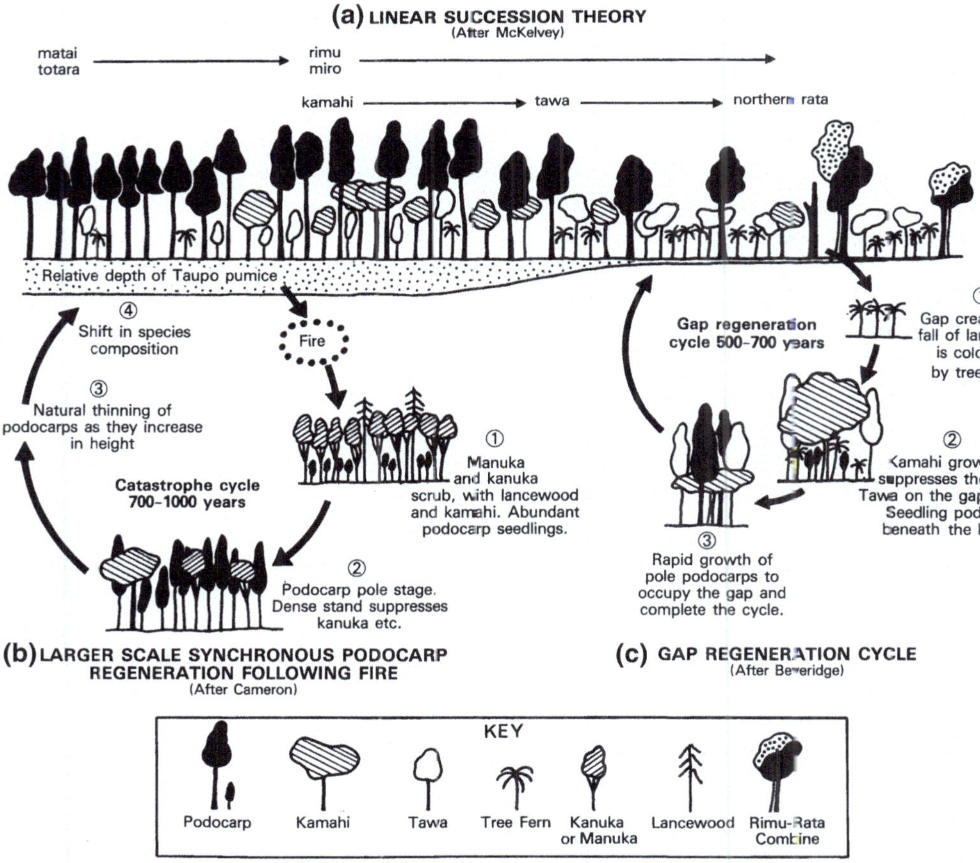

Fig. 2.8 McKelvey's volcanic succession hypothesis explaining the relationship between the diminishing depth of pumice soil, and the consequent changes in forest composition, with distance from Lake Taupo. Parts of this hypothesis are no longer accepted. Kanuka is shortlived even where not suppressed by podocarps. *From* Morton et al. [42: 59], *courtesy David Bateman Ltd*

by species different to those dominant at earlier successional stages. McKelvey's concept is summarised in a lavishly illustrated description of the rather similar pattern at Whirinaki State Forest, by Morton et al. [42].

McKelvey's hypothesis was hugely influential, not only because it claimed largely to explain the existing pattern of the forests. It also supported then prevailing views about the necessity of logging native forests. It offered a broad understanding of the mosaic of wildlife habitats of very high conservation value in PFP, and provided a background for early research on management of native wildlife in PFP [27]. Over time, this influence has been challenged strongly by different researchers producing new information to re-analyse his key assumptions, as follows.

First, while McKelvey assumed that the largest podocarp trees are the oldest, subsequent studies in dense podocarp forests showed that stem diameters are only weakly correlated with tree age [20]. The largest known tree in PFP, the Pouakani totara, is possibly about 1500 years old [5], but most other giant trees are much younger. Unfortunately, most podocarps that reach such a size are hollow at the base, so their age can be estimated only from the section of sound wood that can be sampled.

Second, McKelvey assumed that the existing dense podocarp forests are the first generation of trees to have colonised the original Taupo ignimbrite pumice surface, but more recent evidence demonstrates the rapid recolonisation of the fresh tephra surfaces by forest within 200 years of the Taupo eruption [55]. Hence, at least some sites occupied by dense podocarp stands must have supported two or more generations of trees since the Taupo eruption. McKelvey later agreed that the dense podocarp stage includes several generations of long-lived trees, so could not be a single colonising first crop [40]. Moreover, disruptive events such as fires and wind-throw, which open canopy gaps and reset the succession (Fig. 2.8), have made the existing pattern more a mosaic than a simple linear sequence [15].

Third, McKelvey assumed that the present distributional limits of many podocarp species have been set by the slow processes of recovery after the Taupo eruption. But a numerical study of the distributions of major forest tree species, using forest plot data collected during the National Forest Survey (Chap. 6, Fig. 13.17) linked to climate estimates for individual plots, found strong evidence that these limits are more likely to be set by climate and topography than by history [34].

Fourth, McKelvey assumed that dense podocarp forest is not self-sustaining, but is replaced in time by forest dominated by broadleaved species. This idea proved to be a useful argument for advocates of the logging industry. Canopy collapse or hollow stumps observed in large, long-lived trees implied to some forest managers that the podocarp component of the forest was "overmature" and dying, and therefore must be harvested quickly before it fell down.

Up until the early 1980s, this fear of wasting a valuable resource was used to strengthen the case for logging of old-growth forests both at Pureora and at Whirinaki, 100 km away on the edge of Te Urewera National Park. Indeed, the official management plan for Whirinaki asserted that the giant podocarps were senescent and were not replacing themselves. The plan illustrates the then attitude of NZFS, which saw itself as conducting a salvage operation: "The long term objective of forest management in Whirinaki is to anticipate natural decrement by judicious selection logging" [46: 34].

This justification for continued logging was severely criticised by forest ecologists at the time [42], and later studies of regeneration and stand dynamics of podocarp/broadleaved forests showed that these forests can sustain themselves over multiple generations, at least where tawa is rare or absent [36]. The Buried Forest data support this conclusion [8].

And there is another complication, especially on the eastern slopes of PFP nearest to Lake Taupo. This area was a major centre of Maori settlement (Chap. 3), and forest clearance by fire had a long history there [50]. The repeated post-fire successions that followed have probably produced vegetation patterns that mimic the processes of forest recovery set off by the Taupo eruption (Fig. 2.8).

Overall then, McKelvey's post-volcanic succession hypothesis now seems too simplistic to account for the heterogeneity—in terms of physiography, soils, and climate—of the physical environment and the corresponding diverse patterns of forest disturbance and recovery. Especially, it underestimated the effects of the long-term maturation (podzolisation) of soils under the humid climatic conditions of west Taupo. Recent re-measurement of permanent plots in PFP showed little evidence that tawa, the supposed final dominant in the succession, increases in plots where it was not already prominent, supporting Leathwick and Mitchell's [34] conclusion that most forests in the region have now returned to limits largely determined by climate. McKelvey [40] later agreed that there is no evidence to assume that the west Taupo podocarp forests are senescent or failing to replace themselves.

Interactions Within the Forest Community

The community interactions within a mature forest are nowhere better demonstrated than in the mutual dependence of forest trees and the birds that help to disperse their seeds. NZFS scientist Tony Beveridge [30] documented these effects in podocarp forest near Pureora over a period of 7 years (1958–64). The timing of his work was very significant, because possums were only just arriving in the region and had not yet added to the ongoing disruption of these ancient relationships already being imposed by rats and introduced birds [3]. The study included observations of birds feeding in the crowns of fruiting podocarps, and seed traps set to collect data on the abundance of seeds, their condition and fate.

Podocarp seeds are produced every year, but in huge abundance some years and very few in other years. The seeds often detach from the receptacles in the tree crowns before they fall to the ground. Seeds of green miro fruits are broken open by kaka, and totara seeds by kakariki. Rimu seed is favoured by many seed-eating birds, rodents and insects; chaffinches and other finches feed in tree crowns (Fig. 2.9).

New Zealand pigeons travel long distances on seasonal migrations, and concentrate on trees offering an abundance of food—a fact well known to the Maori (Chap. 3). Their dropping of ingested seed from perches in large broadleaved trees and podocarps has important consequences for the regeneration of mixed podocarp/broadleaved forests. Pigeons have always been the main dispersers of miro, matai and tawa seed, assisted by kokako and moa in earlier times, and still by tui, bellbirds and silvereyes, sometimes in local concentrations [11]. Now, feral pigs may help by grubbing up fallen tawa berries and miro fruits, and depositing the seeds elsewhere along with large amounts of fertiliser.

Killing native birds to get samples for gut analyses is not encouraged, so for one study, the New Zealand Wildlife Service (NZWS) made the most of 144 pigeons confiscated from poachers in central North Island forests. In summer, the succulent berries of species such as fuchsia, wineberry and mahoe were common foods, and tawa berries in late summer [38]. Most hunters targeted the pigeons only between May and August when the birds were feeding on miro berries, so, predictably, the guts of birds collected during the autumn were crammed with miro fruits. At Whirinaki, a year-long study confirmed that pigeons have a strong preference for miro and tawa fruits, but they also need access to a range of fruits, foliage and flowers through the seasons [14].

During the autumn of 1978, PFP produced the best rimu seed crop for 10 years [45]. Most of the sound seed fell, with fleshy ripe receptacles attached, during periods of strong winds in mid-April and early May, but during calm

Fig. 2.9 Seed-eating birds: New Zealand pigeon (kereru) (*David Cook Wildlife Photography*): North Island kaka (*D. Veitch*) yellow-crowned parakeet (kakariki) (*L. Bernard*); introduced chaffinch (*T. Smith*)

periods, finches feeding on the seed contents dropped the ripe receptacles detached from the seed and split seed coats. Whiteheads, silvereyes, bellbirds and tuis dispersed many smaller seeds, ship rats fed on the seeds that had fallen into the counting trays, and possums ate the receptacles, some with seed attached.

In 1978 there was seed enough for all. From counts of the numbers of seeds fallen to the ground under large trees (>5000 sound fallen seeds per m^2), Beveridge estimated that a single mature tree in PFP could produce a crop of 2 million seeds. In other years (monitored 1958–64) the rats and other introduced seed predators might well reduce the number of seeds available to birds [3].

Forest Regeneration

Southern Hemisphere conifers (podocarps) are very long-lived. That means that their population dynamics operate over longer time scales than those of broadleaved angiosperm trees [48]. It also means that they are better adapted to low-frequency, large-scale disturbances than to more stable environments [16].

The age distributions of existing old stands suggest that, after a disturbance such as a canopy gap or a fire, the first cohorts of podocarps establish in a predictable sequence, variable with local microclimate but generally starting with the least shade tolerant: first totara, then matai, then rimu, and finally miro. The reverse sequence tends to follow progressive collapse of an existing canopy [36], except that light-demanding totara does not re-establish, and broadleaved species become increasingly important over time.

Regeneration rates are the key to sustainable management of any kind of resource, so in 1960 and 1976, forest ecologists assessed the regeneration rates of podocarps in second-growth forest in the Waipapa Ecological Area. They marked a 200 m line transect of successive permanently-marked 2 × 2 m plots [2] along a transect extending outwards from the high forest margin through zones of shrub broadleaves, manuka and monoao to open frost flats. Podocarp regeneration was fastest in a narrow belt of large kamahi between tall podocarp/tawa forest and scrub associations. Seedlings grew in patches of broad-leaved shrubs, but not on frost flats.

The regenerating podocarp forest did not spread out uniformly from the high forest margin, but as a mosaic of developing podocarps in patchy broadleaved shrubs on higher ground, only slowly invading frosty depressions. Manuka, fivefinger and lancewood provided shelter for small podocarp seedlings, many of which did not survive unless they reached heights of 2–3 m.

Rimu, totara, matai, kahikatea and tanekaha were all abundant as seedlings, but not miro. Well-established podocarp seedlings added height by only 6–12 cm/year from 1960 to 1976, as new growth can be damaged by frost and cold wind. The figures indicated a substantial increase in densities of established podocarp seedlings in the period observed [2].

On a larger scale, assessments over 7600 ha of secondary vegetation in the Waihaha Block of Tihoi State Forest have provided regeneration data by canopy type. They showed dense pole stands developing around the forest margins, mainly of tanekaha, on 5 % of the area, and a high overall abundance of podocarp regeneration, especially in the general broadleaved shrub cover type (45 % of total area), sufficient to replace trees over the next century. Rimu and totara (mainly Hall's totara) will be significant elements in the future forests of these sites. Significantly, John Herbert [19] added, the forest supports the main prerequisites for podocarp regeneration, which are viable bird populations for seed dispersal, suitable nurse species and adequate seed production.

Native Wildlife

Extinct Vertebrate Species

In the limestone landscape of the southern Waikato, barely 50 km northwest of Pureora Village, the Waitomo Caves preserve a matchless record of the pre-human fauna of the central North Island. Sinkholes and underground caverns have trapped examples of many now-extinct species that were once important food sources, at least for a while, for the earliest Polynesian settlers (Chap. 3).

The remains of these long-vanished species may also be found in places near the west Taupo forests where the earliest human settlers once lived. For example, fragments of tuatara and North Island takahe have been excavated from the lower levels of two archaeological sites on the shores of Lake Taupo (Fig. 2.10: [23]).

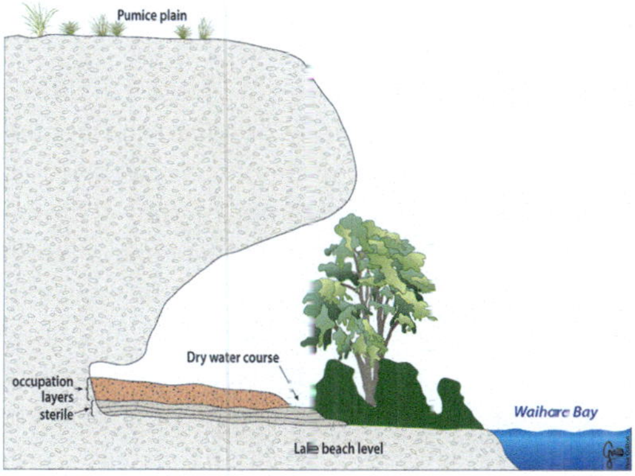

Fig. 2.10 A rock shelter at Waihora Bay, near Taupo. Fragments of tuatara and takahe, once important as food for Maori but now locally extinct, have been recovered from the older archaeological levels. *Modified by Max Oulton from* Hosking and Leahy [23]

By the time of Dieffenbach's visit in 1839, the tuatara was already long gone from the mainland, but he had heard of it, and eventually obtained a live specimen from offshore:

> I had been apprized of the existence of a large lizard, which the natives called Tuatera, or Narara, with a general name, and of which they were much afraid. But although looking for it at the places where it was said to be found, and offering great rewards for a specimen, it was only a few days before my departure from New Zealand that I obtained one, which had been caught at a small rocky islet called Karewa, which is about two miles from the coast, in the Bay of Plenty, and which had been given by the Rev. W. Stack, in Tauranga, to Dr. Johnson, the colonial surgeon. From all that I could gather about this Tuatera, it appears that it was formerly common in the islands; lived in holes, often in sandhills near the sea-shore; and the natives killed it for food. Owing to this latter cause, and no doubt also to the introduction of pigs, it is now very scarce; and many even of the older residents of the islands have never seen it. The specimen from which the description is taken I had alive, and kept for some time in captivity: it was extremely sluggish, and could be handled without any attempt at resistance or biting [12: 205].

According to traditions recorded in the Minute Books of the Native Land Court, the takahe (moho) were once hunted in the Pureora hills as far as Titiraupenga. The North Island species of takahe is extinct, though the South Island species survives under Department of Conservation (DOC) protection in Fiordland [18].

Likewise, the kakapo, a large flightless ground parrot, was once widespread throughout the North Island, and its bones are well represented in the limestone caves of the Waitomo area [57]. It was still present in the central North Island in the early 1800s [18], but is very vulnerable to human hunters and dogs. It is long gone from the North Island mainland, and remains one of New Zealand's most severely threatened species, surviving only on predator-free offshore islands or in captivity. Many other species once common in forests similar to PFP until less than 800 years ago have also disappeared (Figs. 2.11 and 2.12).

Hochstetter [21: 179] was fascinated by stories about the moa, and he concluded from a review of the localities where moa bones had been discovered that those birds were once distributed throughout the North Island. He tried hard to gather a good collection of specimens, but he was too late.

> Upon North Island I had scoured every district, that had been noted for the occurrence of Moa bones, I had ransacked all the so-called Moa caves, but all in vain. The Moa enthusiasts, that had been there before me, had carried off the last fragment of a Moa bone, and the Maoris on having discovered, that they could make some money by it, had gathered whatever there was still to be found, and sold it to European amateurs at enormous prices. The only relic I at least found out, was in the possession of a chief in the Tuhua district, who produced from the dust and rubbish of his raupo-hut an old bone, which he had hidden for a long time, and with which he parted only after lengthy negotiations. It was the pelvis of a small species [21: 184].

In a later chapter he described this incident slightly differently:

> My stay at Katiaho enabled me also to buy from a Maori for the price of one pound Sterling the pelvis of a small Moa which had been found near Teruakuaho a few miles above Katiaho on the Ongaruhe, under a cliff of the Herepu mountain. This was the first Moa relic, that fell into my hands, and I was not little gratified at the lucky circumstance [21: 354].

Living Species of Land Vertebrates

The total list of native land fauna now living in PFP (Box 2.2) includes many endangered, nationally threatened, regionally threatened and rare species (Fig. 2.13). It has been claimed that PFP has the greatest concentration of rare and endangered species of any area in New Zealand [43, 56], hence the great urgency and significance of protecting it.

Fig. 2.11 An imaginary scene in a central North Island forest about 20,000 years ago. *Left* to *right*, *above* red-crowned kakariki, morepork, kereru (native pigeon), **piopio (New Zealand thrush), *North Island saddleback, *North Island kokako, grey warbler, short-tailed bat, harrier hawk, kaka. *Left* to *right*, *below* brown kiwi, **bush wren, *Hochstetter's frog, *kakapo, *snipe, *tuatara, *blue duck, **giant moa, weka, forest gecko, **laughing owl, **small moa, **stout-legged wren, *little spotted kiwi, **owlet-nightjar. Species marked with one asterisk are now rare or threatened, and survive only under DOC protection or on offshore islands; those with two asterisks are totally extinct. *Modified by Ellen Clarkson from a painting by Pauline Morse*

Box 2.2 List of living species of native land vertebrates present in PFP

Scientific names are listed as an Appendix.

Endangered, threatened and rare endemic species

North Island kokako, North Island brown kiwi, blue duck, New Zealand falcon, North Island kaka, yellow-crowned kakariki, rifleman, North Island fernbird, North Island robin, long-tailed cuckoo, lesser short-tailed and long-tailed bats, and Hochstetter's frog.

Non-threatened to common native and self-introduced species

Paradise duck, grey duck, Australasian harrier, *spur–winged plover, pukeko, spotless crake, New Zealand pigeon, shining cuckoo, morepork, kingfisher, grey warbler, tomtit, *welcome swallow, bellbird, tui, fantail, New Zealand pipit, *white-faced heron, whitehead, *silver-eye, forest gecko, striped skink.

*Self-introduced since European settlement.

Fig. 2.12 Endemic species once resident in the central North Island but now totally (North Island takahe), or locally extinct (tuatara, kakapo, North Island saddleback). *Crown Copyright, Department of Conservation Te Papa Atawhai. Photographers; kakapo, G. Martin (2005); saddleback, D. Veitch (1979); takahe, unknown; tuatara, D. Veitch (1980s)*

Introduced species

Californian quail, eastern rosella, skylark, Australian magpie, yellowhammer, goldfinch, greenfinch, chaffinch, redpoll, song thrush, blackbird, house sparrow, dunnock, starling, Indian myna.

Data from Imboden [27], updated from Leigh and Clegg [35], R. Hay and T. Thurley (pers comms).

Hochstetter [21: 179] reported that "the natives speak of sorts of Kiwi, which they distinguish as Kiwi-nui (large Kiwi) and Kiwi-iti (small Kiwi). The Kiwi-nui is said to be found in the Tuhua district, West of Lake Taupo, and is in my opinion *Apteryx Mantelli*." In his time the kiwi was still abundant in mountain country, although "nearly exterminated by men, dogs, and wild cats" in the inhabited areas. Not long afterwards, Kerry-Nicholls agreed that kiwi were still common in the west Taupo forests [28: 364].

Now, brown kiwi still survive in parts of the central North Island, especially under DOC's active protection in Tongariro Forest Kiwi Sanctuary (Fig. 0.1 in Preface, and 12.10). Their prospects for survival in the Hauhungaroas, and in other areas outside sanctuaries with kiwi-focussed pest management, are not good [51].

Bushmen and forest rangers working in the Pureora bush in the 1940s (Chap. 8) remarked on the numbers of kaka, kakariki and pigeons they saw ("Flocks of 50–100 kaka screaming overhead", remembered Ivan Frost; "we used to hear them all the time", added Buster Seager) [54]. By the 1950s, these autumn flocks were down to 20 or 30 birds, but could still be observed in the tops of fruiting miro at Pikiariki, cracking open the green unripened fruit to extract the contents of the hard seed coats [3].

Kakariki (yellow-crowned parakeets), often heard flying over high forest, were common in late summer, feeding on the seed of tanekaha that flanked dense podocarp forest, or in the forest edge around Pureora village (Fig. 9.1). On Pureora Mountain, Tony Beveridge found caches of Hall's totara seed with the fleshy receptacles removed but no seed eaten. Only parakeets extract the contents of totara seed after cracking the seed coats.

The North Island kokako (Fig. 2.13), like many other predator-sensitive endemic species, had been common and widespread throughout the lowland forests of the North Island before the arrival of Europeans and their rats, and in 1944 were still reported as "fairly plentiful in SF 93—say one to every 100 acres" [44]. The unlogged parts of Pouakani SF 93 (Fig. 7.3) were later taken into Pureora SF 96 (Fig. 10.1).

By 1978 there were very few remnant kokako populations left, all apparently declining (Fig. 13.13), and forest staff working all day in the bush

Fig. 2.13 Endemic species still resident in Pureora Forest Park: karearea (falcon), whio (blue duck), brown kiwi, North Island kokako. *Crown Copyright, Department of Conservation Te Papa Atawhai. Photographers: falcon, B.J. Harcourt (1974); whio, T. Smith (2005); kiwi, R. Colbourne (1980s); kokako, unknown*

never saw a kokako. But this beautiful and nationally endangered bird became disproportionately important to the story of the protest action (Chap. 9) and the subsequent creation of PFP (Chap. 11).

One very important and nationally threatened resident of Pureora is the endemic lesser short-tailed bat. The population of bats surviving there is one of the few remnants of a once abundant species found throughout New Zealand [29]. It belongs to a distinctive lineage that originated in Australia, some of whose members flew across the Tasman millions of years ago, when New Zealand was free of all four-footed mammals. At that time, feeding on the ground was a safe and energy-saving strategy, and the new arrivals adapted well to those conditions, in time evolving into the short-tailed bat and its larger cousin the greater short-tailed bat (now extinct). They became the principal pollinators of the endemic *Dactylanthus taylorii*, a flowering root parasite (Fig. 2.14), which produces quantities of nectar on the ground. Both the bats and plants are now severely threatened by rats and possums.

Invertebrates

Many of the invertebrates of PFP's terrestrial and freshwater habitats are unknown, although those that are known are very diverse [25, 26]. Entomologists often complain that forest ecologists know almost nothing of the insect component of the ecosystem on which so much else hinges [25]. In PFP, this is a serious loss: the insects of the central North Island are especially significant, because they offer a window into the very different world of the pre-human past.

The central North Island forests support many insect groups unique to New Zealand. Some of the insects of PFP belong to lineages that could have lived in ancient Zealandia since the Pleistocene or long before, only distantly related to other groups that have disappeared everywhere else in the world, and new species are frequently found [13]. For example, we have hard evidence, from dateable subfossil deposits, of several large, flightless invertebrates that are now extinct. The most important and revealing of such deposits come from two quite different sources.

Fig. 2.14 The endemic New Zealand lesser short-tailed bat behaves much like a mouse, tucking up its wings and feeding on the ground. Pureora retains one of few surviving populations of this important pollinator of the endemic parasitic plant *Dactylanthus taylorii* (as shown here by the pollen covering its snout). *David Mudge, Nga Manu Images*

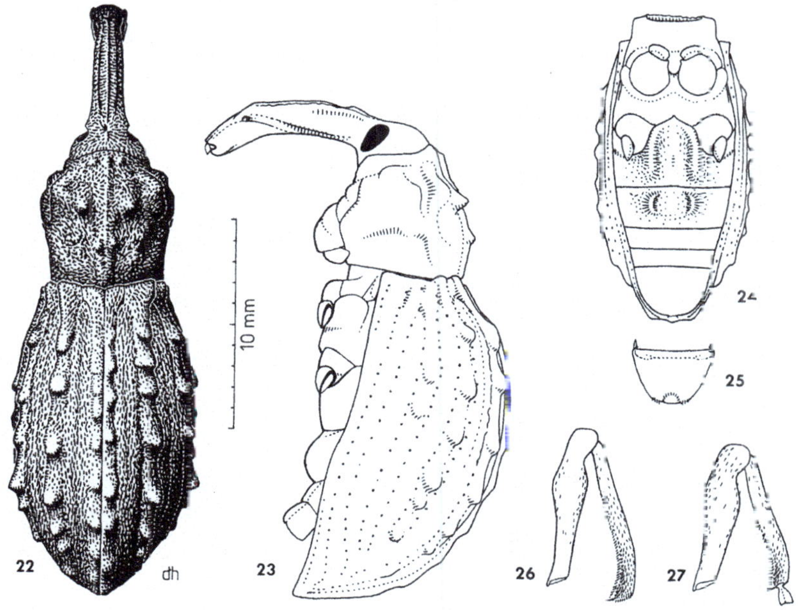

Fig. 2.15 *Tymbopiptus valeas*, an extinct giant flightless weevil. The average body length of the nearest living relatives of *T. valeas* is about the length of the scale bar. *From* Kuschel [31], *courtesy Entomological Society of New Zealand*

First, the buried forest material contained remains of many insects preserved in ideal anaerobic conditions, representing the layer of litter and humus that was on the forest floor and on the surface of acid (pH 3.5–4.9) bogs at the time of the Taupo eruption [17]. Chris Green identified a total of 210 insect species, mainly beetles (Coleoptera)—and 68 species of these were weevils. Other groups represented were mites, bugs, cockroaches, ants and flies.

All except two of the insect species identified in the buried forest material are present in the central North Island today. The two exceptions are both giant, flightless weevils, one a species of *Anagotus*, and the other, *Tymbopiptus valeas*, both now presumed extinct. *T. valeas* belongs to the tribe Phrynixini, represented in Chile by three living genera which have no relatives elsewhere in the Neotropic but many in the Australian Region. That dates the ancestry of the group back to at least the late Cretaceous [31: 21–25].

Second, fragments of insects have been recovered from a limestone shaft at Waitomo (Fig. 0.1 in preface), alongside bones of takahe, kiwi, weka and kakapo in a layer dated to around 200 years after the Taupo eruption [57]. Both *Anagotus* sp. and *Tymbopiptus valeas* were among them. They were both flightless ground-feeding insects, "considerably larger" than any living New Zealand Phrynixini. They survived the eruption in c. AD 232, but not the arrival of Pacific rats after around AD 1280.

T. valeas has only ever been found as a subfossil (Fig. 2.15), and the closest living relatives of the vanished *Anagotus* weevils are now restricted to rodent-free islands offshore, or to high country of >1000 m elevation. Still, Kuschel was an optimist: he coined the name of *T. valeas* from the Greek *tymbos*, a tomb, and *pipto*, to fall, plus *valeas*, a Latin farewell bidding that does not give up hope of an eventual return [31].

The implication of this story, that the arrival of rats was catastrophic for the largest of the endemic flightless invertebrates, is illustrated by contemporary experiments comparing the abundance of insects on islands with and without rat control. The species most often eaten by rats, the large Auckland tree weta (a flightless endemic insect related to the crickets), increased in numbers threefold in areas where rats were held to fewer than four per hectare for three consecutive years [52]. As soon as the rats came back, the weta declined again.

The living insect fauna of PFP is, predictably, very rich [25]. It would be impossible to sample all species equally, so one study concentrated just on the beetles of the Waipapa Ecological Area. Malaise traps, made of fine netting which guides insects passively to collecting jars, are suitable for catching low flying species. Three of these traps were set and cleared at weekly intervals over spring, summer and autumn (September 1983 to March 1984), in shrubland and in the adjacent mature podocarp forest. The results demonstrated many differences in the beetle faunas by habitat, especially in the diverse vegetation protected in Waipapa EA.

Nearly 5000 beetles of 400 species from 50 families were captured and analysed by a classification procedure which grouped the beetles by habitat. About 125 species were restricted to each of the two habitats, and 150 were shared. The shrubland samples were dominated by beetles that feed on live plants, and those from the forest were dominated by beetles that break down debris [25].

References

1. Angas, G. F. (1847). The New Zealanders illustrated. London: Thomas M'Lean.
2. Bell, A. (1976). Scrub associations and podocarp regeneration in the Pouakani Ecotone. Rotorua: New Zealand Forest Service, Forest Research Institute.
3. Beveridge, A. E. (1964). Dispersal and destruction of seed in Central North Island podocarp forests. Proceedings of the New Zealand Ecological Society, 11, 48–55.
4. Beveridge, A. E., & Herbert, J. W. (1989). Pureora buried forest. Rotorua: Forest Research Institute.
5. Burstall, S., & Sale, E. V. (1984). Great trees of New Zealand. Wellington: Reed/New Zealand Forest Service.
6. Clarkson, B. D., Clarkson, B. R., & McGlone, M. S. (1986). Vegetation history of some West Taupo mires. In B. Veale & J. Innes (Ed.), Ecological research in the central North Island Volcanic Plateau region (p. 34–37). Rotorua: New Zealand Forest Service, Forest Research Institute, Rotorua.
7. Clarkson, B. R. (1984). Vegetation of three mountain mires, west Taupo, New Zealand. New Zealand Journal of Botany, 22, 361–375.
8. Clarkson, B. R., Patel, R. N., & Clarkson, B. D. (1988). Composition and structure of forest overwhelmed at Pureora, central North Island, New Zealand, during the Taupo eruption (c. AD 130). Journal of the Royal Society of New Zealand, 18, 417–436.
9. Clarkson, B. R., Clarkson, B. D., & Patel, R. N. (1992). The pre-Taupo eruption (c.AD130) forest of the Benneydale-Pureora District, central North Island, New Zealand. Journal of the Royal Society of New Zealand, 22, 61–76.
10. Clarkson, B. R., McGlone, M., Lowe, D., & Clarkson, B. D. (1995). Macrofossils and pollen representing forests of the pre-Taupo volcanic eruption (c. 1850 yr BP) era at Pureora and Benneydale, central North Island, New Zealand. Journal of the Royal Society of New Zealand, 25, 263–281.
11. Clout, M. N., & Hay, J. R. (1989). The importance of birds as browsers, pollinators and seed dispersers in New Zealand forests. New Zealand Journal of Ecology, 12, 27–33.
12. Dieffenbach, E. (1843). Travels in New Zealand, Volume II. Christchurch: Capper Reprints.
13. Dugdale, J. S. (1986). Insects of the central North Island. In B. Veale & J. G. Innes (Ed.), Ecological research in the central North Island volcanic plateau region (p. 29). Rotorua: Forest Research Institute.
14. Emeny, M. T., Powlesland, R. G., Henderson, I. M., & Fordham, R. A. (2009). Feeding ecology of kererū (*Hemiphaga novaeseelandiae*) in podocarp-hardwood forest, Whirinaki Forest Park, New Zealand. New Zealand Journal of Ecology, 33, 114–124.
15. Enright, N. J., & Hill, R. S. (Eds.) (1995). Ecology of the southern conifers. Carleton, Victoria: Melbourne University Press.
16. Enright, N. J., & Ogden, J. (1995). The southern conifers—A synthesis. In N. J. Enright & R. S. Hill (Ed.), Ecology of the southern conifers (p. 271–287). Melbourne: Melbourne University Press.
17. Green, C. J. (1987). Insects of the buried forest at Pureora. Rotorua: Forest Research Institute
18. Heather, B., & Robertson, H. (1996). The field guide to the birds of New Zealand. Auckland: Viking (Penguin Books).
19. Herbert, J. (1978). Forest pattern and regeneration ecology of the Rangitoto-Hauhungaroa ranges. Wellington: New Zealand Forest Service
20. Herbert, J. W. (1980). Structure and growth of dense podocarp forest at Tihoi, central North Island and the impact of selective logging. New Zealand Journal of Forestry, 25, 44–57.
21. von Hochstetter, F. (1867). New Zealand, its physical geography geology and natural history. Stuttgart: J G Cotta.
22. Hogg, A. G., Lowe, D. J., Palmer, J., Boswijk, G., & Ramsey, C. B. (2012). Revised calendar date for the Taupo eruption derived by ^{14}C wiggle-matching using a New Zealand kauri ^{14}C calibration data set. The Holocene, 22, 439–449.

23. Hosking, T., & Leahy, A. (1982). Investigations at rock shelter N93/5 Waihora Bay, Taupo, New Zealand. Records of the Auckland Institute and Museum, 19, 81–99.
24. Hudspith, V. A., Scott, A. C., Wilson, C. J. N., & Collinson, M. E. (2010). Charring of woods by volcanic processes: An example from the Taupo ignimbrite, New Zealand. Palaeogeography, Palaeoclimatology, Palaeoecology, 291, 40–51.
25. Hutcheson, J. A. (1988). Representative samples from arthropod communities. What's New in Forest Research, 167, 1–4.
26. Hutcheson, J. A. (1990). Characterisation of terrestrial insect communities using quantified, Malaise-trapped Coleoptera. Ecological Entomology, 15, 143–151.
27. Imboden, C. (1978). Wildlife values and wildlife conservation in the Hauhungaroa and Rangitoto ranges [*Transcript of the Taupo seminar*]. Wellington: New Zealand Forest Service
28. Kerry-Nicholls, J. H. (1884). The King Country, or explorations in New Zealand—A narrative of 600 miles of travel through Maoriland. London: Sampson Low & Co.
29. King, C. M. (Ed.) (2005). The handbook of New Zealand mammals, 2nd Edition. Melbourne: Oxford University Press.
30. Klitscher, K. (Ed.) (1997). Characters of FRI. Rotorua: New Zealand Forest Research Institute Ltd.
31. Kuschel, G. (1987). The subfamily Molytinae (Coleoptera: Curculionidae): General notes and descriptions from New Zealand and Chile. New Zealand Entomologist, 9, 11–29.
32. Leathwick, J. R. (1987). Waipapa ecological area: A study of vegetation pattern in a scientific reserve. Forest Research Institute Bulletin 130. Rotorua: New Zealand Forest Service.
33. Leathwick, J. R., Wallace, S. W., & Williams, D. S. (1988). Vegetation of the Pureora Mountain ecological area, west Taupo. New Zealand. New Zealand Journal of Botany, 26, 259–280.
34. Leathwick, J. R., & Mitchell, N. D. (1992). Forest pattern, climate and vulcanism in central North Island, New Zealand. Journal of Vegetation Science, 3, 603–616.
35. Leigh, A. H., & Clegg, S. E. (1989). Maniapoto District wild animal management plan, 1989 to 1999. Hamilton: Department of Conservation.
36. Lusk, C. J., & Ogden, J. (1992). Age structure and dynamics of a podocarp/broadleaved forest in Tongariro National Park, New Zealand. Journal of Ecology, 80, 379–393.
37. McClelland, E., Wilson, C. J. N., & Bardot, L. (2004). Paleotemperature determinations for the 1.8-ka Taupo ignimbrite, New Zealand, and implications for the emplacement history of a high-velocity pyroclastic flow. *Bulletin of Volcanology*, *66*, 492–513.
38. McEwen, W. M. (1978). The food of the New Zealand pigeon (*Hemiphaga novaeseelandiae novaeseelandiae*). New Zealand Journal of Ecology, 1, 99–108.
39. McKelvey, P. J. (1963). The synecology of the West Taupo indigenous forest. New Zealand Forest Service Bulletin, 14, 1–126
40. McKelvey, P. J. (1986). Future research perspectives. In B. Veale & J. Innes (Ed.), Ecological research in the central North Island volcanic plateau region (p. 54–57). Rotorua: New Zealand Forest Service, Forest Research Institute.
41. Meade, H. (1870). A ride through the disturbed districts of New Zealand. London: John Murray.
42. Morton, J., Ogden, J., & Hughes, T. (1984). To save a forest: Whirinaki. Auckland: Bateman.
43. Native Forests Action Council (1977). The Podocarp rainforests of Pureora. Nelson: Native Forests Action Council (NZ).
44. New Zealand Forest Service (1944). Forest journal entry for 21 July 1944. Te Kuiti: New Zealand Forest Service.
45. New Zealand Forest Service (1979). The rimu seed crop in 1978, central North Island forests. Wellington: New Zealand Forest Service.
46. New Zealand Forest Service (1981). Whirinaki state forest park management plan. Rotorua: New Zealand Forest Service, Rotorua Conservancy.
47. Nicholls, J. L. (1976). A revised classification of the North Island indigenous forests. New Zealand Journal of Forestry, 21, 105–132.
48. Ogden, J., & Stewart, G. H. (1995). Community dynamics of the New Zealand conifers. In N. J. Enright & R. S. Hill (Ed.), Ecology of the southern conifers (p. 81–119). Melbourne: Melbourne University Press.
49. Palmer, J. G., Ogden, J., & Patel, R. N. (1988). A 425 floating tree-ring chronology from *Phyllocladus trichomanoides* buried by the Taupo eruption at Pureora, central North Island, New Zealand. Journal of the Royal Society of New Zealand, 18, 407–415.
50. Perry, G. L. W., Wilmshurst, J. M., McGlone, M. S., & Napier, A. (2012). Reconstructing spatial vulnerability to forest loss by fire in pre-historic New Zealand. Global Ecology and Biogeography, 21, 1029–1041.
51. Robertson, H. A., Colbourne, R. M., Graham, P. J., Miller, P. J., & Pierce, R. J. (2011). Experimental management of Brown Kiwi *Apteryx mantelli* in central Northland, New Zealand. Bird Conservation International, 21, 207–220.
52. Ruscoe, W. A., Sweetapple, P. J., Perry, M., & Duncan, R. P. (2013). Effects of spatially extensive control of invasive rats on abundance of native invertebrates in Mainland New Zealand forests. Conservation Biology, 27, 74–82.
53. Tucker, S. (1855). The southern cross and southern crown. London: James Nisbet and Co.

54. Wilkes, O. (2005). Transcripts of oral history recordings. Hamilton: Department of Conservation.
55. Wilmshurst, J. M., & McGlone, M. S. (1996). Forest disturbance in the central North Island, New Zealand, following the 1850 BP Taupo eruption. The Holocene, 6, 399–411.
56. Wilson, R. (1982). From Manapouri to Aramoana: The battle for New Zealand's environment. Auckland: Earthworks Press.
57. Worthy, T. H., & Holdaway, R. N. (2002). The lost world of the moa. Bloomington: Indiana University Press.

The Maori of the Central North Island Before 1860

3

C.M. King and T. Roa

Abstract

This chapter describes the history of the arrival and occupation of the Maori tribes that settled in the Pureora area; their original use of the forest for food resources; some of the most significant kainga (villages); the impacts on them of the European introductions of pigs and potatoes, and the famous feast of 1856 at Pukawa which marked the beginning of the King Movement.

Keywords

Maori traditional tracks and boundaries · Tuwharetoa · Maniapoto · Origins and settlement · Forest fires · Moa · Polynesian dog · Pacific rat · Kiore · *Rattus exulans* · Birdhunting · Kumara · Potatoes · Pigs · The Pukawa hui of 1856

Tribal Boundaries

The central position of Pureora Mountain had a defining influence on early Maori history. The territories of three great tribal groupings met at a point somewhere near it, just south of the Weraroa trig marking where the Hurakia hills joined the Hauhungaroas (Fig. 3.1). This was the common intersection of sub-boundaries radiating out in all directions, the pivot from which territorial boundaries were drawn. Percy Smith [42] drew a map, on which three great tribal groups, and the tracks and rivers marking their borders, are marked.

In pre-European times the land and its resources were not "owned" by Maori in the sense that it was property, a disposable commodity that could be bought and sold. Negotiations between European land agents and local chiefs, as for example described in great detail by Wakefield [48], were constantly dogged by semantic misunderstandings that now read as poignant tragedies.

The original version of this chapter was revised. For detailed information please see Erratum. The Erratum to this chapter is available at 10.1007/978-3-319-18410-4_16

C.M. King (✉)
School of Science, University of Waikato, Private Bag 3105, Hamilton 3240, New Zealand
e-mail: cmking@waikato.ac.nz

T. Roa
School of Maori and Pacific Development, University of Waikato, Private Bag 3105, Hamilton 3240, New Zealand
e-mail: tomroa@waikato.ac.nz

C.M. King et al. (eds.), *The Drama of Conservation*,
DOI 10.1007/978-3-319-18410-4_3

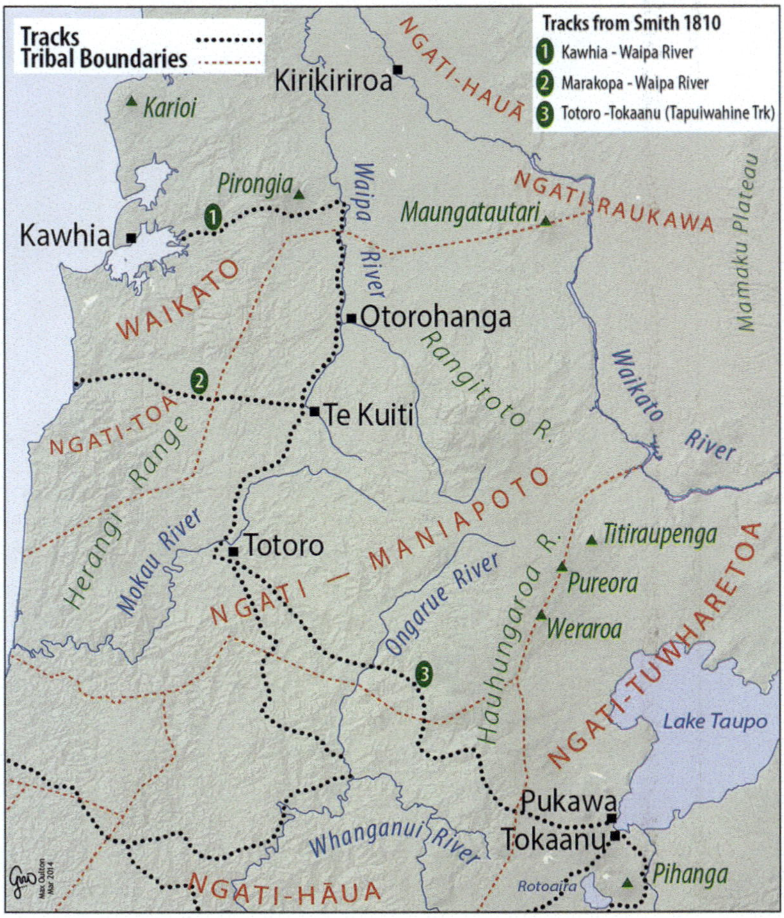

Fig. 3.1 Tribal areas significant to the Pureora story, and their ancient pathways in the western North Island in about 1800. Track No. 3, the Tapuiwahine, was the one used by Hochstetter in 1859 (see Fig. 4.5). *Redrawn by Max Oulton from Smith 1910*

Maori people occupied land in extended kin groups, whanau and hapu, under a system of interlocking and overlapping rights of use. These rights, or take, were derived as follows:

- *Take whenua kite hou*: a right of discovery, such as one related to journeys of occupants of an ancestral canoe.
- *Take tupuna*: an ancestral right derived from continuous occupation, particularly one which could be traced from an ancestral canoe.
- *Take raupatu*: a right obtained by conquest, with displacement or servitude of the original occupants, followed by occupation of the land by the conquering group.
- *Take tuku*: a right by virtue of a gift or exchange awarded in special circumstances such as a marriage or settlement of a dispute.

The principal sources of rights of occupation were based on take tupuna and take raupatu. However, these rights did not stand alone. An important related concept was ahi ka or ahi ka roa, the principle of keeping the fires burning on the land as a symbol of long-standing occupation.

This did not necessarily mean continuous settlement, but it did mean intermittent use, such as seasonal visits for fishing or birding in which temporary encampments might be made. If occupation rights were not maintained, the fires grew cold, and after three or more generations the fires could be regarded as being extinguished [47: Chap. 2]. This principle has an important consequence for the Pureora story (Chap. 10).

Maori Values

The values of Maori society arose from its communal nature. Individuals were seen as the repository or the voice of the group. There was a tapestry of intricate genealogical relationships, committed to memory and illustrated on carvings, and the notion that what affects a part also affects the whole was strongly upheld.

Similarly there was a firm belief that humans are part of nature—the forests, seas and waterways, and especially the mountains. People saw themselves in a sacred relationship with the natural world, and the exploitation of natural resources was conducted under strict regimes of tapu (sacredness) and mana (spiritual authority) administered by tohunga (priests) [36]. Violations of tapu by individuals could bring disaster on the whole tribe, so were severely punished.

Every tribal group had its own sacred mountain, and the significance of the mountains, of the valleys, of the landmarks and the stories the tribe maintained in its relationship to the landmark reflected on the mana (prestige) of the people. Sacred mountains had personalities, and were regarded as ancestors of high rank (tupuna, rangatira) and as extended family (Chap. 9) with their own mauri, essential elements of their own. These were recognised by the people, and ascribed a sacredness in accord with their mana—including the ability to argue and to love, to marry and to produce offspring [26]. In their roles as kaitiaki of the people, and conversely of people as kaitiaki of them, the mana was shared. Hence the names of significant mountains were given respectful personal names, not, as in European usage, preceded by the impersonal word "Mount".

Maori legends explain the locations and origins of geographic features such as the gorge of the Whanganui River and the isolated position of Taranaki, both attributed to a battle between Taranaki and Tongariro over the only female mountain of the group, Pihanga. Taranaki lost the argument and fled westwards, gouging out the Whanganui valley on the way.

Legend also explained the origins of volcanic activity as the arrival of fire from the mythical Maori homeland far to the north, traveling underground and occasionally emerging to form hot springs and geysers before bursting out at the summit of Tongariro [26]. Such stories made clear to visiting geologist Ferdinand von Hochstetter in 1859 that Maori had grasped the connection between volcanism and hydrothermal activity along the NE-SW trending line of the Taupo Volcanic Zone, from Ruapehu to White Island (Fig. 1.3).

A vital concept to Maori, frequently misunderstood by Pakeha, was that of wahi tapu.

> It was not only natural features but also practical knowledge of places where food, fibre and other resources were obtainable that were significant. There were also spiritual qualities of places that had special significance as wahi tapu (sacred places including burial grounds, urupa), places associated with past battles or other incidents involving important ancestors, or places where the mauri (life force or essence of a place and its people) were said to be preserved [47: Chap. 3].

To Maori, every place had meaning, a point that greatly impressed European explorers travelling with them:

> ...they seemed to take a pride in being able to define thoroughly all the natural features of their country. Each mountain and hill had its special name, and every valley and plain and river down to the smallest stream, each being called after some characteristic feature or legendary tale connected with it; whilst every tree, plant, bird, and insect was known by a designation which betokened either its appearance or habits [23: 323].
>
> It is really astonishing to observe how minutely the Maoris know their country, and how they have named not only each plant, bird, or insect, but also nearly every place. Every single cave and cleft, every rock and every hole in these parts has a special name with some legend or other attached to it [18: 340].

Some early European travellers, especially Dieffenbach—a remarkably forward-thinking scientist and ethnologist [6]—had great sympathy for the Maori outlook and traditions. But British military men such as Ensign Best and Herbert Meade had no concept of why some things, quite ordinary to them, had intense spiritual significance to their Maori companions.

> On the way we passed the graves of several chiefs who had fallen in battle; on one of these was a stone head which I alarmed the Mauries very much by touching; one of them declared that the Spirit would certainly rise in the night and avenge itself on me; he appeared to feel the danger I was in very much [43: 299].
>
> The Mauries are terribly afraid of the Lizard I happened to catch a small one which I put in a box and thus armed I kept the whole party at a swinging pace by occasionally producing my box to the hindermost [43: 307].
>
> They are very intelligent and reflective....at the same time they are very superstitious, having a childish fear of darkness and solitude, ascribing omens to the commonest occurrences...[including] the green tree lizard, from which, whether alive or dead, men, women, and children are said to fly in real terror [33: 167].

Georg Angas often found his work seriously inconvenienced and sometimes actually endangered by tapu. During his stay at Taupo, he was not allowed to shelter his portfolio of drawings from rain under the only available roof.

> The people were angry because my portfolio was placed under the cooking shed to preserve it from the rain, for E Pera had told them that it contained the head of Te Heuheu, and as it is sacrilege for him to enter a place appropriated to food, it is equally sacrilege for his portrait to be placed under similar circumstances [3: 133].

On an earlier occasion he almost lost the lot.

> Anything relating to food, if represented with the same pencil that depicted the head of the sacred chief, or put into the same portfolio with it, is considered a sad and fearful sacrilege. The whole of my sketches narrowly escaped being committed to the flames, through the indignation of Ko Tariu; and they were only rescued by the influence of my friend, the chief Te Heuheu. I was obliged in future to make drawings of the patakas, tapu buildings, &c, by stealth [3: 112].

One can sympathise with Angas' frustrations: the gulf of incomprehension was in his time too wide for most Europeans to cross. One who did, F.E. Maning, provides a helpful explanation: because all forms of personal property such as clothes, weapons, ornaments and tools took such great labour to make by hand, and because the fighting men and chiefs were so indispensable to the survival of the tribe, the tapu laws attached to them and their possessions were of real service in protecting them from irretrievable loss [27]. The same applied to the guarding of important communal property such as the fishing and hunting grounds and the gardens that provided their food.

For a pre-literate and warlike society, the spiritual world functioned to preserve cultural memory, law and order as well as do European libraries, museums and police: it was simply indispensable. The effects of the dismissive European attitude to wahi tapu is usually underestimated by non-Maori, but it is important to see how devastating that attitude was and is. The destruction of their cultural history affected nineteenth-century Maori in the same way that the 2003 ransacking of the Iraq Museum in Baghdad affected European archaeologists and historians.

When we understand the attitudes and beliefs of the original owners of Pureora's forests, we can appreciate their grief over what they have lost. During the heyday of the logging boom, the Maori values of the past were thoroughly over-ruled and desecrated, when bush crews could not and did not honour the ancient tradition that required them to pause and ask permission from Tane, the god of the forest, before felling any large tree [36]. Any members of a logging crew who refused to fell trees that they regarded as rangatira (noble, or especially valuable) were threatened with dismissal (Chap. 9).

At least now, twenty-first century Pakeha authorities such as the Department of Conservation take great trouble to respect the ancient Maori views, and train their staff to protect and preserve Maori sites (which includes not disclosing their locations). There is close collaboration with the appropriate tangata whenua, and no work is done on Maori cultural sites without consultation and approval.

Myths and Facts

Tribal traditions recounting the history of exploration and naming of geographical features were for generations handed down in verbal form only, and along independent lines of descent. Unsurprisingly, they have developed various contradictory versions, which make them indispensable and greatly treasured by their current holders but not necessarily literal truths. Like their Biblical equivalents, they were formulated as explanations of how things are and which things matter, in simple terms based on the shared knowledge of the time, vital to social cohesion and understandable by everyone [40]. A similar distinction between knowledge and values is critically important for understanding some contemporary debates, such as the one surrounding the use of 1080 toxin (Chap. 12).

Both Biblical and Maori traditions are enormously important, and they each demand unqualified respect as the highest authoritarian statements within their own traditions on how each culture derives meaning from facts. The ultimate purpose and value of both sets of traditions is to explain the relationships between the people and their worlds, and they can and do achieve that purpose independently of whether or to what degree their stories are literally true (both usually include a mixture).

Sometimes traditional interpretations can be mixed with gentle, affectionate humour. For example, one might ask, why are there so many species of beetles in the Waipapa Ecological Area (Chap. 2)? A famous biologist J.B.S. Haldane, when asked if there is anything to be concluded about the Biblical creator from the study of creation, replied that he must be "inordinately fond of beetles", since at least 300,000 species are known [17]. The abundance of beetles in the Waipapa EA might be taken to show that Tane, the Maori god of the forests and its creatures, shared that view.

The correct and perfectly respectable technical term for both Maori and Biblical traditions is that they are "myths", but in modern parlance that word has been taken over by a different meaning, emphasising lack of literal truth, as if that made them totally meaningless.

That interpretation signals an inappropriate application of scientific standards to pre-scientific data [24]. Such judgements are to be recognised and avoided, because they destroy the value of traditional material carrying meaning of a quite different kind. Both kinds of truth belong here, but each must be interpreted on its own terms.

Origins of the Central North Island Maori

The Maori people of Aotearoa are descendants of Polynesian voyagers who first arrived on the shores of New Zealand in about AD 1280 [54]. While there is still some debate about the precise date and the number of vessels, there is general agreement that during the late 1200s a number of ocean-going waka (canoes) made their way from east Polynesia, to land at various points on the coast of New Zealand. Similarity of place names, languages, and cultures, and DNA analyses shows a close link between the peoples of eastern Polynesia and New Zealand.

According to oral tradition, Whangaparaoa, at the very eastern tip of the Bay of Plenty on the north coast of the North Island, was the first landing place of several canoes, including the famous *Tainui* and *Te Arawa*—the two that are of most concern here. The *Tainui* sailed north, was portaged across the Tamaki isthmus, and made a final landfall on the west coast, at Maketu on the Kawhia Harbour.

Exploration and Settlement

The Tainui peoples are said to have remained near Kawhia Harbour for six or seven generations. During this time, they came to experience a new climate (much colder than they were used to, and with no distinct wet-dry seasons) and new species of flora (such as flax) and fauna (including many unfamiliar birds). These early settlements were often at harbours or the mouths of rivers—close to the sea, with good access to fishing and shellfish grounds. There was extensive hunting of coastal groups of seals, sealions, penguins and many other seabirds at their nesting colonies [12], and of flightless birds across the country, especially the meatiest of all, the moa [56].

The descendants of the colonists, the Ngati Maniapoto, Ngati Toa and Ngati Raukawa, eventually moved from the west coast to the inland hill country that was later called the King Country (Fig. 3.1). The Te Arawa and Ngati Tuwharetoa peoples moved inland from the opposite direction, the Bay of Plenty, to the Rotorua-Taupo basins and the eastern side of the volcanic plateau. By the late 1300s the whole of New Zealand had been explored, and by the time Cook arrived in 1769, every corner of it fell within the interest and influence of a particular Maori tribal or sub-tribal grouping.

The Taupo area had long been the focus for inland settlement (and for European explorers), because the waterways provided by the Waikato, Waipa, Mokau and Whanganui Rivers and by Lake Taupo offered not only relatively easy movement by canoe, but also important food resources.

By contrast, the dense intervening forests and rugged landscape were significant barriers to travel (Chap. 1). They were also cold and wet in winter, so were virtually uninhabited by people. Figure 3.2 (left) shows many dots (each representing 30 people) along major rivers and the shores of Lakes Taupo and Rotorua, but only two on the mountains of the future Pureora Forest Park [32].

The lack of settlements does not mean the mountains of the tribal lands were unimportant. They were highly prized hunting areas [38], regularly visited by foraging parties during the bird-snaring season.

In the 1820s, many North Island villages were decimated by inter-tribal wars set off by the introduction of muskets and gunpowder, deadly European imports that radically altered the balance of power between tribes [35].

> A series of raids by northern tribes from around the Bay of Islands had a ripple effect on tribal alliances, rejuvenated old feuds and started some new ones. Nga Puhi raids into Hauraki and Waikato put pressure on the tribes of the region, who retreated up the Waikato and Waipa valleys in the 1820s. Ngati Maru of Hauraki were expelled from the Maungatautari area in 1830, but before this their war parties had ranged widely in the Taupo district. In the early 1820s, a combined force of Waikato and Maniapoto tribes had expelled Ngati Toa, led by Te Rauparaha, from the Kawhia district… The Pouakani block [Pureora] straddled one of the important routes south from the Waikato via Maungatautari to Taupo and many war parties passed through [47: 4.1].

Maori used a wide range of locally available resources, including birds, fish, and plants, but large areas of the region had few food resources and were little inhabited until the 19th century [32]. Walton's analysis of many different sources suggests that 1600 is a likely estimate for the number of Maori people living in the vicinity of Lake Taupo in 1840 [49]. A detailed map by Ward [50] shows some 65 settlements inhabited from 1830 to 1880 (Fig. 3.2, right).

> The map is no doubt incomplete, but sufficient sites have been located to show the basic pattern of distribution. In virtually every case settlements were located around the shores of the lakes, on the Waikato River, or close to the edges of the bush [50].

At the time of Dieffenbach's visit in 1840, Maori settlements around the edge of Lake Taupo were still quite small and widely scattered, and apparently declining [13]. They lived in many small kaingas (villages) and hunting camps, and moved frequently between them.

> The evidence given to the Native Land Court also suggests a good deal of mobility in settlement patterns. Te Waiti Hohaia commented that "we had so many kaingas we travelled from place to place" [47: 3.3].
>
> The Waikato river was a food resource, but was not a place for permanent settlement. There were no tuturu kainga [permanent settlements] on the Waikato River… the houses were only temporary, used when fishing and catching birds. …Along the river bank were koura [fresh water crayfish] fisheries and duck snares… They… belonged to our

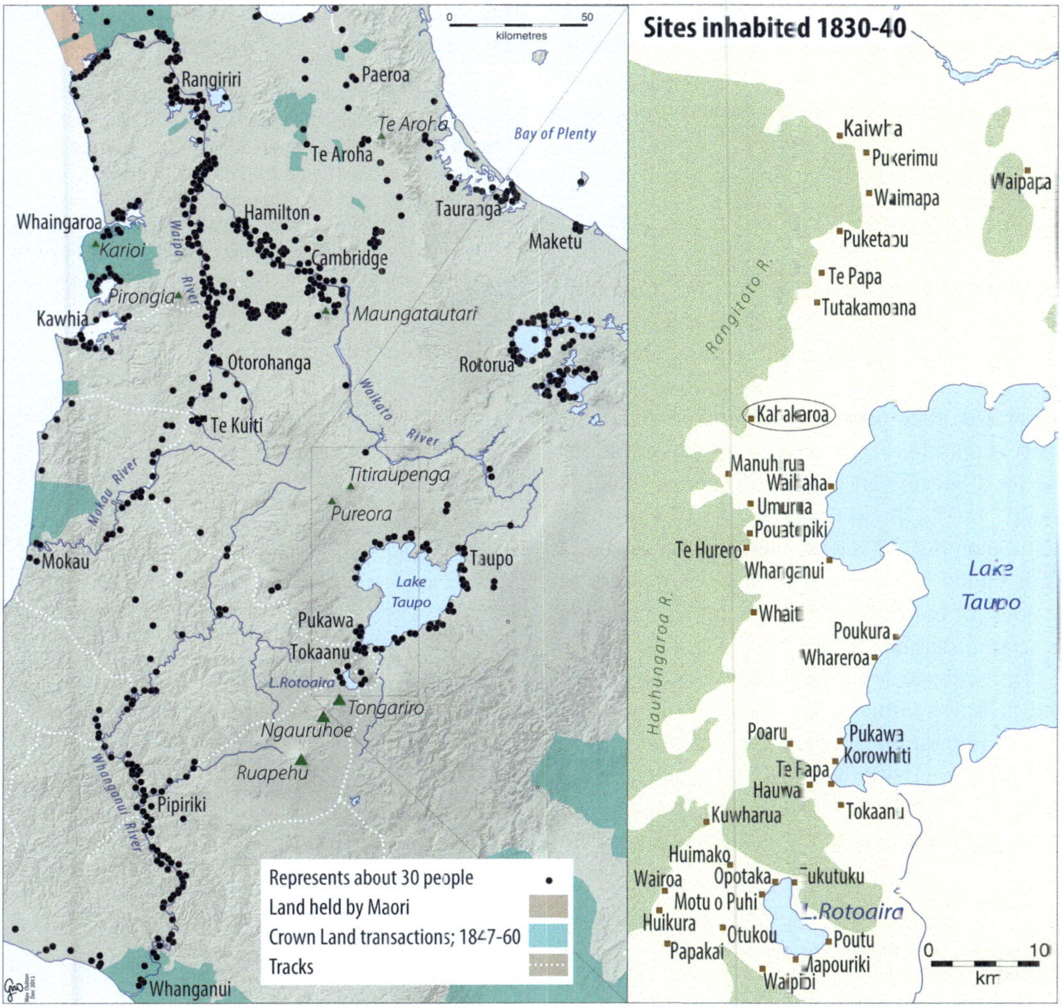

Fig. 3.2 *Left* Distribution of Maori settlements in the central North Island in 1853. All the land shown except the small blocks in *green* was held by Maori. *Redrawn from McKinnon (Ed.) (1997: Plate 31)*. *Right* Maori settlement sites between the west Taupo forests and the lake in 1830–1880. Kahakaroa pa was probably the one visited by Kerry-Nicholls in 1882 (named by him Kahakaharoa). *Redrawn by Max Oulton from Ward (1956), courtesy Journal of the Polynesian Society*

> matuas and tupunas [parents and ancestors] down to ourselves. No cultivations were along the river bank, the plantations were all near the bush away from the river [47: 3.3].

Counting a shifting population is always difficult, but Dieffenbach was right that the local population was already declining, and not only because of warfare and the stress of dealing with the Pakeha. Well after Dieffenbach's and Angas' relatively short visits, missionary sources reported widespread epidemics of influenza in the Taupo area in 1853. As is typical for influenza, the heaviest mortality fell on the very young and the very old. By the late 1850s, numbers in the Taupo area could have been down to about 1100 [52]. Against that, the significant new food supplies brought in by the Pakeha (pork and potatoes) certainly improved life for the survivors.

Fire

Forest clearing by fire, deliberate or accidental, had been advantageous and already widely practiced when the main subsistence crop had been fern root [30], but the process rapidly accelerated after potatoes suddenly entered the Maori economy. From as soon as this nutritious new food became an important item of trade, in around 1805, Maori cleared forest lands for potato planting. Maori farmers in the northern North Island became the principal suppliers of potatoes for both Maori and Pakeha consumption for several decades (Chap. 4).

The wide-scale destruction of the forest (Fig. 3.3) was witnessed and criticised by many of the early European travellers reaching inland areas in the 1840s and 1850s. Their earnest warnings of the inevitable consequences for forest fauna and long-term resource management were, like those of the conservationists of the next century (Chap. 9), simply ignored.

> It is evident that forest has at some former period covered a greater extent of land in the neighbourhood of Taupo than it now does; it does not appear to have been destroyed by volcanic eruptions, but by the fires kindled by the natives in order to clear the ground for the purpose of cultivation [13: 366].
>
> Large tracts of fern in this dreary region have been devastated by fire, and add a further gloominess to the scene [3: 133].

R.G. Ward explains further:

> Fires started in connection with the cultivations destroyed the bush in many parts of the region and it seems that repeated burning prevented regeneration. In order to get early crops of potatoes the tubers were sometimes planted in the midst of the scrub which was then felled….. Temporary huts were erected but these plantations were rarely used for more than one season… After repeated burning, vegetation on pumice country tends to degenerate to tussock and any accumulation of organic matter is rapidly destroyed. It is probably that in causing frequent fires the Maori was responsible for the modification of the vegetation over large areas and it is likely that as the forest was pushed back older settlements would be abandoned and new ones established closer to the bush edge. The pattern of settlements indicated in [Fig. 3.2] may represent the last stage in this trend

Fig. 3.3 Fierce flames sweep across the forest. This drawing, although made in 1919, gives a clear idea of the effects of early fires on the native flora and fauna. *Alexander Turnbull Library, Wellington, New Zealand. MNZ 1108*

> prior to the interruption caused by the introduction of European ways of life. In the last seventy to eighty years, that is following the abandonment of many of the old kainga and consequent upon the occurrence of fewer fires, manuka and heavy second growth have replaced the tussock over large areas [50].

Soil profiles and pollen analyses from peat beds laid down since the Taupo eruption confirm that there were large areas devoid of forest in 1840, and some surviving monoao-dominated frost-flats were originally induced by uncontrolled fires. On the lower ground, most surviving dense podocarp forests in PFP show signs of advanced regeneration developed during the last 100 years. Much of the fire history of the western shore of Lake Taupo (the eastern edge of PFP) is reconstructed by Perry et al. [37].

Food Resources

Moa

J.A. Wilson landed at the Bay of Islands with his family in 1833. In Wilson's time, most people agreed with the theory proposed by Julius von Haast, that an earlier and culturally separate group, the so-called "moa-hunters" or "aborigines", had preceded (and were destroyed by) the ancestors of the classic Maori (named by von Haast the "Hawaikians"), but that theory has long since been discredited [7: 237, 16, 25].

One of Wilson's informants summarised some Maori memories of moa hunting, as passed down to their descendants.

> The ancient inhabitants hunted the moa until it became extinct. The last bird was killed with a taiaha by a man at Tarawera. The habits of the moa are described as solitary, living in pairs in secluded valleys in the depths of the forest near a running stream. It fed on shoots, roots, and berries, and was particularly fond of nikau and tree fern. It was supposed to feed at night, for it was never seen to eat in the daytime. … The moa had a plume of feathers on its head. In the depths of the Motu forest there is a mountain called Moanui, where, no doubt, the bird was killed by the people of Rotonui-a-wai and Wharikiri, for their descendants knew fifty years ago that their forefathers had slain the moa [55: 128].

There is little evidence of moa-hunting in the central North Island, but the Maori tribes of the inland mountain country were well aware of the loss of the moa. They speculated on their own decline in proverbs, like the one quoted by the missionary Richard Taylor:

> *Ka ngaro a moa te iwi nei* (This tribe will become extinct like the moa). The moa… is now supposed to be extinct, the bones only having been discovered. It would be thus with the tribe alluded to; the people would all die, and their skeletons only would remain to show that they had been [44: 132].

Happily for us all, this gloomy prediction never came to pass. The Maori did indeed decline to very low numbers by the 1890s, but after the turn of the century they began an impressive renaissance. Now, a hundred years later, the Waitangi Tribunal has restored much of their lost mana, and addressed many old grievances over lost lands, including some of their ancient hunting grounds in the central North Island [47]. The ancient doomsayers were wrong; the Maori are not exactly what they were, but the worst fears of their ancestors were not realised.

The moa are indeed extinct, but interest in them, and in their relationship with the kiwi, has never been stronger. Arguments about how and when the endemic flightless birds arrived in Zealandia, and about the number and distribution of species that evolved here, rage all the louder now that studies of genes and geology come to different conclusions [1, 11, 34].

Kuri

The kuri was a short-legged breed of the domestic dog, with a bushy tail and pointed ears, brought to New Zealand by the ancestors of the

Fig. 3.4 Te Maioha, a cousin of the Maori King Te Wherowhero, wearing a parawai (dogskin cloak), in 1844. *From Angas, The New Zealanders 1847 p. 82 Plate XXXV, courtesy University of Auckland Libraries, Early New Zealand Books Collection*

Maori from East Polynesia. It was the largest mammalian species available to the Maori from which they could harvest meat and skins (Fig. 3.4), so it was a valuable asset. "In former times the natives prized both flesh and skin; and a dog-skin mat was a robe of state", wrote Charles Hursthouse [21]. Warriors decorated their spears (huata) with long tufts of dog's hair bound with flax [23: 292]. Understandably, Maori families kept their kuri in the villages, and cared for them well.

The usefulness of kuri as effective hunting companions was observed by early visitors.

> The kakapo betrayed itself at night time by its cry. With the assistance of a dog it was easily caught. Only within the present century did it become extinct, through constant hunting. Its loss as a source of food, was very much felt by the Maoris [55: 128].

> Every pah abounds with dogs, which are used principally to hunt the wild pigs that run loose in the woods [2: 337].

Pure-bred kuri did not bark, but howled like a jackal. In time, the distinctive characters of the kuri's separate lineage were overwhelmed by crossbreeding with European dogs. By 1843, Dieffenbach commented that "It is now rarely met with, as almost the whole race of the island has become a mongrel breed" [14: 184].

There was a time when lost dogs, supposedly kuri-European cross-breeds, roamed the back country in wild packs, including in parts of the Hauhungaroa Range. Stories about them are hard to confirm, but Jenks [22] reckons that some at least could have been true. True or not, the possibility of meeting a pack of wild dogs gave one gang of construction workers, surveying the central North Island for a power line in about 1946–47, a good bargaining tool. They refused to go into the bush without any protection against a pack of fierce dogs that might attack on sight. The men insisted on being issued with rifles before the work could go ahead. Even if they never met a dog, the rifles were handy to have next time the gang met a pig. Even during the 1980s, a pack of wild dogs roamed the Hauhungaroas, led by a white Alsatian [51].

Rats

The kiore or Polynesian rat was the first four-footed small mammal to live wild in the forests of New Zealand. It is the smallest of the three species of rats to have arrived (Chap. 13). It had the islands to itself from about AD 1280 [53], when the first Polynesian rats jumped ashore from voyaging canoes hauled up on the unexplored beaches, to the late eighteenth to early nineteenth centuries, when the much larger Norway rats arrived.

Genetic analyses show that the kiore is most closely related to Pacific rats from the southern Cook and Society Islands, and they have

Fig. 3.5 In the ancestral meeting house Wāhiao of the iwi Tūhourangi-Ngāti Wāhiao, on the marae Te Pākira–Whakarewarewa in Rotorua, is a carved panel (a poupou) showing an ancestor holding a rat (kiore) affectionately in his hands. It is regarded as the most sacred poupou in this particular meeting house because the tupapaku (deceased) of the tribe lie at its feet during iwi and tribal tangihanga (funeral ceremonies). *Image reproduced by permission of the elders of Tūhourangi Ngāti Wāhiao of Te Pakira Marae. C.M. King (2003)*

colonised New Zealand from there many times independently [29]. They spread inland throughout both main islands, including the Taupo region, but disappeared from there soon after the arrival of the aggressive Norway rats.

The little kiore was both a familiar companion, in life and on carvings (Fig. 3.5), and an important item of food for inland Maori tribes.

> Though now nearly extinct and seldom seen, [the kiore] was formerly so numerous as to form a considerable article of food. It was taken by an ingenious kind of trap, which somewhat resembles our common mole-trap. These were set on lines of road, which had been made expressly for this purpose in the forest; and they generally succeeded in taking sufficient at once to feed the whole pa [44: 380].
>
> The rat was, perhaps, the most valued kind of Maori game; when in season the flesh was greatly relished. They were kept in rat runs or preserves, which no stranger would venture to poach upon [55: 128].

Elsdon Best, an early ethnographer deeply interested in pre-contact Maori life, described in great detail the rigid customs and rituals involved in the trapping, cooking and preserving of rats during the hunting season in his classic work *Forest Lore of the Maori* [8].

He records a discussion with Tutakangahau, a Tuhoe chief, about the food-gathering customs of his tribe. Tutakangahau was born about 1830, and remembered that when he was young, he saw the kiore die out as the newly introduced Norway rat, the pou-o-hawaiki, appeared. He recalled a particular expedition to collect food (muttonbirds, titi chicks) from the main range at Maungapohatu, which at that time still supported a breeding colony of titi, only to find that the new species of rat had eaten all the young birds [20].

Kiore are mainly nocturnal, and were considered to be harmless vegetarians until the early

1970s, when evidence emerged that populations of small native species such as lizards, juvenile tuatara and weta could not survive on islands with kiore. More recently, detailed records of the benefits of kiore eradications from offshore islands give us some idea of the damage that must have been done in the mainland forests over centuries of unhindered occupation by these innocent-looking little rodents [45]. Far from being merely a clean and useful food resource for people, kiore were the beginning of the end for many of the small and defenceless members of the native New Zealand fauna that once lived safely on the ground, yet kiore were by no means the worst of it: that was still to come (Chap. 13).

Kiore eat a wide range of plant and animal foods, including tree seeds and seedlings, caterpillars, weta, cicadas, land-snails, native frogs, tuatara, skinks, geckoes and small birds and their eggs. The full consequences of the combination of the effects of kiore on forest regeneration and on the density and distribution of small and vulnerable fauna are unknown, but certainly very substantial [56]. Kiore still survive on a handful of offshore islands, and in parts of Fiordland.

Beech forests (Nothofagaceae) are virtually absent from the Pureora area (Chap. 2), so the Maori living there could not benefit from the massive irruptions in numbers of kiore that predictably followed a heavy fall of beech seed. On the other hand, maybe kiore in podocarp forest responded to a good season for berries and fruits the way the birds did. If so, rat irruptions would have been, at least until the late eighteenth century, a reliable sign of an excellent rat-trapping season to come.

Although the kiore had in its time been a valuable resource, and its disappearance a cause for great regret, it had also been a nuisance.

> It is customary amongst the New Zealanders to erect within their Pahs, or about their Kaingas and plantations, storehouses for the reception of food, and the preservation of maize, kumeras, and other seeds and roots: these storehouses are universally elevated from the ground by one or more posts, in order to preserve their contents from the destructive attacks of the native rat, which is extremely numerous in some parts of the country; they are termed "whata" in the northern parts of the Island, whilst on the west coast, and about Taupo, they are more commonly styled "pataka" [4: 72]

It is just as well that the Maori had already developed a method of keeping their food safe from marauding rodents (Fig. 3.6), well before two species of larger European rats arrived (Chap. 13).

The Maori were and are good observers of nature, and well aware of the difference between what they thought of as a native rat, called kiore maori, and the European species, which they called kiore pakeha, or the stranger rat. Worse, they witnessed the new arrivals turning into an unstoppable predatory tide that over-ran the country with drastic consequences, not only for the kiore but also for the people. As Angas observed, the Maori were not slow to notice that they themselves were in the same situation as the kiore.

> In former times they used the "kiore maori" for food in large numbers, but latterly it has become very scarce, owing to the warfare carried on against it by the European rat. It is a favourite theme with the New Zealanders to speculate on their own extermination by the Europeans, in the same manner as the stranger rat has superseded their native one [5: 55–56].

Hochstetter [18: 195] quotes (in a footnote to a discussion about the extinction of the moa from over-hunting), that:

> They [the moa] succumbed,—the larger the species, the sooner,—to the same fate, that is gradually sweeping the Kiwi, the Kakapo and the rat Kiore[1] in a similar manner, and before our eyes, from the face of the land.

The chief's request was not necessary, as the European rats were running ashore with or without any human permission, but they proved to be an unacceptable substitute for the kiore. F. E. Maning, an Irish adventurer who bought Maori land in 1833, describes the consequences for trading negotiations. He had to pay

[1]A chief, on observing the large European rats on board one of the vessels, entreated the captain, to let those rats run ashore, and thus enable the raising of some new and larger game [18: 195].

Fig. 3.6 Part of a Maori village painted by Georg Angas in 1844, showing four carved pataka (food stores on poles), a tame kaka (used as a decoy when bird-hunting), a low-roofed wharepuni (sleeping house), and several pigs. *From Angas (1847) The New Zealanders, Plate XXX p. 71, courtesy University of Auckland Libraries, Early New Zealand Books Collection*

...about fifty different claimants... One man required payment because his ancestors, as he affirmed, had exercised the right of catching rats on it, but which he (the claimant) had never done, for the best of reasons, i.e., there were no rats to catch, except indeed pakeha rats, which were plenty enough, but this variety of rodent was not counted as game [27: 77].

Birds

As the early Maori moved inland over several generations, they encountered the great forests, and their culture moved from being largely maritime to one which, in certain places including Pureora, was strongly influenced by trees and bird life.

One of the clearest and most relevant descriptions of the dependence of inland Maori on birds is given in the Pouakani Report to the Waitangi Tribunal.

A variety of birds were snared both in the swamps and forests. The most important forest species were kereru (pigeons) and kaka (red parrots). The main methods of catching birds included use of a bird spear from a perch in a tree, fixed snares, or a running noose at the end of a long rod held in the hand. Certain species of tree were known to be favourites for birds in the fruit season, including tii (cabbage tree), miro, kahikatea, tawa. The miro was a particular favourite of pigeons and had the added quality of making them very thirsty. One technique was to provide a bird trough, waka manu or waka kereru, a wooden vessel up to 1.5 m long and sometimes carved. Either snares were set over it or the snarer would conceal himself nearby with his tahere, noose, on a rod. The term waitahere was used to describe either a patch of water or a bird trough above which this method of snaring was used. A waituhi was a pool of water or bird trough with fixed snares over it. Because both techniques were often used in the same place both terms were used. Miro trees were scattered in the bush and did not grow in single stands, but were sometimes located at intervals along a ridge. A series of bird troughs would be set up, known as ara waka, a path or route of bird snaring troughs, or ara waituhi. The process of preserving birds was known as huahua manu, and preserved birds were often described simply as huahua [47: Sect. 15.2].

Birds were a significant resource and rights to bird catching places were guarded jealously. ... Each person named [see Box 3.1] would have been regarded as the kaitiaki, guardian of that place, who controlled access and therefore conservation of bird resources [47: Sect. 15.2].

In a paper describing and illustrating the technical skills of pre-contact Maori in bird-snaring, Downes comments:

The Maori of pre-European days was a singularly efficient craftsman, fowler, and warrior. As a food-seeker he secured the eel on its way to the ocean, caught the less known but more highly esteemed piharau (lamprey) on its way up stream, trapped the kiore or rat in the forest, and most ingeniously snared the feathered creatures of Tane-i-te-rere. In bush-lore he had little or nothing to learn from the white man. ...the Maori, with great skill, made river, sea, and bush minister to his necessities. He lived well where the average European would have starved [15].

The Maori had no way to kill birds at any distance (they had no equivalent to the bow and arrow), but their own methods (Fig. 3.7, Box 3.1) had other advantages:

[Snaring with a noose on the end of a rod] is a weapon...much more destructive in the hands of a skilful person than a gun, as it causes no alarm among the birds, which will even continue to hop about in the same tree after several have been taken. When a bird is caught, the rod is gently withdrawn, and, the bird being secured, is again elevated to its former position ready to deceive another [41: 213–14].

Box 3.1 Traditional bird hunting sites in the west Taupo forests

Since the end of the nineteenth century, timber milling and the subsequent clearing of much of the bush has made it almost impossible to identify places where birds were once snared by Maori hunters. But there is some very interesting general information in the public domain. Witnesses who gave their submissions to nineteenth-century sessions of the Native Land Court certainly wanted to put on

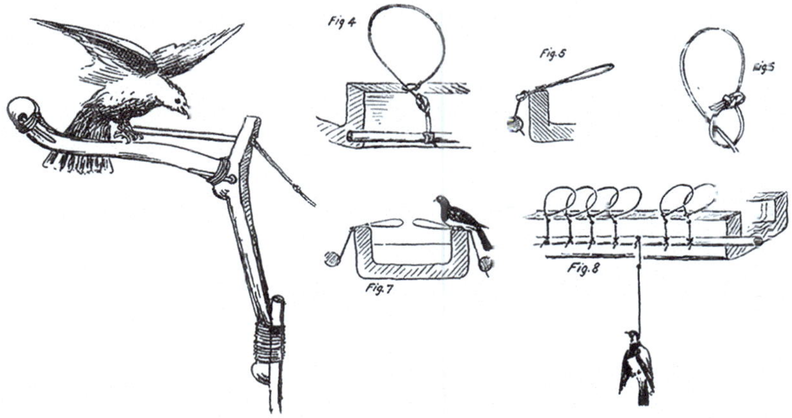

Fig. 3.7 *Left* a snare set for kaka. *Right* construction details of snares set on waka kereru (bird troughs) to catch native pigeons during the tree-fruit season. *From Downes (1928), courtesy Journal of the Polynesian Society*

open record the basis of their personal claims of ownership and former use of the forests in pre-European days.

The report of the Treaty of Waitangi hearing on the (much later) Pouakani claim cites several Native Land Court witnesses who had very clear memories of important birding places in the forests of Pureora and Titiraupenga [47: Chap. 15].

Hitiri Te Paerata described the bird snaring places in the northern part of Tuaropaki bush within the Pouakani block.

> The hapus to which I belonged hunted and caught birds in the forests. Paiakapuru, a Rimu tree at Hapotea, was one of snaring trees, it belonged to Ngakao. I have seen this tree. Te Punapuna was a bird trough (waitahere manu) belonging to Te Paerata. Another bird trough was near Mokaiteure, it belonged to Makawaiatemomo. At Paengawhakarau were other bird troughs elevated on trees belonging to Wereta Te Hikapai. A great snaring place (waitahere manu) was at Otanepai, this belonged to Te Arawaere and Ngahiku and Hoani Karapehi. At Moanui was a tutu manu, this belonged also to Te Atawaere and Hoani Karapehi. Te Tarata on a ridge was also a tutu (bird snaring) owned by Te Arawaere. Te Matai snaring place belonged to Rota, and Taurakumekume to Te Oneroa. Te Aramahoe (a tutu manu) belonged to Te Paerata. Te Puwharawhara tutu belonged to Ngakao. The last four were all on Te Tarata ridge. I know of another tutu at Waitutu called Te Whakapahi, this belonged to Rota. A bird snaring water (waitahere manu) below Moanui was called Kopuatahi, was the property of Te Awaiti, Te Haeana, Te Arawaere and Hoani Karapehi [47: 15.2].

Significant places, such as good sites for bird snaring (Fig. 3.7) and kainga mahinga manu (bird snaring camps), were given distinctive names. Hitiri Te Paerata described the kainga on the Pouakani block, beginning with the settlements below Titiraupenga:

> Owairenga was a kainga and cultivations… the principal houses were at Pukerimu a short distance from Owairenga the largest house was Tataurangi …….There is one large wharepuni at Kaiwha besides many smaller ones….Te Hapainga was an ancient kainga… from this settlement people went to catch birds…. Whatapo was another settlement inhabited at the time of our ancestors.
>
> Te Waimahana… is situated on both banks of the Waikato River. I lived there and my father before me. The houses of this settlement were not wharepunis but wharetoetoe [i.e. not substantial buildings but thatched huts], it was merely a kainga for cultivation purposes….. at this settlement crops of potatoes were planted and birds were snared….

> Horaaruhe was a kainga and a pa... At this kainga were two large wharepunis one of which was called Wairangi....In connection with this settlement were extensive plantations....The bird snaring places of this kainga were at Waipapa. Ngawhakawhitiwhiti, a Matai tree, was owned by Te Paerata Kaiawha. Ngamataiturua, two Matai trees, belonged to my father. Hamutira a waitahere manu (bird snaring water) belonged to my father. Te Waipopotea belonged to Paora Ngamotu [47].
>
> One of the seasonal hunting camps, Tuturuwhakamate, was about 500 m north of the present Ngaherenga camping ground at Pureora [38].

A good early summary of the animal resources of the North Island forest was published by Richard Taylor [44]. Much of it would have applied to the first foragers in Pureora Forest Park.

> The kiwi was hunted at night with dogs. The natives can so closely imitate the cry of this bird, that they soon draw to them all which may be in that part of the forest. ...The kiwi is still abundant in some places.
>
> The kakapo, or ground parrot, is a gregarious bird, larger than the common fowl: it was hunted with dogs... at night; it is now all but extinct in the northern island, though it is said to be plentiful in the southern one.
>
> The weka, or large rail, is still found in the interior of the north island. The natives imitate its cry so exactly that it readily approaches them. This bird is so pugnacious, that if a bit of red cloth, or other rag, be tied to a stick, it flies at it immediately, and is thus easily caught by a noose held in the other hand.
>
> The kereru, or wood-pigeon, is a very fine bird, but very stupid. It is frequently taken by placing a pole near the water's edge, where it is accustomed to drink. When it has quenched its thirst, it alights upon this, which is completely covered with snares made of flax, where it soon gets its legs entangled, and is thus secured. The rimu and kahikatea pine trees, when in fruit, are also thickly set with snares, by which means numbers are taken. The natives have also a long bird-spear, often from twenty to thirty feet in length, armed with a sharp barb of bone. With this they silently approach the tree on which the bird has alighted, and it is, generally, so stupid as to stay and be speared, although it sees its enemy approach. If the tree be one to which the pigeon is accustomed to resort, on account of the abundance of fruit, the natives then construct a ladder by ingeniously binding two young trees together, which may be growing near, and use them for that purpose, by which means they can approach the bird with the greater facility. Great numbers of pigeons are thus captured. The natives frequently extract all their bones, and, when cooked, place them in a large papa–a vessel made of the totara bark; thus preserved in their own fat, they will keep many months. The tui, when in season, is very fat, and is also preserved in a similar way. It is a most lively bird, and can only be taken by snares.
>
> The kaka, or great brown parrot, one of the largest of its family, is also eaten. It is generally caught by means of a tame parrot (Fig. 3.6), which is used as a decoy. A pole is stuck in the ground, in an inclined position, in some shady part of the forest, on which the tame parrot is placed. The native forms a little arbor with a few large leaves of the fern-tree, in which he sits concealed with a small stick in his hand. The call of the tame bird soon attracts some of its wild companions, which, when they alight on the pole, are enticed still lower, until the fowler either seizes them with his hand, or knocks them down with his stick.
>
> The kakariki, or small green parrot, is taken by snares; it is a delicious bird, and very abundant.
>
> These are the principal birds upon which the New Zealanders lived... The water-fowl, the wild duck, whio, pukeko, and several others–were more rarely captured, and can scarcely be said to have contributed much to their support [44: 279–282].

The huge forests of the central North Island were favoured hunting grounds. Surplus birds captured during a good hunting season were preserved in their own fat, especially the larger species such as pigeons and kaka. They were kept for leaner times, or as stock to give away in barter or to welcome honoured guests.

> The inhabitants of the villages on the upper part of the river Wanganui are celebrated parrot catchers, and keep great numbers of tamed birds to be used as decoys. About the month of June, a great part of the population migrate to the immense forests lying between their river and the more central parts of the island, for the express purpose of catching parrots. Every evening, the birds taken during the day are roasted over fires, and then potted in calabashes in their grease, for they are very fat. Thus preserved, parrots and other birds are considered a delicacy, and are sent as presents to parts of the country, where they are scarce: and in due time a return present of dried fish, or something else not to be obtained easily in an inland country, is received [41: 214].

The seasonal migrations of important bird species were well known, and turned to the advantage of the Maori hunters. By about March of each year, great flocks of pigeons were moving south as far as Pirongia and Te Aroha; then to Rangitoto, Ranginui and Titiraupenga by May and June; then to Pureora and Ketemaringi by July and August. Hunting camps in the forest were established for occupation during the snaring season. At each stage,

> …a great multitude [of birds] was to be found, feasting upon the abundant miro berries. It seems that the Creator designed and developed the native forests so that, at every season, some tree or shrub…produced food for the pigeon, as the Maori gratefully realised to his profit [38].

This seasonal harvest within the boundaries of a tribe's own lands was fiercely defended against intruders from other tribes. The last great battle on the northern and western slopes of Pureora was in about 1807, when a war party of Ngati Maniapoto and Ngati Matakore drove out Ngati Haua invaders (Fig. 3.1). By contrast, the Ngati Tuwharetoa to the east were linked with the Tainui by marriage, so relations between them were more cordial [38].

Kumara

Increasingly, Maori developed horticulture. With careful techniques, often involving the use of shelter walls of stone, they succeeded in cultivating several tropical plants brought from the Pacific islands by their Polynesian ancestors, especially the kumara (sweet potato) (Fig. 3.8).

> Their ancestors brought the kumara, or sweet potato–the taro–an arum–and the hue, or calabash, with them from Hawaiki: these were the only vegetables they possessed, and they carefully cultivated them in large quantities, until the arrival of Europeans, who gave them the potato, the value of which was so soon discovered, that now it may be said to be their staple article of food. It is far more universally cultivated than the kumara, from its taking less labour in planting, and yielding a more certain and larger return.
>
> The kumara requires not only a warm aspect, but also, in general, an artificial soil; sand or gravel being laid on the ground to the depth of six inches. So also the taro, which needs the aid of bush screens and other expedients to make it flourish. … The kumara, taro, and even potato grounds, are generally selected on the sides of hills, having a northern aspect; by this declivity towards the sun, they gain an increased degree of heat [44: 378].

The central North Island climate was generally too cool for kumara, but local tribes made good use of sites where geothermal activity warmed the soil—as directly observed by two early visitors to Taupo.

> A space of about 10 acres on either side of the Tokaanu stream is perforated with holes and cavities of various sizes, from which steam issues in large quantities. Some part of this space is barren

Fig. 3.8 A Maori garden with kumara plants and a pataka, as recreated in authentic detail in the Te Parapara Garden, Hamilton. The garden is named Nga Pukeahu o Puna (Puna was the god of the kumara). *C.M. King (2014)*

> and whitened by the sulphureous exhalations from the hot springs; but in other places, manuka and rich grass grow to the very edge of a boiling cavity… close to the mouth of the stream, on the flat, which is here perhaps a mile broad…extensive patches, sown with the kumera, are neatly fenced in and cultivated [48: 102–103].

> [We came to] a settlement of the Taupo natives: some were busy at work in their little corn and kumera beds, whilst others were cooking food. Their maize plantations were exceedingly neat, and the light pumice soil was turned up into little heaps very carefully where their taro was planted [3: 108].

Another such favourable place was probably Waimahana, at the eastern end of the Pouakani block. Werohia Te Hiko told the Waitangi Tribunal that a kainga mahinga kumara [a settlement for kumara cultivation] had been planted by his father and all Ngati Wairangi ….before Te Ariki [1851 or 1852]. "Perhaps", commented the author of the report, "the hot springs at Waimahana provided sufficient warmth, a local micro climate which allowed kumara growing here but not elsewhere. This site is now flooded by Whakamaru hydro lake" [47: Sect. 3.3].

Pigs

The first two pigs the Maori ever saw were gifted to them by French explorer Captain Jean François Marie de Surville at Doubtless Bay in 1769. Nothing is known of their fate, but they were probably not the ancestors of any later population [31: 336].

During Cook's second (1773) and third (1776) voyages, a number of boars and sows were released, mostly in Queen Charlotte Sound, and two breeding pairs were given to the Hawke's Bay chief Tuanui.

In 1795, Lieutenant-Governor King of Norfolk Island established contact with the northern chief Te Pahi, and sent him a total of 56 pigs in three shipments in 1804 and 1805. It is probably from these, and from pigs gifted between tribes, that pigs became established in the North Island. By 1805, Maori living near the coast were selling pigs and potatoes to Europeans, and providing pork and potato dinners for explorers.

Maori adapted very quickly to these significant new resources, not only as a food for their own people, but also as valuable trading items. Especially in the 1820s and 30s, European ship captains depended on Maori to supply their crews with fresh food (Fig. 3.9), and Maori depended on European captains, despite the objections of the missionaries, to supply muskets, gunpowder and bullets [39].

> The domestic animals reared by the New Zealanders have been introduced at various times by the Europeans who have visited their coasts. The pig… is bred in great numbers throughout the country [2: 337]
>
> [At Puhanga, SW of Piopio, in 1859] We evidently created quite a sensation; the children seemed for the first time to have seen white men; yet we were received with a cordial welcome, and the women at once proceeded to prepare a meal, cooking it by means of heated stones in holes dug in the ground. It was set before us in newly made baskets—potatoes and pork, as ever and everywhere [18: 348].

By the 1850s the forest in many parts of New Zealand teemed with pigs, providing a rich new food resource for Maori hunters. Pigs were prized for their meat and fat, and were kept in a semi-domestic state around the kainga (Fig. 3.6). Precious vegetable gardens had to be protected from marauding pigs enclosing the plants inside fences (Fig. 5.6), which took a lot of effort to build, but there was no attempt to construct European-style pig sties.

> The natives about here [Oruanui, about 11 miles from Taupo] keep neither pigs, poultry nor livestock of any kind [33: 67–70].

One might wonder why not, since hunting takes more effort than farming. The explanation is supplied by Hursthouse: the only pigs available were the wild and unpredictable descendants of the gentle domesticated porkers gifted to Maori decades ago, so were impossible to manage in captivity.

> In the old "war-times," when the Tribes were constantly besieging and destroying each other's village strong-holds, slaves and pigs formed the conquerors' chief booty–many of each were killed

Fig. 3.9 Maori bargaining with a Pakeha, 1845 or 1846. Pigs and potatoes were eagerly traded for money to buy muskets in the Bay of Islands at that time. The social disruptions caused by the northern tribes that first obtained muskets had serious consequences for others much further south, including those in the central North Island. *Painting by John Williams, Alexander Turnbull Library, Wellington, New Zealand. ATL A-079-017*

> and eaten at the victors' feasts; but, during the siege and melee, some of the pigs would escape into the forests. Here, they ran wild and bred; and having a fine climate, abundance of fern root and other food, millions of acres of close covert, and no animal enemies to molest them, they spread over both islands, and may now be numbered by thousands. They roam about in little herds of a dozen or so; but, keeping in the deepest recesses of the forest, are seldom or ever seen except by bush travellers or pig-hunters. They are small fleet animals, with coarse skins, weighing about 100 lbs. The young pork is excellent; but it is a costly mistake to put one of these forest rangers into the sty: evidently ignorant that the duty of a good pig is to get fat and be killed, they chafe and fret in their prison, invariably catch and devour any intrusive duck which may chance to look in, slight [reject] barley meal, and grow thinner on every meal [21: 124–125].

Potatoes and Other European Vegetables

Early European explorers introduced the Maori to many new food plants. In December 1769, wheat, peas, and rice were left in New Zealand by de Surville. Marc Joseph Marion du Fresne planted wheat, maize, potatoes and various kinds of nuts on Moturua Island in the Bay of Islands in 1772. Lieutenant-Governor Philip King sent potatoes and maize from Norfolk Island in 1793.

> The melon and pumpkin are now also cultivated, as well as the cabbage and turnip, which grow wild, having been introduced by Cook; maize and wheat have been more recently raised, but are now grown in large quantities [44].

These new crops and many others transformed Maori domestic life from at least the turn of the nineteenth century. For example, maize was well established in the Bay of Islands by around 1816, and the cobs were roasted in embers, or fermented and made into cakes.

But by far the most significant of the new food resources were the potatoes. For the first time in their history, the Maori had a crop that could be grown easily and stored for long periods with minimal deterioration—in short, a staple food crop and a reliable resource for feasts and barter [10].

Potatoes were grown at Thames as early as 1801, and traded in the Bay of Islands by 1805. Potatoes quickly became a valuable item of trade with visiting Europeans (Fig. 3.9), so, unsurprisingly, cultivation of the potato stimulated a massive expansion of Maori traditional shifting agriculture, with extensive and irreversible consequences [10]. In fact, R. J. Cameron estimates that by the 1850s, Maori potato growers were clearing the forest at least as fast as European settlers continued to clear it after the 1860s.

Potatoes were also planted in standing bush, for two reasons.

> As we passed through the woods, we found two plantations of potatoes, which would have rendered our bringing any quite unnecessary had we known of them. As my natives never seemed to consider that these kind of plantations belonged to any body, we always used to help ourselves when we came to any of them without compunction. In fact, I suppose that these patches must have been planted by some of the mission-natives, on purpose to save trouble when they went their journeys between the two stations [9: 15].
>
> Were the contrary not well known, the potato might be taken for an indigenous plant, as it is impossible to go anywhere without finding it growing wild. As we know it has not been introduced more than fifty years, this diffusion of the root may be considered wonderful. We may be led to suppose, from this circumstance, that the climate is exceedingly favourable to the growth of the potato [9: 27].
>
> The natives plant all their potatoes in the woods where the soil is much richer [33: 70].

European vegetables were an important component of the Maori diet, although not always with results appetising to Europeans.

> The introduction of potatoes has also wrought a great change in their diet, probably for the worse. Potatoes [and corn]… form almost the entire food of the natives; those [living]…along the banks of the rivers, add fish. They eat their potatoes without salt; and many, who subsist exclusively on them, do not take sufficient exercise to render such a diet wholesome. Fevers, too, are frequent, from the too abundant use of putrid corn; the natives steeping the ears of maize in water for several weeks, to render them soft, until they become perfectly rotten, and give forth a most offensive odour [21].
>
> We were almost suffocated by the violent stench of kaanga or stinking corn, arising from a large pot over the fire in the yard, filled with a sort of gruel prepared by boiling the putrescent maize. Two slave women sat stirring it round with sticks, inhaling with evident delight the odour that to us was indescribably disgusting [3: 64–5].

Kainga

The Maori lived in villages (kainga) located close to important resource areas including forest, water, cultivable land and geothermal heat. Permanent kainga were often fenced (Fig. 5.6], to keep out the wind and the wild pigs [47]. Important food sources such as the Waikato River and the shores of Lake Taupo, hunted for koura (crayfish) and ducks, were dotted with kainga and fishing places [13: 360], but the population seemed sparse and not permanent. There was no cultivation in these areas. Where natural resources were less reliable or only seasonal, the scattered kainga were periodically abandoned and later rebuilt.

A wharepuni was a sleeping house, built of wood rather than raupo (reeds), and roofed with totara bark or thatch (Fig. 3.6). To maximise insulation, wharepuni were semi-subterranean, set so deep in the ground that only the roof showed, and even that was often covered in soil to a thickness of a foot or more. Because there was no window and no entry for fresh air but a close-fitting narrow door, the atmosphere inside was, to some travellers unused to it, "hot, stifling and abominably unwholesome" [33].

Hochstetter's experience at the Maori settlement Katiaho [Ongarue] in 1859, where he arrived late at night, cold and tired, was quite different.

> The dogs hailed our arrival with a perfect jackal's howl; the pigs, roused from their repose, were running to and fro; but human voices also became audible, and at length some persons came up to us in the dark, who conducted us to a large house, of which only the roof seemed to protrude from the ground. One after another we slipped in through a low square-hole, and found ourselves in a spacious apartment lit up by two blazing fires and heated to an almost tropical heat (85 °F), in which we were most cordially received by the chief of the place, Taonui, with the surnames Tekohue and Hepahapa, and by the whole people gathered about him, all expressing their unfeigned surprise at being honoured yet so late at night, and in such a weather, with a visit from Pakehas. There might have been twenty or thirty persons in the hut, which number was almost doubled by the addition of our party. The hut in which we found ourselves was a so-called Wharepuni, a conversation and

sleeping-room, such as existed in former times in every Maori village; which, however, have fallen more and more into disuse, owing to the influence of the missionaries, who opposed the sleeping together of old and young, of boys and girls. This Wharepuni was quite new; it had been but recently erected on the occasion of a visit from a neighbouring tribe. It was a real palace in comparison with the miserable raupo huts in other Kaingas. The side-walls were artificially wrought of plaited reeds and rushes; the ground-floor was covered with neat mats, and a row of carved columns supporting the roof divided the large room into two halves. The right side, according to Maori custom was assigned to the guests; and all of us strangers having arrived in a most deplorable plight, wet to the skin, and tired to death, we could well congratulate ourselves on having found so excellent a shelter. We divested ourselves of our dripping garments, and wrapped ourselves after Maori fashion in woollen blankets. Outside, in the cook-house, the meal was prepared, and after supper we chatted together till late in the night [18: 351–2].

Among the least welcome inhabitants of Maori villages were the massive infestations of fleas.

[We passed] a cold and wretched night in a windy shed they called their "ware karakia," or chapel, which was inhabited by immense fleas, whose size did not render them less nimble or ravenous [3: 132].

Returning to the village [Kaihu], I found my tent pitched not far from the low rush building used as a chapel; and in which I would have met the natives, but was warned not to enter on account of the fleas… When ready to leave them, cloak in hand, I ventured to go near the front of the chapel to look in. A moment's standing in the spot was quite enough. The hopping hosts could be distinctly heard among the dry rushes and litter which strewed the ground. My light trowsers were literally covered; and Paukena to whom I had turned over my cloak to clear it, significantly said on bringing it back, "Tenei ano tou kete puruhi;" –Here is your basket of fleas [46: 59–60].

Horses

Horses arrived with the earliest missionaries from 1814 onwards, and though they were welcome as potential draught animals capable of moving heavy loads, life was difficult for them at first. Horses are strictly grazing animals, so a country that is naturally covered mostly in forest had to be modified to produce grass and hay for them, and the owners of working horses also needed access to the various ancilliary skills of the farrier, saddler and vet.

For early travellers in the interior, roads suitable for riding on were few and short, and cross-country travel was often impossible for horses because it involved many daunting obstacles described in chilling detail by Kerry-Nicholls [23]—such as struggling through impenetrable forest thick with supplejack, and frequent wading through bogs and dangerous rivers.

The arrival of such an astonishing animal amazed the Maori population. Angas described the "extraordinary excitement produced by the arrival of so large and singular an animal" caused by the first horse to appear in the Taupo area—a gift to Te Heuheu Mananui's son Tamati in 1844 [3: 112].

By the late 1850s, horses were the main form of overland transport, used by Maori as well as by settlers, military forces and traders. Escapes and strays were inevitable, including from the Armed Constabulary camps of the 1860s. Some stray horses became truly feral, while others were simply permitted to wander and were rounded up when needed, or, especially stallions, shot as pests disturbing domestic mares. In 1882, herds of wild horses ranged the western tableland of Lake Taupo [23: 307]

The Great Feast of 1856

Any large and important hui or tribal gathering had to be accompanied by a great feast, which in turn meant that the local tribe had to find a huge amount of food. One of the best recorded of these feasts took place on the shores of Lake Taupo in 1856 (Chap. 4).

The great hui known as Hinana, called by Te Heuheu at Pukawa in 1856, put considerable strain on the bird resources of Titiraupenga, Tuaropaki and other bush areas. The numbers of birds to be snared seem to have been related to production of berries from certain trees, and such trees did not always produce fruit every season. Various interpretations were also made about the effectiveness of the karakia, rituals used in bird snaring.

> Takiwa Te Momo stated that, after the Hinana meeting, birds were scarce at Tuaropaki and Tirohanga. …The famine still continues at Tirohanga. I heard that Te Wharepapa cursed those places and the trees not bearing fruit caused the places to be deserted by game.
>
> Te Ahitahu Taiawhio knew of only one bird snaring, when the birds were taken to pay for gun powder, since Hinana. He said: "I don't know how long after Hinana this bird snaring was, but it was before the war [1863–64]. There has been no bird snaring, food preserving, on a large scale since, as the birds are much less numerous. It was Hitau who destroyed the birds". [47: Chap. 15].

The context and significance of Te Wharepapa's "curse" is not explained, but for some reason, birds did not return to Tirohanga bush as quickly as expected. Naturally, the local people interpreted this problem in terms of how they understood the over-riding influence of spiritual powers on the natural world around them.

In our time, it would be easy to blame the excessive harvest required to host the huge 1856 feast for the initial decline [47], but the continued loss of traditional bird-snaring over time throughout the Pureora area is a different matter. For an explanation of the long-term failure of the bird populations, we need to look elsewhere.

One possibility is that, around that time, the widespread use of firearms was beginning to make it easier to over-harvest birds by shooting rather than by snaring. As Ferdinand von Hochstetter (one of the first Europeans to explore the upper Mokau valley) reported in 1859:

> [We] entered a dark, stately forest. We made a hard shift to dig our way through, over the smooth texture of roots in the sombre twilight of the virgin forest, when suddenly close by us a shot was fired, and from behind a gigantic Kahikatea-trunk a human figure stepped forth a double barrelled gun in the hand, and …yonder lay the whole band encamped around a brightly blazing fire, all armed with guns. [We were frightened at first, but they were] a peaceable Maori-party shooting pigeons. We saluted each other with a friendly "tena koutou," exchanged tobacco for some of the finest wood-pigeons,—a capital dish for our dinner,—and passed on [18: 347].

One of the witnesses giving evidence to the Native Land Court made the same point about hunting with guns, when asked to define a tokatoka. He said it is a rock with a kaka-snaring post on it, and that he used one during the year of the Hinana feast. "It is not used now as people kill birds with guns, not snare them as they used to do". When asked in what year did snares cease to be used, when guns were used instead, he replied "At the year of the Hinana Feast" [28].

A later hearing of the Court was told that it was now impossible to stop people poaching birds with guns. The switch in hunting methods probably started as soon as guns and gunpowder were available. As Hodgskin [19] observed, referring mainly to the tribes of the Bay of Islands,

> They are excellent marksmen, and almost every NZer has his fowling-piece [p. 13]… Wood pigeons are found in abundance every where–much larger, fatter, and more beautiful in plumage than our English pigeons. The flesh is delicious… These birds are easily shot, for they are so tame as to allow you to approach within a few yards. The natives shoot hundreds with small pebbles, which are used as a substitute for shot [19: 28].

With hindsight, a third and more compelling reason for the decline of the birds might be suggested—one that is just as tragic as any supernatural fault (karakia failure or enemy malice) or overhunting with firearms, and only indirectly related to human activity. Whereas Norway rats had been abundant in the bush for decades, and kiore for centuries, the real catastrophe for tree-nesting birds in the forests of the North Island began after about 1860, with the arrival of a voracious, abundant and agile tree-climbing predator, the ship rat (Chap. 13). On the other hand, although the kiwi is certainly not immune to the presence of rats, it is less affected by them than by stoats, which had not yet arrived anywhere in New Zealand. In 1882, when Kerry-Nicholls and his companions explored the King Country, they reported that:

> Here, besides the usual diet of pork and potatoes, we were treated to roast kiwi. This bird …is the only remaining representative of the great family of New Zealand Struthionidae. It is a dwarf form of the moa, not larger than a fair-sized hen… These

birds, which live in pairs, are still very plentiful in the dense, unfrequented ranges of the King Country [23: 316–18].

Alas, that is no longer true (Chap. 2).

References

1. Allentoft, M. E., & Rawlence, N. J. (2011). Moa's Ark or volant ghosts of Gondwana? Insights from nineteen years of ancient DNA research on the extinct moa (Aves: Dinornithiformes) of New Zealand. Annals of Anatomy—Anatomischer Anzeiger, 194, 36–51.
2. Angas, G. F. (1847). Savage life and scenes in Australia and New Zealand, Vol 1. London: Smith, Elder and Co.
3. Angas, G. F. (1847). Savage life and scenes in Australia and New Zealand, Vol 2. London: Smith, Elder and Co.
4. Angas, G. F. (1847). The New Zealanders illustrated. London: Thomas M'Lean.
5. Angas, G. F. (1866). Polynesia. London: Society for Promoting Christian Knowledge.
6. Bell, G. (1976). Ernst Dieffenbach: Rebel and humanist. Palmerston North: Dunmore Press.
7. Berentson, Q. (2012). Moa: The life and death of New Zealand's legendary bird. Nelson: Craig Potton Publishing.
8. Best, E. (1942). Forest lore of the Maori. Wellington: Government Printer.
9. Bidwill, J. C. (1841). Rambles in New Zealand. Christchurch: Capper Reprints.
10. Cameron, R. J. (1964). Destruction of the indigenous forests for Maori agriculture during the nineteenth century. New Zealand Journal of Forestry, 9, 98–109.
11. Campbell, H. J. (2013). The New Zealand drowning debate: Did New Zealand sink beneath the waves? Wellington: Bridget Williams Books.
12. Collins, C. J., Rawlence, N. J., Prost, S., Anderson, C. N. K., Knapp, M., Scofield, R. P., Robertson, B. C., Smith, I., Matisoo-Smith, E. A., Chilvers, B. L., & Waters, J. M. (2014). Extinction and recolonization of coastal megafauna following human arrival in New Zealand. Proceedings of the Royal Society B, 281, doi:10.1098/rspb.2014.0097.
13. Dieffenbach, E. (1843). Travels in New Zealand: with contributions to the geography, botany, and natural history of that country, Vol. I. Christchurch: Capper Reprints.
14. Dieffenbach, E. (1843). Travels in New Zealand, Volume II. Christchurch: Capper Reprints.
15. Downes, T. W. (1928). Bird-snaring, etc., in the Whanganui River District. Journal of the Polynesian Society, 37, 1–29.
16. Duff, R. (1977). The moa-hunter period of Maori culture. Wellington: Government Printer.
17. Gould, S. J. (1993). A special fondness for beetles. In S. J. Gould (Ed.), Dinosaur in a haystack: Reflections in Natural History (p. 377–387).
18. von Hochstetter, F. (1867). New Zealand, its physical geography geology and natural history. Stuttgart: J G Cotta.
19. Hodgskin, R. J. (1841). A narrative of eight months sojourn in New Zealand, with: A description of the habits, customs, and character of the islanders; the climate, soil, and productions of the country, including timber for ship-building; with a brief account of birds, fishes, etc., in a series of letters. Coleraine: S Hart
20. Holman, J. P. (2010). Best of both worlds: The story of Elsdon Best and Tutakangahau. Auckland: Penguin.
21. Hursthouse, C. (1857). New Zealand, or Zealandia, the Britain of the South, Vol 1. London: Edward Stanford.
22. Jenks, H. J. (2002). The wind in the tussocks. Wellington: New Zealand Institute of Surveyors.
23. Kerry-Nicholls, J. H. (1884). The King Country, or explorations in New Zealand—A narrative of 600 miles of travel through Maoriland. London: Sampson Low & Co.
24. King, C. M. (2002). Habitat of grace: Biology, religion and the global environmental crisis. Adelaide: Australian Theological Forum.
25. King, M. (2007). The Penguin history of New Zealand illustrated. Auckland: Penguin Books.
26. Lowe, D. J., Newnham, R. M., & McCraw, J. D. (2002). Volcanism and early Maori Society in New Zealand. In R. Torrence & J. Grattan (Ed.), Natural disasters and cultural change (p. 126–161). London: Routledge.
27. Maning, F. E. (1863). Old New Zealand. Auckland: Creighton & Scales.
28. Maori Land Court (1897). Tihoi. Taupo minute book, 11, 26.
29. Matisoo-Smith, E., & Robins, J. (2009). Mitochondrial DNA evidence for the spread of Pacific rats through Oceania. Biological Invasions, 11, 1521–1527.
30. McGlone, M. S., Wilmshurst, J. M., & Leach, H. M. (2005). An ecological and historical review of bracken (*Pteridium esculentum*) in New Zealand, and its cultural significance. New Zealand Journal of Ecology, 29, 165–184.
31. McIlroy, J. C. (2005). Feral Pig. In C. M. King (Ed.), The handbook of New Zealand mammals (p. 334–345). Melbourne: Oxford University Press.
32. McKinnon, M. (Ed.) (1997). Bateman's New Zealand historical Atlas. Wellington: David Bateman.
33. Meade, H. (1870). A ride through the disturbed districts of New Zealand. London: John Murray.
34. Mitchell, K. J., Llamas, B., Soubrier, J., Rawlence, N. J., Worthy, T. H., Wood, J., Lee, M. S. Y., & Cooper, A. (2014). Ancient DNA reveals elephant birds and kiwi are sister taxa and clarifies ratite bird evolution. Science, 344, 898–900.

35. Moon, P. (2012). A savage country: The untold story of New Zealand in the 1820s. Auckland: Penguin Books.
36. Orbell, M., & Moon, G. (1985). The natural world of the Maori. Auckland: David Bateman.
37. Perry, G. L. W., Wilmshurst, J. M., McGlone, M. S., & Napier, A. (2012). Reconstructing spatial vulnerability to forest loss by fire in pre-historic New Zealand. Global Ecology and Biogeography, 21, 1029–1041.
38. Phillips, F. L. (1989). Landmarks of Tainui Nga tohu a Tainui: A geographical record of Tainui traditional history. Otorohanga: Tohu Publishers.
39. Polack, J. S. (1838). New Zealand: Being a narrative of travels and adventures during a residence in that country between the years 1831 and 1837. London: Richard Bentley.
40. Rue, L. D. (1989). Amythia: Crisis in the natural history of western culture. Tuscaloosa, Alabama: University of Alabama Press.
41. Shortland, E. (1856). Traditions and superstitions of the New Zealanders, 2nd edition. London: Longman & Co.
42. Smith, S. P. (1910). History and traditions of the Maoris of the West Coast North Island of New Zealand prior to 1840. New Plymouth: Memoirs of the Polynesian Society.
43. Taylor, N. M. (Ed.). (1966). The journal of ensign best 1837–1842. Wellington Government Printer.
44. Taylor, R. (1855). Te Ika a Maui, or, New Zealand and its inhabitants. London: Wertheim and MacIntosh.
45. Towns, D. R. (2009). Eradications as reverse invasions: Lessons from Pacific rat (*Rattus exulans*) removals on New Zealand islands. Biological Invasions, 11, 1719–1733.
46. Wade, W. A. (1842). A journey in the Northern Island of New Zealand. Hobart: George Rolwegan.
47. Waitangi Tribunal (1993). The Pouakani report. Wellington: Waitangi Tribunal.
48. Wakefield, E. J. (1845). Adventure in New Zealand from 1839 to 1844. With some account of the beginning of the British colonisation of the islands. Vol II. London: John Murray.
49. Walton, A. (1986). The population of the Lake Taupo region, New Zealand, 1839–1859. New Zealand Journal of Archaeology, 8, 73–88.
50. Ward, R. G. (1956). Maori settlement in the Taupo Country, 1830–1880. Journal of the Polynesian Society, 65, 41–44.
51. Wilkes, O., Gaukrodger, J., & Ritchie, N. (2011). Chronology of Pureora. Hamilton: Department of Conservation.
52. Williams, A., & Walton, T. (2003). Early landuse patterns in the Lake Taupo area. Science for Conservation, 222, 1–28.
53. Wilmshurst, J. M., & Higham, T. F. G. (2004). Using rat-gnawed seed to independently date the arrival of Pacific rats and humans in New Zealand. The Holocene, 14, 801–806.
54. Wilmshurst, J. M., Anderson, A. J., Higham, T. F. G., & Worthy, T. H. (2008). Dating the late prehistoric dispersal of Polynesians to New Zealand using the commensal Pacific rat. Proceedings of the National Academy of Sciences of the United States of America, 105, 7676–7680.
55. Wilson, J. A. (1907). The story of Te Waharoa: Sketches of ancient Maori life and history. Christchurch: Whitcombe and Tombs.
56. Worthy, T. H., & Holdaway, R. N. (2002). The lost world of the moa. Bloomington: Indiana University Press.

The European Impact: Exploration to Conflict, 1840–1890

4

C.M. King and N.A. Ritchie

Abstract

This chapter describes the impressions of the country and of the last years of the classic Maori culture recorded in the mid-nineteenth century by Bidwill, Dieffenbach, Ensign Best, Angas, Smith, Hochstetter and Kerry-Nicholls, mostly in their own words. Then it covers the origins of the King Movement, the local consequences of the Waikato War of the 1860s, and the closing of the King Country to Europeans until the early 1880s.

Keywords

Bidwill · Dieffenbach · Ensign Best · Angas · Hochstetter · Percy Smith · Kerry-Nicholls · Thomas Grace · Te Heuheu Mananui at Pukawa · Waihi landslip · Origins of the King Movement · Te Wherowhero · Tawhiao · The Waikato War · Orakau · Te Kooti · Te Rohe Potae · Miringa Te Kakara

The earliest European visitors to New Zealand found a world very different from the comfortable, prosperous and optimistic society they had left.

> It would be an understatement to call New Zealand in 1839 an "undeveloped country". For as yet there was no development at all and its total population of Europeans at that time can only be guessed; estimates vary between 150 (in 1831–32) to 2,000 in the mid-thirties [4: 34].

By the 1860s, much of the interior of the North Island had been explored by Europeans. Among the earliest of those who ventured inland as far as the edges of the Pureora area were J.C. Bidwill (1839), Ernst Dieffenbach and Ensign Best (1841), Georg Angas (1844), Percy Smith and Hursthouse (1858), Ferdinand von Hochstetter (1859) and Herbert Meade (1864). (Another well-known explorer, William Colenso,

The original version of this chapter was revised. For detailed information please see Erratum. The Erratum to this chapter is available at 10.1007/978-3-319-18410-4_16

C.M. King (✉)
School of Science, University of Waikato, Private Bag 3105, Hamilton 3240, New Zealand
e-mail: cmking@waikato.ac.nz

N.A. Ritchie
Department of Conservation, Te Papa Atawhai, Hamilton Office, Private Bag 3072, Hamilton 3240, New Zealand
e-mail: nritchie@doc.govt.nz

C.M. King et al. (eds.), *The Drama of Conservation*,
DOI 10.1007/978-3-319-18410-4_4

did not visit Pureora, as his route of 1847 approached Lake Taupo from Napier and returned eastwards via Waiouru). They all left written accounts of their journeys, which provide detailed descriptions of the land and the people during a critical period of cultural change. After 1864, the west Taupo area was closed to Europeans for 20 years (Chap. 5).

The Europeans did not, of course, really explore the land in the sense of being the first visitors, since most of it had already long been known to Maori. Throughout this period, European explorers used Maori routes and drew up maps from their descriptions. They nearly always used Maori guides and porters, followed Maori tracks (Fig. 3.1), and were fed, canoed and even carried, by Maori [22].

The detailed accounts they published give us a glimpse into a vanished world, written from an outsider's point of view often steeped in European romantic literature (Byron, Scott, Longfellow, Wordsworth), interpreted through the lenses of their personal interests, and formed by the assumptions of their times.

They tended to value the manly, public-school hardiness ("muscular Christianity") that made light of sore feet, rough food, hard beds, biting insects and rats, all the while "singing the praises of life in the wilds, of a bed in the bush with stars to see, and a weka roasted on a supplejack" [22: xiii].

> With the last ray of the sun the sandflies disappear entirely, so that at night at last one is rid of that plague. But, sometimes, certain other still more unwelcome guests intrude at night – rats. They are found even in quite uninhabited countries, and gather after the very first night around the camp. To their running at night leisurely over his head and body, the traveller will easily become used; but eatables must be carefully kept out of their reach by hanging them upon poles [12: 287–8].

Inland from Kawhia in 1882, Andreas Reishek camped one night, watching as

> .....small owls kept flittering round the fire, and rats nibbled away at the fragments of my meal or fought with one another [19: 172].

Occasionally, the rats provided some mild entertainment, as reported by the surveyor John Rochfort:

> Another great evil is the immense number of rats, which destroy corn and everything eatable. They are almost a match for a cat; in fact, I have known six cats turned out of a house by them in a single night. At the same house one of my companions missed a woollen stocking on rising in the morning; after a long search, a small portion of it was found sticking out of a rat's hole in the corner of the room. The officers of the garrison amuse themselves, when indulging in bed in the morning, by practising pistol-shooting on them.... by a little stretch of imagination [they can] practise bush-fighting from behind the bedposts [20: 26].

These authors do not specify whether they were referring to Norway or ship rats. Norway rats were the first to arrive and were the larger and more aggressive of the two species (Chap. 13), so Rochfort's rats sound more like Norways, but in the North Island bush after around 1859–1869, Hochstetter and Reischek could have encountered either species. Both were unwelcome, to residents and visitors alike.

Even at so early a stage in settlement history, some of the visiting authors included pointed criticisms of the ransacking of New Zealand timber and marine resources (whales and seals) by the Pakeha. It is a curious fact, remarks David Young [26: 62], that three of the most trenchant early critics were all German: two (Dieffenbach, Hochstetter) were visiting trained naturalists, and a third (in the far south, Johann Wohlers) a missionary. Their outsiders' perspective contrasted pointedly with that of the mostly British immigrants, who came to settle into a new life.

Travellers' Tales

Because there were no roads over any useful distance outside settlements, the only ways to get about were on foot, on horseback, or by boat or canoe. Settlements built along the main rivers or on the shores of Lake Taupo were relatively accessible (Chap. 3), but other places with few tracks, many steep gorges and fast, deeply incised streams with no bridges, including much of the interior of the west Taupo forests, were seldom visited by Europeans.

Nevertheless, European influence had already significantly changed the way of life of the people living in even the most remote places, long before they ever met a Pakeha. The widespread adoption of pigs, potatoes and corn as staple foods (Chap. 3), plus metal tools with which to manage them, had already greatly improved the nutrition and survival of Maori, especially of children. The opposite effect was achieved by muskets and axes, dangerous weapons that in the early 19th century were swiftly adopted by dominant tribes.

The first Pakeha travellers to venture inland found the going rather hard, and the resident population widely scattered. As the Waitangi Tribunal Report dryly comments,

> It was not an empty region. Although large areas appeared desolate, there were many clusters of kainga, small settlements of perhaps up to 20 or 30 people, scattered along the bush margins, by lake shore or river, and associated with areas of surface geothermal activity. Unlike the lowland forests of the North Island, there were large expanses of fern, scrub and tussock on the Volcanic Plateau. . . The ranges to the west of the lake were clothed with dense podocarp forest [23: Chap. 3].

The Austrian geologist Ferdinand von Hochstetter's vivid first-hand descriptions (Fig. 4.1) summarised the conditions met by most European travellers venturing into the forest in the mid nineteenth century:

> The slender paths of the natives lead over hills and mountains in steep ascent and descent, rarely in the valley, nearly always along the ridge of mountain-heights. Where they cross the bush the clearing is just broad enough for one man to wind himself through. [A horseman] used to European paths will scarcely recognise those Maori-trails, and man and beast would be in continual danger upon them – the horse, in danger of sinking into the deep holes between the roots of trees, and of breaking its legs; the rider, of being caught among the branches, or strangled among the loops of the "supple jack." Hence there is no other choice left but to travel on foot; and it requires full, unimpaired bodily strength, and sound health, to pass uninjured through the inevitable hardships of a longer pedestrian journey through the New

Fig. 4.1 Hochstetter's sketch of himself and his companions on Ngariha (a peak of 560 m on the western edge of the Hauhungaroa Range, northeast of Ongarue) in 1856. *From Hochstetter 1867 p. 354. courtesy University of Auckland Libraries, Early New Zealand Books Collection*

Zealand bush, over fern-clad hills, over steep and broken headlands, through the swampy plains and cold mountain-streams of the country. Whatever the traveller needs for his individual wants, he must carry with him, and therefore must be limited to the most necessary articles. Now and then, a solitary European squatter may be met with; and more frequently still, a Mission station. On all these occasions the traveller will meet with a cordial welcome, and hospitable treatment, and transiently he will enjoy even the comforts of civilized life; but, taken as a general thing, he must resign them all; he must learn to find pleasure in living in the open air with the skies for a canopy and the earth for his table and bed [12: 283].

Bidwill, Dieffenbach and Ensign Best

In the 1820s and 1830s, visiting botanists Allan Cunningham and J.C. Bidwill sent collections of plants to London's Royal Botanic Gardens at Kew. Bidwill headed inland from the Bay of Plenty to Rotorua in 1839, crossed Lake Taupo in a canoe and then, ignoring the Maori tapu protecting the sacred mountains of the Tuwharetoa, climbed Ngauruhoe. He was lucky to get away with it because even decades later, when Kerry-Nicholls and his companion climbed Ruapehu in 1882, the tapu was still strong enough to put the Europeans in real fear of being killed if they had been caught trespassing on the sacred mountain [14]. At that time, Pureora was also protected by tapu, but less than 100 years later, trampers were allowed free access to the summit, and public use of this privilege was rapidly increasing by 1971 [24].

Bidwill described the landscape around Lake Taupo in the late 1830s. The forest and the population depending on it surrounded the western shores of the lake and extended into the deep sheltered valleys inland, while the higher-altitude mountains supported neither forest nor permanent populations (Fig. 3.2).

Ernst Dieffenbach spent two years (1839–41) exploring much of the North Island on behalf of the New Zealand Company, which was promoting the country for British settlement. His prime task was to look for agricultural land and minerals, which probably coloured his views of the difficulty posed to settlement by the dense forest.

In April 1841 Dieffenbach travelled up the Waikato River with Ensign Best, a naval officer on leave from his ship, and from a hill near Mangakino, Dieffenbach described his first view of the central North Island:

> The forest was interwoven with creepers, and the road very tedious. We encamped about a mile from the left shore of the river Waikato. On ascending the hill which separated us from it, a novel scene opened before me. Looking to the eastward the land appeared as if the waves of the sea had suddenly become petrified: ....We also saw Titiraupena, a pyramidical mountain, with naked black rocks heaped upon its pointed summit, and bearing S. 20° E. The Waipapa presented a very wild scene. The river, here about forty yards broad, lost itself in successive falls in a deep fissure which it had corroded out of the soft rock...... Of animal life nothing was visible, with the exception of the *Cigale Zelandica* [cicada], which filled the air with its chirping note, and a brown ground-lark very common in New Zealand. .... Here and there I found pieces of obsidian, and everything proved that we were fast approaching a great centre of volcanic action [10: 323–5].

Dieffenbach was interested in all aspects of natural science, and was an indefatigable collector of botanical and geological specimens. His Maori porters could not understand his interests at all, and sometimes quietly dropped the rock specimens he had given them to carry, substituting others collected along the path just before they reached the day's camp [4: 63]. Dieffenbach gave a detailed account of the fauna, flora, landscape and Maori inhabitants living in New Zealand as he observed them in 1839–41.

Dieffenbach endured with minimal comment all the hardships of an inhospitable environment:

> .....[sandflies] are perhaps the most bloodthirsty animal that exist, attacking all the exposed parts of the body. With the last ray of the sun they all disappear, but are immediately replaced by the musquittos.... We had taken our abode in an old house, where the rats ran over us all night, and two species of smaller animals, not to be named to ears polite [10: 145].

Other, less inhibited authors such as Polack [18] did not hesitate to name the offending two smaller species as fleas and lice, both of which over-ran the houses and the persons of the local people. Explorers soon found that sleeping in their own tents was a safe precaution, regardless of the weather. Angas and his companions

> pitched our tent, overlooking the broad surface of the Waikato, at about half a dozen yards from its brink. The fear of too many visitations from that active parasite, the flea (cleverly styled "e pakea nohinohi," or "the little stranger," by the natives, who say it was introduced by the Europeans), prevented our encamping within the enclosure of the pah. [2: 20].

On the other hand, Dieffenbach was an expert observer, and relished the bird life of the forest:

> The sonorous fluting call of the large parrots, varied by their harsh scream when, on a sudden alarm, they started over the tops of the hills, and then returned to rest, were the only sounds that broke the deep silence.....Sometimes a parrot would perch on one of the trees embowering our huts, as if curious to ascertain who had ventured to disturb his repose. During the night a solitary cry from one of these birds might be heard from time to time, after which everything again became quiet. The sweet song of the mako-mako [bellbird], which I can only compare to that of our nightingale, although I must confess that the former is simpler, and therefore more impressive, and the warbling of the tui, whose note resembles that of our thrush or blackbird, cease at the setting of the sun; but in the morning, before he is above the horizon, the little songsters renew their music with increased vivacity, and their combined tunes form a pleasing concert [10: 117–8].

He attempted to compile a national inventory of animals, the very first 'Fauna of New Zealand'. Among his finds were the freshwater crayfish, the green puriri moth, and the giant weta. He was also instrumental in locating several species of skinks and geckos.

Dieffenbach was unlike many other travellers, in that some of his strong opinions were well in advance of his age. He loved New Zealand and its people, and was never scathing about them (at that time, other Europeans usually called them "savages"), or the conditions in which they lived.

He stressed repeatedly that Europeans were neither culturally nor morally superior to other peoples, which would have been a novel idea to many Victorians, especially the missionaries. He even recognised the potential of the Taupo district as a future tourist attraction (Fig. 4.2):

> The scenery of Taupo lake, the whole character of the landscape, the freshness and peculiarity of the vegetation, with the white smoke rising around

Fig. 4.2 View northwards across Taupo from Te Rapa (marked on Fig. 3.2), with Tauhara Mountain in the distance. At the time of Dieffenbach's visit, Te Rapa was the site of Te Heuheu's pa. *From Dieffenbach (1843), Travels in New Zealand, Vol 1, courtesy University of Auckland Libraries, Early New Zealand Books Collection*

from so many hot-springs, are singularly beautiful, and well calculated to attract visitors from all parts of the world. The excellent disposition of the natives will ensure every one a good reception who does not come with the arrogant and ridiculous prejudices which are too frequently characteristic of a European traveller [10: 363].

Dieffenbach wanted to stay on in New Zealand after his contract expired, and might have had an interesting influence on the processes of colonialism just starting, but his work was not seen as essential, so was not further funded. He reluctantly left in 1841 [4].

Angas

Georg Frederic Angas was an English explorer and artist, and a tireless traveller. His journals and drawings are important first-hand observations of the places and people he visited, although his comments were, understandably, coloured by comparisons with his previous experiences. More than that, all visiting artists interpreted what they saw, and drafted their artworks, in terms of their own ideas about what constitutes the sublime and the picturesque [16].

Early in the spring of 1844, Angas set out on a journey of upwards of eight hundred miles, on foot, to explore the interior of the North Island.

Travelling in New Zealand is very different from travelling in Australia, where the open nature of the country enables one to ride for hundreds of miles in almost any direction: in New Zealand the traveller must go on foot, and so dense and extensive are many of the mountain forests, that he has to cut or force his way through them; whilst the frequent precipices, swamps, and rivers, offer obstacles to his progress that require some ingenuity to overcome [2: 1].

Angas approached the Taupo region by the same route later travelled by Hochstetter and Smith (Fig. 3.1), across the steeply dissected mountain forests of the southern Hauhungaroas. They were all following part of an established Maori trail, apparently the Tapuiwahine track that ran between Totoro and Tokaanu.

Angas leaves no doubt that even well-known Maori trails were not for the unfit or faint-hearted.

We struck over some fern hills into the most awful and almost impenetrable forest and jungle we had yet encountered. Supple jacks, fallen trees, and masses of decayed vegetable matter, impeded our progress; and to surmount these obstructions we were obliged to creep on our hands and knees through tangled brakes, jump over trunks of trees, slide down precipitous banks of slippery roots, and endure all manner of horrors and abominations. On, and on, and on, we toiled– wading, creeping, jumping, sliding, and scrambling– till sunset, when we reached a few deserted huts in an old potato-clearing upon the slope of a hill amidst the forest, beside a stream of water embowered with beautiful fuschias in full blossom. [Next day] we proceeded through [dense forest] for eight weary hours without finding a single opening, and during this time we had frequently to cut or force our way through the tangled overgrowth of vegetation. In these primeval and all but impenetrable forests, the birds are so tame that, on resting and imitating their various notes, we frequently brought round us a flight of little songsters, that approached without the slightest manifestation of fear. Amongst the smaller varieties, I observed the white-headed manakin, a black and yellow fly-catcher, and an extremely diminutive wren. At intervals, in the silent and gloomy forest, one passes an old shed constructed of bark, or the leaves of the nikau-palm, where the remains of fires bespeak the resting-place of native travellers [2: 104–106].

Whiteheads and fantails still live in the Pureora Forest (Box 2.2), but the bush wren is extinct, and beautiful fuchsias in full blossom have become a rare sight since the arrival of the introduced browsing mammals (Chap. 12).

After three days, Angas reached Lake Taupo in late October 1844 [3], and made the customary courtesy visit to Te Heuheu Mananui, paramount chief of Tuwharetoa at his lakeside pa at Te Rapa (Fig. 4.3).

Te Heuheu was superintending his people, who were at work in the potato grounds; but he at last arrived, and saluted me by pressing noses. After sitting down again in silence for some time, I delivered to him a letter of introduction, which I had brought from Te Whero Whero, the principal chief of Waikato.

Te Heuheu is a fine old man; he stands nearly seven feet high, and is very corpulent. His hair is

Fig. 4.3 Te Heuheu Mananui (seated) and his brother Iwikau. *From Angas (1847) The New Zealanders p. 126, courtesy University of Auckland Libraries, Early New Zealand Books Collection*

silvery white, and his people compare it to the snowy head of the sacred Tongariro; there being no object, except this tapu mountain, of equal sanctity to permit of its being mentioned in connection with the head of their chief. ....

After Te Heuheu had heard the contents of the letter, which was read to him by one of his grandchildren, he immediately ordered a large pig to be killed for us; in the mean time I was fed, much against my inclination, with potted pigeons, boiled down in their own fat and kept in a gourd until perfectly rancid; for no salt is used in preparing them: this is a delicacy reserved for visitors and state occasions [2: 109–110].

After some days enjoying Te Heuheu's hospitality, Angas and his party reluctantly set off again into the "volcanic wilderness", which he described in less than flattering terms while at the same time delighting in new species of plants and animals.

We started in the rain and wind, through a dismal and desolate country, composed of broken hills, ravines, and rocky masses of pumice, intersected by swamps. Amongst the wiry grass, or wiwi, of these volcanic districts, two very beautiful species of moss occur in considerable abundance: the one is a cup-moss, with brilliant scarlet sealing-wax-like tips and edges; the other resembles bushes of white coral. Larks (*Alauda Novae Selandiae*) are abundant here, and amongst the grass I captured a new species of butterfly, belonging to the genus *Polyommatus*; the latter has since been described and figured from my specimens by my friend Edward Doubleday, Esq., of the British Museum.

We crossed a foaming cataract by jumping across a chasm of rocks, which was our only means of passing a broad and swiftly-flowing river: it was a dreadful leap, and had either of us slipped we should have been dashed to pieces in the raging cauldron of rock-beaten surf below. A flower, which I have observed in no other locality, somewhat like the Christmas primrose, grew on the banks of this river [2: 132–3].

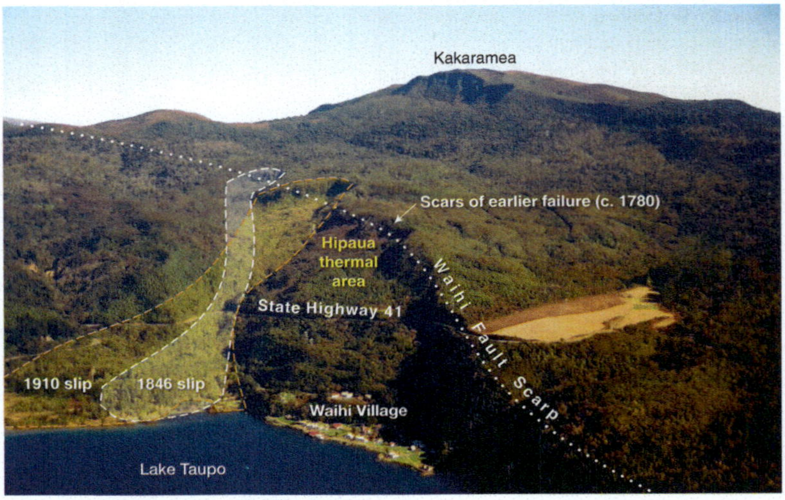

Fig. 4.4 Position of the unstable thermal area above the southwest shore of Lake Taupo. Te Heuheu's village of Te Rapa was buried under the 1846 slip. Both SH 41 and the modern village of Waihi are still at risk. *Data from Geological Survey NZ, image from G.T. Hancox, GNS Science, Lower Hutt, in Crozier* et al. *(2008: 199)*

Raging rivers and impenetrable bush were not the only hazards of travel, and some of them are still familiar to modern campers.

> We were….. annoyed by the mosquitoes during the night; and no sooner had the sun risen, and we issued from our tent to wash by the river side, than those peculiarly vexatious pests, the sand-flies (namu), commenced their attacks on our bare hands and feet. The sand-fly is a small black insect, and swarms in such myriads, that one is never free from their vengeance, if remaining for a single instant in the same position: whilst sketching, my hands are frequently covered with blood, and their numbers being inexhaustible, one at last gets weary of killing them [2: 20–21].

Smith

Percy Smith, then only 18 (and described by historian Nancy Taylor as a "young colonial savage"), with four companions set out from Taranaki on 4 January 1858, via the Mokau River (Fig. 3.1) and across country as Angas had done, making comparable comments on the hazards of travelling on foot through forest and swamps [21].

On 17th the party reached Pukawa, and stayed until 21st. Smith was unimpressed by the state of the village, which, he saw,

> …..may have been a large and populous place at one time, but now is fast sinking into decay. There are not more than 15 houses inhabited, if so many, and we scarcely saw anyone but Te Heuheu and his women. Not far from his house is a splendid pataka, or store house, and it is elevated on four red posts, in fact the whole concern is painted red. The doorway end is beautifully carved, the carvers seeming to have vied with each other in making the ugliest faces and letting in the biggest bottoms of glass bottles for eyes. I believe that at the meeting called together to erect this, the Maori King movement was first mooted. In it Te Heuheu keeps all his household goods, such as plates, chairs, dishes, basins, wheat etc. [21: 365].

Smith was shown the site of the pa at Te Rapa (Fig. 4.4),

> …where in May 1846, the former Te Heuheu [Mananui], brother of the present one, was killed, with 60 of his people. It appears that the hot springs above had loosened the earth, stones etc., which, rushing down the course of a stream, had overwhelmed the pa in the night, covering the houses, fences, and everything in its way. The position of the pa is now occupied by a mass of yellow hard mud, with the tops of some houses just showing

> above. One of the houses had been dug out to get at the body of Te Heuheu…….It is believed he would have been saved, but an old woman (whom we saw at Pukawa, bent nearly double with age) called out to him that his child was in the house, when he ran back and was caught by the avalanche. The stream at present runs over the tops of some of the houses [21: 367–368].

Modern studies of the volcanic activity in the area above the village [7] show why this was not the first such catastrophic landslide in that area, nor would it be the last.

In 1931, two generations later, Ted Lattey, the grandson of one of Smith's companions, Charles Hursthouse, attempted to re-enact the central part of Smith's journey from Ongarue to Pukawa. Like Smith, Lattey did the trip the hard way—on foot and with four companions from Taranaki.

They could follow the Ellis and Burnand sawmill tramway (Chap. 5) up the Mangakahu Valley at first (16 km), but after that, getting through the bush was just as tough as Smith found it, especially where there were no wild cattle to make tracks. They missed Percy Smith's route, suggesting that the old Maori tracks were by then invisible. Relevant to this story is the report that the inland country was still sufficiently unknown that, on their return, the newly formed State Forest Service asked Lattey to describe the timber resources he had seen [6].

Hochstetter

In March 1859, Ferdinand von Hochstetter set off from Auckland on a long surveying trip into the interior, which he stated was at that time "topographically almost unknown" to Europeans. He was not the first to venture there, but Hochstetter was a skilled geologist, and he compiled a detailed and relevant account of the central North Island in the mid nineteenth century. He published an invaluable and detailed map on returning to Auckland (Fig. 4.5). It marks the Rangitoto Ranges, Titiraupenga and Pureora (labelled with its original name, "Hurakia"), as he observed them from neighbouring mountains to the west and south.

The published version of the map was accompanied by some notes explaining his intentions and his route.

> Since Dieffenbach's memorable travels in New Zealand…., no naturalist has visited the southern part of the Province of Auckland, so justly celebrated for its grand volcanic phenomena. …I started with a numerous suite, well provided for a long journey on foot, and for a campaign in that thinly inhabited country. On the 6th of March, 1859, near Maungatautari, I reached the main stream of the Waikato, flowing from the heart of the Island.
>
> I travelled along this river in the canoes of the natives…… and entering the Waipa, visited the Mission Station, and took a tour to the West to visit the Harbours of Whaingaroa, Aotea, and Kawhia. All those localities are of geological importance on account of the numerous localities for petrefactions. At Kawhia, I found, besides Belemnites, the first Ammonites found in New Zealand.
>
> From Kawhia, I took a tour inland to the Mokau district. Penetrating through numerous primitive forests, and traversing large mountain chains, I passed the springs of the Wanganui River in the Tuhua district, and on the 14th of April, our party arrived at the majestic Lake Taupo [11].

This laconic summary of the route gives no clue as to how challenging such a journey was at that time. But Hochstetter's full account goes into much greater detail, leaving no doubt about the effort required and the dangers he faced. Some sections of it are very valuable and relevant to our understanding of the appearance and history of the Pureora district only 160 years ago, starting with his first sight of the Rangitoto Range.

> [At] Orahiri,…. the river [Waipa] takes a south-easterly direction towards the Rangitoto-range. The valley here changes to an extensive plain covered with Kahikatea-forests; in the background rises the wood-clad Rangitoto-range forming the water-shed between the Waipa on one hand, and the Mokau and Wanganui rivers on the other. The river-pebbles lead us to suppose, that the Rangitoto-range consists of shale, sandstone and marl, and consequently is not of a volcanic origin [12: 335–36].

Fig. 4.5 Part of the map drawn by Hochstetter in 1859, showing his route from Kawhia to Pukawa. An "Overland Mail Track" is marked running north of Titiraupenga. *From Hochstetter (1867), Map 2, last (un-numbered) page*

He was right about that (Chap. 1). A week later, on the evening of April 8th, Hochstetter and his companions reached the Ongarue valley.

> Two valleys, bordered by picturesque mountains, meet together at Katiaho [near the present town of Ongarue], the Ongaruhe valley from the North, and the Mangakahu valley from the East. Between the two, opposite the settlement, arises the Ngariha mountain. The Ongaruhe is the main river; the Mangakahu only a small tributary; and Katiaho lies just opposite the junction of the two rivers on the right bank of the Ongaruhe….
>
> The terrace formation, so remarkable on the Waipa, is still more marked in …. all the valleys of the Upper Wanganui district. There are here three terraces in the valley and as many on the declivities of the bordering hills. The former are cut into thick beds of pumicestone gravel, which fill the

> bottom of the valley; the latter into trachyte-tuff, composing the hills and mountains on the sides of the valley. To the terraces on the sides of the valley correspond farther up the valley extensive table-lands covered with pumicestone, and everything indicates to the observer that he is drawing nearer and nearer to a powerful volcanic hearth, from which those huge masses of pumice-stone and trachyte-tuff are originating [12: 352–3].

He was absolutely right about that too, as modern geologists can explain in great detail (Box 1.1).

Hochstetter and his companions climbed Ngariha on 10 April 1856 to get a view of the volcanoes. He wrote a lyrical description of the view, and made a neat drawing of himself and his companions admiring it (Fig. 4.1).

Next day the party walked another 16 miles (26 km) to reach the western borders of the present Pureora Forest Park.

> At the mouth of the Mangakahu we crossed the Ongaruhe by means of canoes, and followed the river up the valley in an easterly direction on a miserable road, which, in a manner very fatiguing to the traveller, led continually up and down over the terraces, through swamps, over numerous small creeks, and several times across the river itself. .... At a distance of three miles, the valley turns South-East towards the Tuhua-mountain, the most prominent point in the whole country, about 3400 feet high with a broad platform and a steep descent on its South side ....
>
> ... The chief ornament of this landscape is the Hikurangi (i.e. ascending towards heaven), at the right between the Piaua and the Ongaruhe rivers, a volcanic cone of a very regular shape, rising from a very gently sloping base steeper and steeper to a height of 800–1000 feet, the top appearing as if cut smooth with a knife (Fig. 1.7). Dark woods cover its declivities. The top is said to contain a water basin at the bottom of a funnel-shaped crater, and on the South-Eastside two powerful mineral springs (probably chalybeate) bubble out. [we came to Petania]... a Maori village ... situated at the southern foot of the Tuhua-mountain, 754 feet above the level of the sea [12: 354–6].

The last leg of the journey to reach the lake took longer than expected, but despite the miserable conditions Hochstetter kept detailed notes and later wrote the first comprehensive European description of the unexplored forest country west of the lake, skirting the southern edge of what is now Pureora Forest Park.

> The distance from Petania to Lake Taupo is estimated a two days' journey. The road, however, is extremely difficult; it leads up and down from valley to valley, from mountain to mountain, across the ridges springing from the Tuhua-mountain in a southerly and southwesterly direction, and through dusky primeval forests. It traverses the sources of the Wanganui, and, ascending higher and higher, it finally reaches the watershed between the Wanganui and Lake Taupo. We were three whole days passing over this route. On the first day, after a most fatiguing passage through deep ravines cut into pumicestone gravel, we crossed the Takapuiraha range (1534 feet high), and encamped on the left bank of the Pungapunga river upon a beautiful grass-plain, called te Patate, 897 feet above the level of the sea.
>
> April 13.– We had now to scale the Puketapu, ...and thus gained an interesting view of the sources of the Wanganui, over a sombre mountain-country and wood-landscape, in the back-ground of which the Ruapahu loomed up in all its majesty, its peak wrapt in clouds. Southwest of the Ruapahu another volcanic cone, 3000 feet high, was visible; it was pointed out to me as Hauhanga. To the Northwest and West the Tuhua-mountain and the Hikurangi-cone were the most prominent points. In clear weather, Mt. Egmont also is said to be visible from here. ...
>
> After leaving the Puketapu we were continually in the bush; it seemed as though it would never come to an end. Up and down, from ridge to ridge, from dale to dale; we passed the Waipari, then the Waione, cold creeks, the water of which showed a temperature as low as 50° F. Again we had to climb up-hill, over roots and logs in the sombre dusk of the bush, the huge crowns of the tall forest trees shutting out the light of day, and the sky being moreover veiled by dark, dismal clouds of rain. The magnificent fern *Leptopteris hymenophylloides* grows in those damp woods with a extraordinary luxuriance, in the shape of the variety called *superba.*
>
> At length we came to a small creek flowing in a direction different from that of all the other creeks we had hitherto passed; it was the source of the Kuratao, running in a N.E. direction towards Lake Taupo, a sign that we had crossed the water-shed, and we hailed with joy the first indication of our having come quite close to our long looked-for destination. .... Of the lake, however, the sight of which we had expected to greet our longing eyes, there was as yet nothing to be seen But in its place two beautiful mountain-cones, the Kuharua and the Kakaramea, rose before us. We had reached a

> Pumicestone-plateau, called by the natives Moerangi, and I was greatly surprised at finding the result of my barometrical observations to show a height of 2188 feet. We pitched our tents by the banks of the Kuratao river, which, cutting through the pumicestone, forms a ravine about 100 feet deep with triple terraces.
>
> April 14.– The distance from the lake was greater than we had supposed….we struck once more the Kuratao valley at Whakaironui, a potato-plantation at the margin of the plateau. …. after Poaru, we plodded along through marshy woods, we came to an open eminence, whence we had the first view of the lake. Like a sea it lay there spread out in the distance, without our being able to discern the opposite shore in consequence of the murky weather. Gently sloping down-hill, and passing along the foot of the Kuharua… to an elevation, from whence we saw the celebrated pah Pukawa, the residence of the great Maori-Chieftain Te Heuheu [Iwikau], situated beneath our feet at the margin of the lake [12: 357–359].

Hochstetter lodged at the missionary station run by Rev. Thomas Grace near Pukawa. He described the scene at the southern end of the lake:

> The West shore of the lake is formed by vertical bluffs of rocks, which near Karangahape, at a promontory projecting far into the lake, attain a height of more than 1000 feet. Upon that side of the lake a landing is practicable only at the few points where little rivers empty into the lake. The long-stretched wooded ridges of the Rangitoto and Tuhua [= Hauhungaroa] mountains, rising to a height of 3000 feet above the level of the sea, shut out the horizon in a northwesterly direction, and only one point attracts the attention by its rather singular form,—I am speaking of the Titiraupenga mountain, from the summit of which a bare pyramid towers up, resembling a ruined castle [12: 366].

Hochstetter spent five days at Pukawa, sketching a detailed map of the local geology and benefitting from Grace's "exact local knowledge" on expeditions, both for mapping field work and for visiting the pa at Te Rapa. Hochstetter was deeply impressed by the mission station and its Pakeha inhabitants.

> The generous hospitality of the Rev. Mr. Grace and his amiable family…..[and] the arrangements of the excellent lady of the house made us utterly forget, that we were sojourning in the remotest interior of New Zealand. The picture of that happy family circle, blessed with a number of blooming children, was truly calculated to awaken the most grateful emotions [12: 360].

One of his expeditions took him along the lake shore to Te Rapa, the geothermally active area in the hills above Te Heuheu's original village, and the site of the natural disaster that had destroyed it and killed his brother Te Heuheu Mananui in 1846 (Fig. 4.4).

> East of Pukawa, in the rear of a steep promontory, a small bay extends south. The western-shore of this cove is formed by vertical bluffs consisting of alternating horizontal banks of trachyte, trachytic conglomerate and tuff. A small creek, the Waihi, plunges quite close to the South-end of the cove, in a magnificent fall about 150 feet high over this bluff of rocks. At this cascade the mountains recede somewhat from the lake; and here already, from the conglomerate-layers forming the beach, hot water, of 125–153 °F., is seen bubbling forth. By conducting this water into artificial basins, the natives have prepared several bathing-places, the water in which showed a temperature of 93° F. … Above these springs on the side of the mountain, probably 500 feet above the lake, steam issues from innumerable places. The whole Northside of the Kakaramea mountain seems to have been boiled soft by hot steam, and to be on the point of falling in. From every crack and cleft on that side of the mountain hot steam and boiling water are streaming forth with a continual fizzing noise, as though hundreds of steam-engines were in motion. Those steaming fissures in the mountain-side, upon which every stone is decomposed into reddish clay, the natives call Hipaoa, i.e. the chimneys, and it was at the foot of that mountain-side, that in the year 1846 the village Te Rapa was overwhelmed by an avalanche of mud, and the great [Mananui] Te Heuheu perished. [12: 367–8]

The Mission House still stands just outside Pukawa, now restored and privately owned. The cliffs above the modern village of Waihi still steam and gently shudder with geothermal activity (Fig. 4.4).

The landslide mentioned by both Smith and Hochstetter was in fact only one of three recorded there (the other two were in about 1780 and in 1910), which between them killed 200 people. The most recent alarm was on 29 June 2009, when the village was evacuated after a series of small earthquakes led to fears of another landslide, but within days the residents were allowed to return [17].

Hochstetter's map, "The Southern Part of the Province of Auckland" (Fig. 4.5), labels the area between Titiraupenga and the Waikato River as "Volcanic Tableland 2000 feet high consisting of trachytic rocks thickly covered with forests and unexplored" [11]. A track is marked with the inscription "Overland mail track" between the Rangitoto Ranges and the Waikato River. This was the route of the Maori track by which local Maori carried the mail between 1857 and 1863, until the service was stopped by war in the Waikato.

Hochstetter was also an observant naturalist, and he made large collections of plants and animals representing the unfamiliar species he found along the way. He found a unique species of frog (now named Hochstetter's frog) and a giant snail. Unlike some other explorers, he described the birds as very abundant in some places:

> The country here [Ohura] seems to abound in birds; for thousands of Tuis *(Prosthemadera Novae Zelandiae)*, which had perched themselves upon a group of Kahikatea pines, gave here a concert, such as we hear in Germany from the starlings, when they visit the vineyards in autumn [12: 350].

Even as late as 1883, this area was still largely unknown to Pakeha and, some claimed, to Maori also. On the other hand, local Maori at Taupo may well have had other motives for telling Pakeha enquirers "that it was covered in dense bush, and that it would be impossible to travel through it for any distance, and especially on account of the numerous rivers and creeks that would have to be crossed" [23: Chap. 3].

Kerry-Nicholls

In 1882 J.H. Kerry-Nicholls described the landscape along his route north-west from the Waihaha area of Lake Taupo. He was one of the first to attempt an exploration of the recently opened King Country, which had been closed to Europeans for the previous 20 years, so his account is much later than those of the first European explorers. The much diminished local population still offered the travellers traditional Maori hospitality.

> At about a mile distant from the Waihaha River.... we arrived at Kahakaharoa [Fig. 3.2, right], a small pa situated on a winding mountain stream called Te Pikopiko. At one time there had been a considerable native settlement here, but now the whole place was nearly abandoned [14 318].
>
> Here, besides the usual diet of pork and potatoes, we were treated to roast kiwi....these birds... are still very plentiful in the dense, unfrequented ranges of the King Country [14: 319–320].

He was one of the few who attempted the journey on horseback, but even by then, riding was not easy.

> Journeying still further on, we crossed the Te Tihoi Plains, a fine tract of open country extending around the mountains of Titiraupenga as far north as the banks of the Waikato River, and thence north-westerly to the Te Toto [Rangitoto] Ranges. ....To the north-east high, forest-clad mountains rose up one above the other in the direction of Ouranui [Oruanui] and the valley of the Waikato, while to the west were rugged, forest-clad ranges, crowned by the towering form of Titiraupenga.
>
> This magnificent mountain, which is one of the highest peaks in the northern portion of the King Country, rises to an altitude of some 4000 feet above the level of the sea. It assumes in general outline the formation of an extensive cone, with a broad base and long, sweeping sides, while its summit is surmounted by a gigantic pinnacle of rock, of a pointed form, and which serves with the great mountain as a conspicuous landmark all over the surrounding country. It is covered from base to summit with dense forests and its enormous gorges and deep ravines give rise to many streams and rivers [14: 319–320].

In one of the ravines flowed the Mangakowiriwiri, "a tremendous gorge of the mountain, flanked on either side by tall precipices of rock.... Looking down into the deep fissure we could just see the silver streak of water foaming nearly a hundred feet below". The stream was crossed "by means of a very narrow and very primitive footway, which the natives told us was known as the 'Bridge of God'" [14: 321].

Between the Mangakowiriwiri and Mangakino Rivers was "open undulating country covered with a luxuriant growth of tussock and other native grasses". The Mangakino was crossed by swimming the horses across a ford "like ducks", and the travellers continued on to cross the Waipapa River and Rangitoto ranges into the Waipa valley:

> We gained the crossing place by a steep winding descent, the mountains with their rocky bluffs on the opposite side of the river being clothed with a dense vegetation of giant trees, while to the right of the track by which we had to descend was a small mountain forming a complete cone, and which was clothed from base to summit with a luxuriant growth of fern and tall manuka. ....It took us several hours to traverse the Te Toto [Rangitoto] ranges, the track winding about in every direction with deep ravines on either side. Here the vegetation was of the most luxuriant and varied order, but the enormous roots of the great trees made riding very difficult [14: 324–5].

Te Heuheu Iwikau and the King Movement

Maori tribes were generally independent, but they also had a form of loose association of groups linked by ancestry and marriage.

> By 1840, the region around Lake Taupo was peopled by a number of different hapu led by chiefs who operated independently of one another, but not in total isolation. There was a form of confederation of the various hapu whose lineages could be traced to Tuwharetoa.
>
> It was the custom with Ngati Tuwharetoa, from the time of Turangitukua until the close of the nineteenth century, to select from a panel of high-born men the paramount chief and war-leader of the tribe. This rank was not necessarily a hereditary right. It was conferred by the tribe on the most suitable man, irrespective of seniority. The ariki of the tribe were the direct descendants of the senior male line from the tribal ancestor Tuwharetoa himself, and it was the senior ariki's prerogative on behalf of the people to install the paramount chief. The rank of ariki could not be removed as was the case of the paramount chieftainship [23: Chap. 2].

After the death of Mananui in 1846, his younger brother Te Heuheu Iwikau succeeded him and led Ngati Tuwharetoa until his death in 1862. Iwikau went to live at Pukawa, in the territory of Ngati Manunui, the hapu of his senior wife Ruingarangi (or Morunga). Pukawa became his principal pa, where he was visited by several prominent European explorers. Iwikau had long requested a missionary to live there, and in April 1855 Thomas Grace settled with his family at Pukawa, under the protection of Iwikau.

According to the geologist Ferdinand von Hochstetter, who met Iwikau a few years later, the Ngati Tuwharetoa leader was 'not averse to Christianity' but feared that baptism would bring about a loss of his influence and authority. He would also be obliged to give up several of his wives before baptism. Although he never became a Christian, Iwikau attended church services regularly, and Grace respected him as 'friend and protector'.

Te Heuheu Iwikau had signed the Treaty of Waitangi in 1840, although his older brother Mananui had not done so. He had consented to the Pakeha coming to New Zealand, but he objected to land sales to Europeans, and therefore lent his significant authority to the King Movement.

Hochstetter was impressed by Te Heuheu during his visit in 1859:

> Long ago I had heard of the great and mighty Te Heuheu Iwikau, residing in Pukawa at Lake Taupo. His name is known wherever the Maori language is spoken; for he belongs to one of the oldest and most renowned noble families of the country; and is numbered among the heroes and demigods of his people. He had been pictured to me as a man of considerable talents, as the best and worst fellow at the same time; as proud, shrewd, generous; as a mysterious medley of modern civilization and ancient heathenism [12: 361].

In 1850 Governor George Grey had visited Pukawa, admired Te Heuheu's richly carved food storehouse, and remarked that all chiefs should have such storehouses as a sign of their standing and generosity. When later this storehouse was destroyed by fire, Iwikau, remembering Grey's words, set out to build another one, larger and more elaborately ornamented.

After 4 years of building, it was completed in 1855, and named Hinana. To demonstrate his mana, which was being challenged by some of Ngati Te Aho, Iwikau invited people from all the major tribes to its opening in November 1856 [9].

Under the increasing impact of Pakeha intrusion during the 1850s, there was growing Maori concern about the alienation of land and the effects of the advance of Pakeha settlement. Resistance to land sales was building up in Taranaki, Waikato and elsewhere. Discussion during a series of tribal gatherings, from 1853 on, led to the idea of some sort of great confederation of tribes to protect their lands from further alienation.

Iwikau had been seeking both to restrain Maori protest and to support the growing grievances over the loss of land. He used the opening of Hinana to convene a meeting at Pukawa to debate the idea of electing a Maori king. Te Heuheu was an ardent proponent of Maori nationalism, and he used his influence to encourage the King Movement.

Through the later 1850s there was intense debate about who should lead such a confederation, and about what form of Maori local government should be set up. These ideas developed into what became the Kingitanga, or King Movement. There was keen support for the concept throughout the central North Island, although less elsewhere.

At the large gathering at Pukawa in 1856 (Chap. 3), nearly every tribal group in the North Island was represented. It was resolved that Taupo would be in the centre of a district extending to Waikato and Hauraki in the north, and Taranaki, Whanganui and Rangitikei in the south, in which no land sales would be allowed.

Tokena Kerehi gave a first-hand description of the hui to a sitting of the Native Land Court, as recorded in the Waikato Minute Book and cited by the Waitangi Tribunal:

> I was present at Hinana meeting, that was our meeting, all of Taupo. All the Taupo hapus, including N'Wairangi prepared food for it. The object of this meeting was to elect Potatau King. The Arawa wished Te Heuheu Iwikau to take that position, N'Tuwharetoa and N'Raukawa concurred [23: 4.2].

So there was strong inter-tribal support for Te Heuheu Iwikau to become King. He had keenly supported earlier discussions about a confederation of tribes.

But this was a difficult situation for him, because Te Heuheu also realised it might not be in the best interests of Tuwharetoa for him to accept such a position. He refused, and instead supported the installation of the paramount chief of Tainui, Potatau Te Wherowhero, as the first Maori king. Te Wherewhero, equally cautious, consulted with his Maniapoto relatives at a special meeting held in 1857 at Haurua, a marae just south from Otorohanga. They gave their approval, and a memorial besides the road records the decision. At Ngaruawahia in 1858, Te Wherowhero became the recognised first head of the Kingitanga movement until his death in 1860, when he was succeeded by Tawhiao (Fig. 4.6).

War

Within 2 years after the pivotal 1856 meeting in his home village of Pukawa, Thomas Grace was becoming apprehensive about the effects of the developing King Movement on native life. He wrote from Pukawa to the Church Missionary Society in London, referring to the Constitution Act 1852:

> The constitution, which has been given to this country, has placed the Maoris in a far worse position than they were, seeing they have no share in any of the representation.
>
> Here in New Zealand we have about four-fifths of the population, British subjects and lords of the soil, and paying the greatest portion of the revenue, cut off from all shares in the representation of the country, either in person or by proxy.

Fig. 4.6 *Left* Te Wherowhero was the first Maori King, from 1858 to 1860. The meeting to confirm his nomination was held at Haurua marae in 1857, near this roadside memorial beside SH 4 south of Otorohanga. *C.M. King (2014). Right* Tawhiao, the second Maori King. Photo by Pulman, E. (1882), taken in Jan 1882 when Tawhiao visited Auckland. *Auckland War Memorial Museum—Tāmaki Paenga Hira. PH-RES-425*

> Surely this is a strange state of things to exist. If a separate house were formed for Maori representatives, there is little doubt that, with a few official leaders appointed direct from home as protectors, the Maori chiefs would be found quite able to take their full share in the representation.
>
> If we deny them the right of British subjects, and thereby ourselves break the Treaty of Waitangi, we should not be astonished if they seek protection for themselves [by setting up a Maori King] [23: 4.4].

Grace could see that a confrontation between the Kingitanga and the government was becoming inevitable. When it happened, he commented in a letter to the Church Missionary Society in London:

> The real cause of the war is, without doubt, the constant coercion that the Maoris have been subjected to in order to induce them to part with their lands. The Government professes not to buy lands, the ownership of which is in dispute, yet nearly all the wars and quarrels that of late years have taken place, have been on this very subject [23: 4.4].

Relationships between Pakeha settlers and the tribes of the Kingitanga confederation became more strained. War erupted in Taranaki in 1860 over a different but related issue, the government purchase of a block of land at Waitara.

Te Heuheu Iwikau had maintained a policy of neutrality in the Taranaki fighting, in which Waikato and Maniapoto tribes participated. He

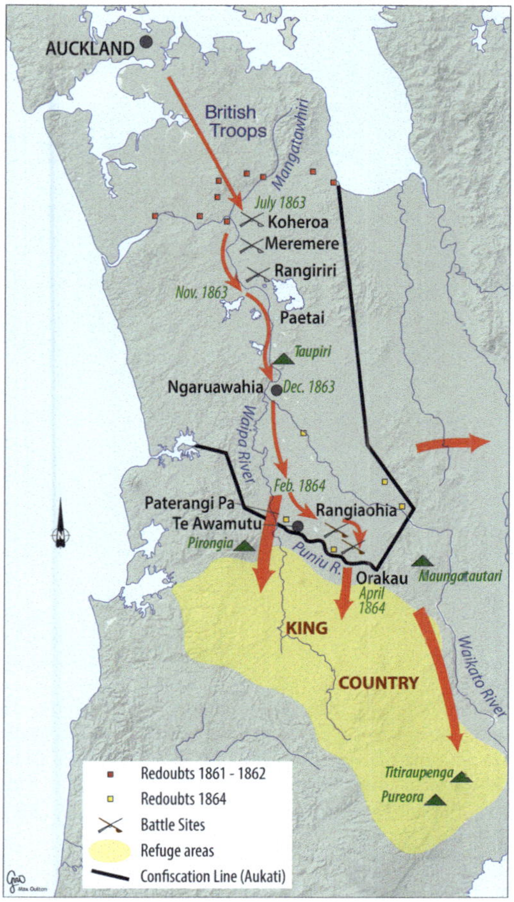

Fig. 4.7 The invasion of the Waikato by British troops in 1863–64, and the battle sites of the Waikato war. After the deciding battle at Orakau, the survivors and their whanau retreated into the King Country, while Waikato lands within the Aukati line were confiscated. *Redrawn by Max Oulton from Waitangi Tribunal* (*1993: Fig. 4.2*)

tried to prevent Ngati Tuwharetoa from joining, fearing that the tribe's lands would be threatened. But in 1862 he died, to be succeeded by Te Heuheu Horonuku, his nephew and son of Te Heuheu Mananui who had been killed in the landslide of 1846. Horonuku later took the name Te Heuheu Tukino.

In July 1863 British imperial troops, led by Lieutenant-General Duncan Cameron, crossed the Mangatawhiri Stream (Fig. 4.7), a tributary of the Waikato River which had, since 1858, denoted the northern boundary of the Kingitanga. During the following "Waikato Campaign", there was a series of battles as the invading troops progressed up the river, reaching Ngaruawahia in early December. Waikato tribes and their allies retreated to Maungatautari and the Waipa valley.

In late 1863, Te Heuheu Horonuku Tukino gathered a force of over 200 men to go to the assistance of Waikato, a change from the neutral policy hitherto maintained by Ngati Tuwharetoa under Iwikau. Another force, mainly from Ngati Te Kohera and Ngati Parekawa of the Tihoi-Pouakani area, led by Te Paerata and Te Kohika, joined with Ngati Maniapoto and others in the final stages of the Waikato war.

The British troops had advanced up the Waipa River, by-passed the pa at Paterangi and attacked Rangiaohia. By the end of March 1864 the troops were in occupation of the Te Awamutu area, while many Waikato and Maniapoto retreated south of the Puniu River. As a last-ditch stand, they

> ….decided to build a pa at Orakau. The construction work was observed by the British and an attack begun. By the time Horonuku and his party arrived they could do nothing but look on during the three day battle (31 March–2 April 1864) that followed [23: 4.3].

At one point, a cease fire was called by Cameron, and the defenders were asked if they would surrender. The response from Rewi Maniapoto was that they would fight on in spite of the lack of water and limited supply of food and ammunition, "ake, ake, ake" (for ever and ever and ever).

The firing began again, and the British troops advanced on the pa The defenders formed themselves into a tight group with the women and children in the middle, broke through part of the earth works and rushed out. The firing continued as British troops stormed the pa.

> The Maori survivors made their way through the swamp below, sheltered by scrub, and retreated south of the Puniu river…The Raukawa and Tuwharetoa survivors, along with Tuhoe people, retreated up the Waikato valley towards Titiraupenga…..There had been no fighting on Tuwharetoa lands, but west Taupo and the upper Waipa valley became refuge areas for dispossessed tribes [23: 4.3].

After Orakau, Te Arawa saw their interests lay in some form of co-operation. Ngati Tuwharetoa had supported the concept of the Kingitanga, but had not participated in the fighting until the final stages at Orakau. The tribes fought to protect their land, but did not see themselves as "rebels" in the Pakeha sense. Few Pakeha saw the situation as the missionary Thomas Grace saw it, or were willing to acknowledge Maori concerns and attitudes.

Te Rohe Potae

Cameron's invasion of the Waikato was more or less complete after the fall of Orakau. The Government then confiscated some 2.1 million acres (849,870 ha) of Waikato land from its defeated but still aggrieved Maori defenders. To prevent further bloodshed and confiscations, the Government agreed to an aukati (boundary) line along the Puniu River (south of Te Awamutu, between Pirongia and Maungatautari), dividing the confiscated Waikato land, the Raupatu, from Ngati Maniapoto land under the control of King Tawhiao, where he and his supporters were free to administer their own affairs. Pakeha were forbidden to travel south of the aukati. The Government further agreed not to pursue King Tawhiao or the defeated warriors beyond it, and no land within Ngati Maniapoto territory was confiscated.

The line of the boundary was supposedly determined by an incident during a meeting in 1864 between King Tawhiao and Governor George Grey. They could not decide how best to separate Maori and European claimants to the land, until King Tawhiao threw his hat down on a map. They agreed that Ngati Maniapoto should be allowed to assert Maori sovereignty over all the area covered by the crown and brims of the hat, Te Rohe Potae. The name loosely translates as 'the area of the hat'; it is known to Europeans as the King Country, centred on Taumarunui, where a memorial to the agreement stands, topped by a hat (Fig. 4.8). Te Rohe Potae became a de facto state within a state, effectively out of bounds to Europeans.

Fig. 4.8 The original outline of the King Country, Te Rohe Potae, was supposed to have been decided from the area of a map covered by King Tawhaio's hat. (Te Rohe Potae means literally "the boundary of the hat"). This memorial in Taumarunui records the event. *C.M. King (2014)*

Te Rohe Potae originally stretched from the west coast eastward to the middle of Lake Taupo, and south from the Puniu River to beyond Taumarunui, although its boundaries changed over the years (Chap. 5). The west Taupo ranges were in the middle of it. The heavy casualties among Tihoi and Pouakani warriors participating in the defence of Orakau, and the immigration of the defeated survivors escaping retribution, had severe social consequences for the people of the King Country after 1864. Their only compensation was that these lands were effectively a self-governing province, largely protected from the government and settlers until the early 1880s.

The agreement ended the war, but did not guarantee peace instead. On the one hand, the whole concept of the Rohe Potae, a separate system of Maori government in a district outside of the control of British colonial administrators and military, was unacceptable to the majority of Pakeha. The Government became exasperated by endless disputes over local boundaries which impeded land sales, so it insisted that traditional collective ownership of land had to be transferred into individual legal titles recognisable by the Crown. That would enable undeniable legal purchases of Maori land, after investigation and survey of land holdings by the newly granted authority of the Native Land Court established in 1865 (the name was changed in 1954 to the Maori Land Court).

On the other hand, although the operations of the Native Land Court were supposed to act in the interests of Maori, to define who actually was entitled to profit from selling which areas, the Court often enforced legislation that served only to separate Maori from their land. To defend their shared ownership of lands, tribal elders had to attend prolonged Native Land Court hearings at locations far from their homes, and pay for costly surveys and legal proceedings unfamiliar to them. These led to arguments over mana and territory, social disruption, massive debts, costly mistakes on survey boundaries and expensive litigation. Some of these mistakes later had a serious effect on the development of Pureora Forest Park (Chap. 11).

There was nothing in the Treaty of Waitangi which required the transmuting of traditional forms of land tenure into titles recognisable in British law. By imposing requirements for survey and associated costs, fees for investigation of title in the Native Land Court, and other costs such as food and accommodation while attending lengthy court sittings, Maori were forced into debt. When the debts were called in, Maori paid in land [23: 307–8].

The tribes supporting the King Movement hated this process, and were determined to stop settlers and Pakeha speculators buying up any of their land. For a while, Te Rohe Potae allowed them to settle their own disputes without interference from the Native Land Court, but it did not last. In the early 1870s, the Government resumed purchases of Maori land, and the original boundaries of Te Rohe Potae rapidly contracted (see Fig. 5.1).

Continuing conflicts of loyalties between those supporting the Government answerable to Queen Victoria ("Queenites") and those supporting the King Movement ("Kingites") disrupted traditional hunting and gardening, and forced the emigration of young people in search of employment. A third group, the basically pacifist Pai Marire (the "Hauhau rebels") had an important cultural centre near Pureora at Tiroa (Box 4.1),

The breakdown of traditional food gathering and social structures were beginning to cause obvious depopulation of central North Island Maori by the late 1850s (Chap. 3). War and disease accelerated the losses, so that later visitors often found formerly occupied areas deserted.

> Their cultivations and enclosures and the settlements generally have a neglected appearance, and one meets everywhere with strong proofs that the population is very rapidly decreasing [8].

The King Movement protected the Pureora district from European settlement and exploitation for almost 20 years after the 1863–64 war. But it never represented all Maori, and over time it lost its mana as Maori lost effective control of the political process. Then, they could no longer defend their traditional lands in Te Rohe Potae from newcomers who knew little of its history and significance to them (Chap. 5).

Te Kooti

In 1869 government troops came into the Taupo district in pursuit of Te Kooti Rikirangi. This mission-educated leader, prophet and founder of the Ringatu church, had been arrested on suspicion of sympathy with "Hauhau rebels" in the Gisborne district in 1865 and transported without trial, with other Maori prisoners, to the Chatham Islands. In July 1868 he and a group of followers captured a ship, the *Rifleman*, and now well armed, sailed back to Poverty Bay. There followed a series of raids on local settlers, and then Te Kooti retreated into rugged bush country, where he was sheltered by the Tuhoe of Te Urewera.

During 1869 Te Kooti was pursued by government troops through Te Urewera. He came out of the bush to cross the southern Kaingaroa plains. On 7 June, he destroyed a small garrison at Opepe [on the present Napier-Taupo road, where a DOC sign marks this historic site] and carried on to eastern and southern Taupo. He stayed some time at Tokaanu and then went on to Te Kuiti, returning to Tokaanu in September. There were more skirmishes in the Rotoaira area in September, and an engagement at Te Porere pa on the lower slopes of Tongariro on 30 October. Te Kooti retreated into the west Taupo bush, where he remained for several weeks (Box 4.1).

Box 4.1 Miringa Te Kakara Te Tiroa
Miringa Te Kakara Te Tiroa, a magnificent sacred house at Tiroa, was of considerable historic value to Maori [13], especially Pai Marire and Te Kooti. The main building was constructed, entirely without nails, in cruciform shape 54 feet (16.5 m) square, whose four arms were oriented north, south, east and west. Unsawn split timber provided the main supports for the walls, and totara bark filled the spaces between them and covered the roof (Fig. 4.9).

In the 1950s the posts marking the four points of the compass around the old building still had legible writing on them. The writing used the Ringatu language, which has extra letters and is part of a faith and culture associated with Te Kooti, who took refuge from British soldiers in the Pureora area in 1869. Next to it was an eating house in octagonal form, with walls made of closely packed punga (tree-fern) logs.

In the 1930s, renovations were done to Miringa Te Kakara and a huge celebration was organised with the support of local sawmilling company Ellis and Burnand, whose Directors were always generous with help for community projects [1: 193].

When Ivan Frost was OC at Pureora, more work was needed, and Ivan gave permission for the main building to be re-roofed with totara bark from a current Pureora logging site. Sadly, the main building was burned down in 1983. But the octagonal eating house survives, and has recently been fully restored by the Rereahu people [1: 193].

The troops pursued him across the Tihoi area and camped at Waimahana, a kainga on the Waikato River north of Mokai. A pa "called Tewe, apparently near Tihoi" was attacked by British troops. Te Kooti escaped his pursuers, and eventually returned to Te Urewera. Te Kooti had no further association with the Pureora area, but continued to elude capture. In the 1870s he returned to live in the north of Te Rohe Potae, at Otewa, beside the Waipa near Otorohanga, under the protection of King Tawhiao. In 1883 he was granted a pardon, and became much more co-operative with the Government. When the surveyor Charles Hursthouse was kidnapped by Maori opposed to the construction of the North Island Main Trunk Railway, Te Kooti helped to rescue him (Chap. 5).

The Beginnings of European Settlement

Until well after the end of the Waikato war, Pakeha settlers were few and far between in the Taupo district. The Grace family had abandoned their mission station at Pukawa in October 1863 [5]. When Meade visited the area in 1864 [15], he found a government medical officer, Dr. Hooper, at Oruanui, north of Taupo, who said he had not seen another European in the district for 2 years.

From the late 1860s, a few government and private land purchase officers appeared, and on 28 October 1867 the first sitting of the Native

Fig. 4.9 Miringa Te Kakara Te Tiroa was an important meeting house and spiritual home for the Rereahu people. It stood at Tiroa between Pureora and Benneydale, and was visited by Te Kooti while he was a fugitive from British troops in 1869. In the foreground stands a niu pole (a Hauhau ceremonial boundary marker) 16 July 1977. *Graeme Reinhardt*

Land Court was held at Oruanui. Negotiations for sale of Ngati Tuwharetoa lands had already begun [23: Chaps. 4 and 5].

The pursuit of Te Kooti in 1869 led to the establishment of an armed constabulary station at Taupo, and the construction of redoubts. In 1870 there were 30 men at Taupo, 180 at Opepe and 40 at Runanga on the Taupo-Napier route. The redoubt built at Taupo became known as Tapuaeharuru (the name of Poihipi Tukairangi's village on the opposite bank of the Waikato River) and was the base for military activities and European government in the region.

The armed constabulary provided the focus for a small Pakeha settlement and associated activities. They planted gardens, repaired the mill at Tokaanu and traded in oats and potatoes with the Tokaanu Maori. Taupo was also a strategic staging post on the mail route and road which was constructed in the 1870s from the Waikato via Atiamuri to Napier.

In due course, Pakeha businesses followed the military families, and by 1878 the number of Europeans living in Taupo county had reached 95, compared with (as near as could be determined) at least 805 Maori [25]. None of these represented any serious incursion of Pakeha into the forested western hill country of the future Pureora Forest Park. But eventually the isolationist policy of the King Movement had to end.

References

1. Anderson, K. (2008). Maoriland sawmillers. Ellis & Burnand Ltd, sawmillers and timber merchants of Mangapehi. Manurewa: Ken Anderson Family.
2. Angas, G. F. (1847). Savage life and scenes in Australia and New Zealand, Vol. 2. London: Smith, Elder and Co.
3. Angas, G. F. (1847). The New Zealanders illustrated. London: Thomas M'Lean.
4. Bell, G. (1976). Ernst Dieffenbach: Rebel and humanist. Palmerston North: Dunmore Press.
5. Brittan, S. J., Grace, G. F., Grace, C. W., & Grace, A. V. (Eds.) (1928). A pioneer missionary among the Maoris 1850–1879: Being letters and journals of Thomas Samuel Grace. Palmerston North: G. H. Bennett & Co.
6. Cooke, R. (1991). Ted's tough trek to Tokaanu. Roll Back the Years, 7, 789–797.
7. Crozier, M., Hancox, G., Dellow, G., & Ferrin, N. (2008). Slip sliding away. In I. J. Graham (Ed.), A continent on the move: New Zealand geoscience into

the 21st Century (p. 198–201). Geological Society of New Zealand, Miscellaneous Publication 24.
8. Cussen, L. (1884). Mr Lawrence Cussen to the Assistant Surveyor-General. Appendices to the Journals of the House of Representatives, 1884 II C1, 29–32.
9. Dictionary of New Zealand biography. Biography of Te Heuheu Tukino III, Iwikau. Last updated 18 July 2012. Available from: http://www.TeAra.govt.nz/en/biographies/1t32/1.
10. Dieffenbach, E. (1843). Travels in New Zealand, Vol. I, With contributions to the geography, botany, and natural history of that country. Christchurch: Capper Reprints.
11. von Hochstetter, F. (1864). The geology of New Zealand: Explanation of the maps. Auckland: T. Delattre.
12. von Hochstetter, F. (1867). New Zealand, its physical geography geology and natural history. Stuttgart: J G Cotta.
13. Hunt, C. G. (1959). Maringa Te Kakara village. Journal of the Polynesian Society, 68, 1–7.
14. Kerry-Nicholls, J. H. (1884). The King Country, or explorations in New Zealand—A narrative of 600 miles of travel through Maoriland. London: Sampson Low & Co.
15. Meade, H. (1870). A ride through the disturbed districts of New Zealand. London: John Murray.
16. Moon, P. (2014). The voyagers: Remarkable European explorations of New Zealand. Auckland: Penguin.
17. New Zealand Herald, 2 July 2009.
18. Polack, J. S. (1838). New Zealand: Being a narrative of travels and adventures during a residence in that country between the years 1831 and 1837. London: Richard Bentley.
19. Reischek, A. (1930). Yesterdays in Maoriland. Auckland: Wilson and Horton.
20. Rochfort, J. (1853). The adventures of a surveyor in New Zealand. London: David Bogue.
21. Smith, P. (1858). Notes of a journey from Taranaki to Mokau, Taupo, Rotomahana, Tarawera, and Rangitikei, 1858. In N. M. Taylor (Ed.), Early travellers in New Zealand (p. 351–386). Oxford: Clarendon Press.
22. Taylor, N. M. (1959). Early travellers in New Zealand. Oxford: Clarendon Press.
23. Waitangi Tribunal (1993). The Pouakani report. Wellington: Waitangi Tribunal.
24. Wilkes, O., Gaukrodger, J., & Ritchie, N. (2011). Chronology of Pureora. Hamilton: Department of Conservation.
25. Williams, A. (1988). Land and lake: Taupo Maori economy to 1860. MA thesis, University of Auckland.
26. Young, D. (2004). Our islands, our selves. Dunedin: University of Otago Press.

Opening the Way for Logging, 1880–1945

5

C.M. King and N.A. Ritchie

Abstract

This chapter covers the opening of the King Country; the completion of the North Island Main Trunk Railway along the western side of the future park in 1908, and the effect the railway had on stimulating the clearing of forest and the development of logging and farming by opening access to markets for timber and farm produce. It describes the histories of some of the earliest sawmills; the development of the milling industry, and how the logging tramlines gradually brought the forest edge closer and closer to the site of the future Pureora village. It summarises the pre-war histories of the local mills significant to the later history of Pureora, and the rise and fall in their productivity.

Keywords

The end of Te Rohe Potae • Tongariro National Park • North Island Main Trunk Railway • Cussen • Hursthouse • Rochfort • Early land surveys • Timber mills • Ongarue • Mangapehi • Bush tramways • Steam lokeys • Piropiro, Maraeroa C, A3B2 • Sawmillers Ellis and Burnand, Marton Sash and Door • Waikato coal seams • Benneydale coal mines

Negotiations

From the first days of European settlement, but particularly after 1870, there was a growing demand for timber and building materials to meet the construction needs of a developing country. New Zealand's railway system began in 1863, but it was not until 1870, when the Immigration and Public Works Act proposed by Colonial Treasurer Julius Vogel was enacted, that real

C.M. King (✉)
School of Science, University of Waikato, Private Bag 3105, Hamilton 3240, New Zealand
e-mail: cmking@waikato.ac.nz

N.A. Ritchie
Department of Conservation, Te Papa Atawhai, Hamilton Office, Private Bag 3072, Hamilton 3240, New Zealand
e-mail: nritchie@doc.govt.nz

C.M. King et al. (eds.), *The Drama of Conservation*,
DOI 10.1007/978-3-319-18410-4_5

progress was made towards accessing the huge timber resources of remote areas. Serious railway works began in Auckland in 1872 [11].

The Public Works Act was part of a large-scale scheme to construct roads and railways to bring settlement, agriculture and industry into the more remote 'unproductive' parts of the country and in the process to 'civilise' and counterbalance rather than defeat the Maori [6, 24].

Vogel's policy involved overseas borrowing of £10 million ($20 M) over a period of ten years, to be spent on assisted immigration and a communications network throughout the country. The Public Works Department (PWD) was established, and after the abolition of the provinces in 1876, the entire national railway network came under the operational control of the PWD. The programme expanded rapidly, and by 1880, the railheads gradually extending the North Island Main Trunk Railway (NIMTR) had reached southward to Te Awamutu from Auckland, and northwards to Manawatu from Wellington.

In between was Te Rohe Potae, a rugged, heavily forested land under the control of Maori, and out of bounds to Europeans. One hundred miles of track from Auckland to Te Awamutu had been constructed over the eight years 1872–80, but "not another inch was to be built southwards for the next five years while the Native Minister was knocking on the door of the King Country" [11: 73].

Over time, some Ngati Maniapoto leaders were becoming convinced that there were benefits for their people in allowing the King Country to be surveyed and settled by Europeans. King Tawhiao and his people were totally opposed to the idea, but the Government sidestepped the King and began direct negotiations with the tribes who actually owned the land needed for the railway [13]. Te Heuheu Horonuku Tukino was a powerful influence on local politics and on negotiations with the Pakeha.

Ngati Tuwharetoa, Ngati Maniapoto, Ngati Raukawa and the Whanganui tribes banded together to protect their interests in the western hill country. They agreed to a survey of about 1.5 million hectares of land south of the Puniu River (then the northern boundary of Te Rohe Potae), and also allowed the prospecting survey for the proposed section of the Main Trunk Railway through Te Rohe Potae to proceed. In return they wanted the government to agree to certain conditions.

Their proposals were listed in an 1883 petition presented by Ngati Maniapoto chiefs Wahanui, Rewi Maniapoto and other tribal leaders at a meeting with John Bryce, Minister of Native Affairs. The petition requested a general amnesty for refugees from the recent war, and specifically for Te Kooti; that the Native Land Court be prohibited from operating in Te Rohe Potae; that Parliament pass a special law to prevent the King Country from ever being sold out of Maori ownership, and another law to ban the sale of alcohol there; and that Maori be allowed to fix the boundaries of—and intra-tribal boundaries within—the area described in the petition, about 142,000 ha (Fig. 5.1; [21, 34]).

Some agreements were reached. First, the region was indeed made 'dry'—the sale of liquor was prohibited throughout the King Country for almost 70 years from 1884, with significant consequences for the residents of Pureora village (Chap. 8). Second, Maori fugitives sheltering within Te Rohe Potae were pardoned—including, despite Bryce's furious objections, Te Kooti. The government refused other requests, but a compromise deal was signed in December 1883 allowing surveys to proceed.

Trouble arose between different iwi affiliations within this area who could not agree on exactly where one iwi's land ended and another's began, so the court split the petition area to create the Aotea Block, a smaller stretch of country of 78,000 ha belonging to Ngati Maniapoto alone, and determined its boundaries.

The other three iwi and the remaining land were dealt with separately. When the court subdivided the Aotea block in 1888, and defined the Pouakani block over the objections of Ngati Raukawa and their allies, serious grievances were created, not only against the Pakeha system but also between tribes. The Government dismissed their concerns, and from then on had sole purchasing rights. But the grievances festered on for decades [34], and in due course they led to

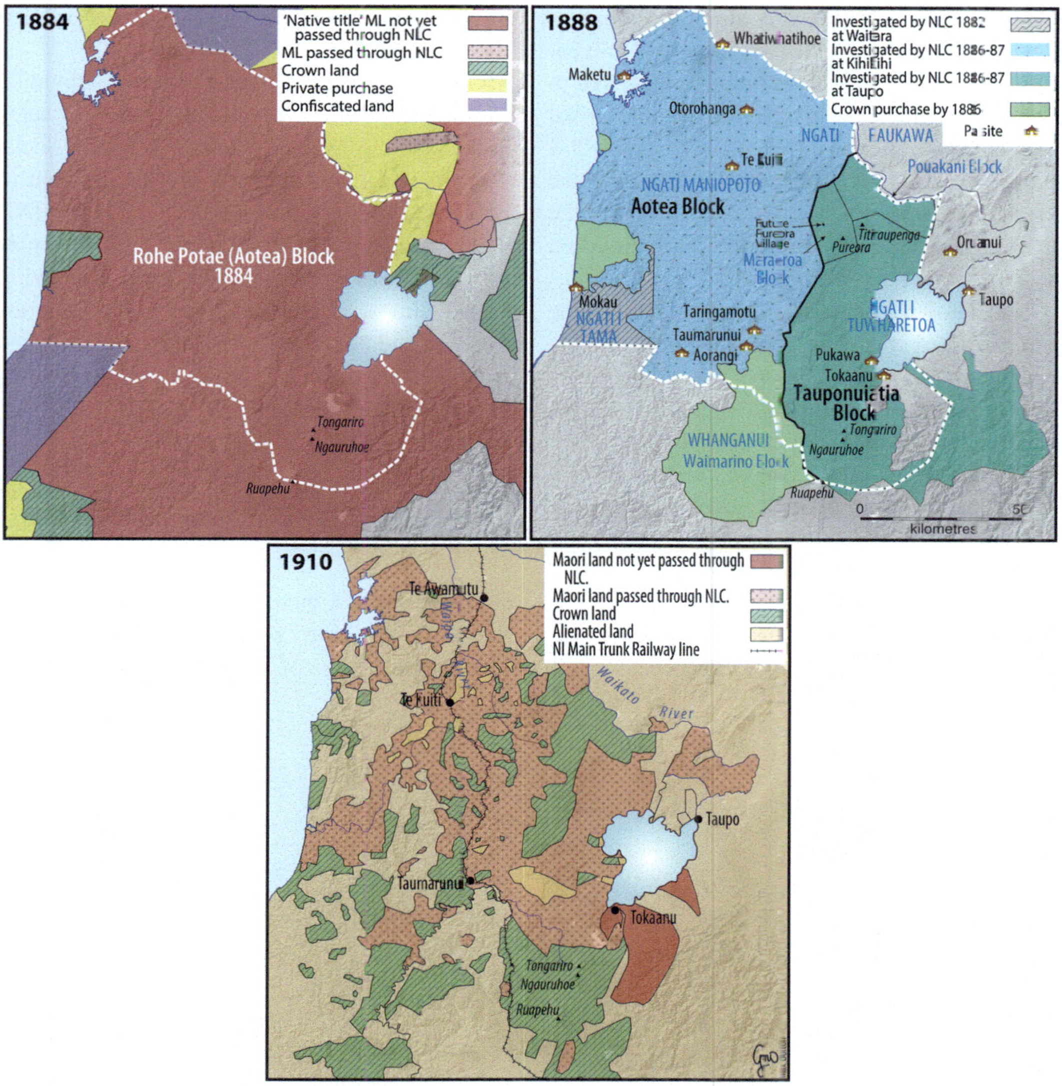

Fig. 5.1 Boundaries of the King Country, Te Rohe Potae, in 1884; after subdivision into the Aotea and Tauponuiatia blocks in 1888, following disagreement over the boundaries between the tribes resident in the two areas; and after 1910, when the North Island Main Trunk Railway was complete, and most of the King Country had passed through the Native Land Court into the hands of Crown or private purchasers. *NLC* Native Land Court. *ML* Maori Land. "Confiscated" land was taken from tribes involved in the Waikato War, including from some who had not fought against the Government but held fertile land coveted by settlers. *Redrawn by Max Oulton from* McKinnon (1997: *Plate 84) and* Waitangi Tribunal (1993: *figs. 7.1 and 7.2*)

court action that influenced the history of Pureora Forest Park (Chap. 11).

Wahanui, the paramount chief and principal negotiator for the Ngati Maniapoto, had seen his tribe's consent to the railway passing through their lands as allowing them, not the Government, to control Pakeha settlement in Te Rohe Potae. The Government interpreted the agreement as the first stage in transferring control from the Kingitanga to the Government.

If Maori expected any advantages to be brought into their homelands by new capital and settlers, they were disappointed. The 1883 compromise agreement over the railway surveys was achieved at the cost of a very significant concession by the Maori leaders. They could not prevent the Native Land Court from being empowered to convert the King Country from traditional tribal (i.e., collective) ownership to individual titles recognisable under Pakeha law. Under the Native Land Alienation Restriction Act of 1884 the Crown's right of pre-emption was re-imposed on the whole of Te Rohe Potae.

The Crown's intention was to serve the national interest by keeping out private land speculators and opening up the King Country for farm development [34: 181]. Individual claimants began making applications to have their rights registered with the Native Land Court, and the first court hearing in Te Rohe Potae was held in 1886.

Over the long term, this process did indeed bring about the catastrophic effect for Maori that their leaders had feared. They paid a disproportionate cost for Pakeha settlement, because little provision was made for them to benefit from it [34: 307–308].

Government agents started buying individual land shares in 1890. Tens of thousands of acres of land in the Maraeroa and Pouakani Blocks were divided into numbered blocks and sold to the Crown for two shillings and six pence (NZ $0.25) per acre [34: 192] (Fig. 5.2). Disputes over boundaries and mistakes in surveying caused almost a hundred years of arguments, litigation and petitions to Parliament, all analysed in meticulous detail by the Waitangi Tribunal's comprehensive report of 1993.

Tongariro National Park

By the late 1880s, there was clearly a risk that the Native Land Court would extend its brief to deciding the fates of the central North Island

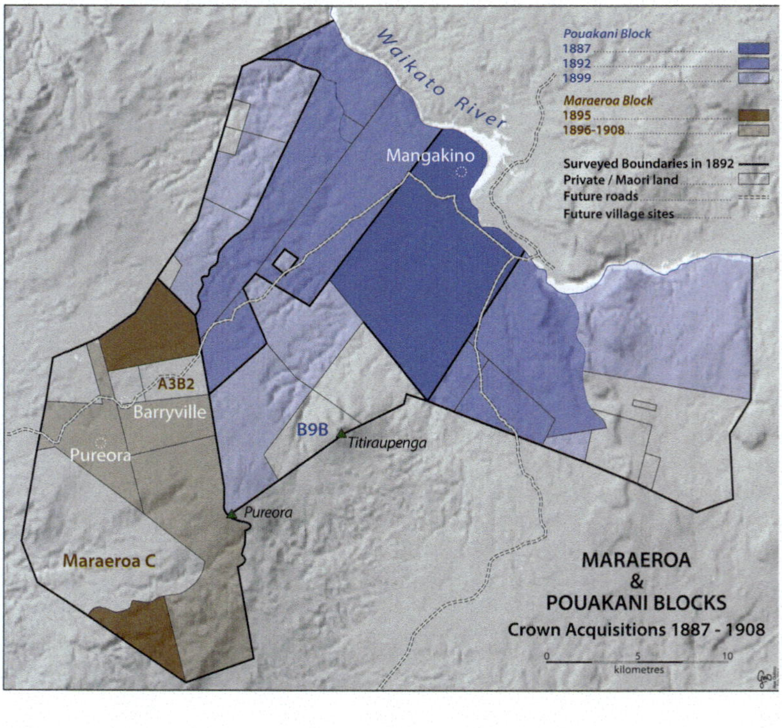

Fig. 5.2 Alienation of subdivisions of the Maraeroa and Pouakani Blocks, 1887 to 1908. Only those blocks significant to the Pureora story are labelled. At that time, there were no roads, and the settlements of Pureora, Barryville and Mangakino did not exist, but their future positions are shown. Pouakani B9B was included in the land sales of the time, but the boundaries were not correctly surveyed, so it was later returned to Maori. *Redrawn by Max Oulton from* Waitangi Tribunal (1993: *fig. 11.2*)

volcanoes. Different tribes were disputing ownership of Ruapehu (Fig. 5.3), Ngauruhoe and Tongariro, and there was a danger that the peaks would be divided and sold. Because Maori regarded the mountains as their tribal ancestors (Chap. 3), the prospect of losing them was appalling. Tuwharetoa chief Te Heuheu Horonuku (Fig. 5.4) agonised over it. He said,

> If our mountains of Tongariro are included in the blocks passed through the Court in the ordinary way, what will become of them? They will be cut up and perhaps sold, a piece going to one Pakeha and a piece to another. They will become of no account, for the tapu will be gone. Tongariro is my ancestor, my tupuna; it is my head; my mana centres around Tongariro [7].

Te Heuheu was attending a sitting of the Native Land Court at Taupo in 1887, along with Lawrence Grace, son of Thomas Grace the Pukawa missionary (Chap. 4), who was acting as Te Heuheu's adviser. Grace suggested that Horonuku could solve the problem by gifting the central North Island volcanoes to the Crown for a national park. Te Heuheu agreed, and his action ensured that Tuwharetoa's sacred mountains would remain protected from Pakeha land developers, even if they were no longer directly controlled by the tribe. The government bought more Maori land to supplement the gift, and Tongariro National Park was established by an 1894 Act.

The North Island Main Trunk Railway

There was considerable debate between 1884 and 1900 about how to choose the optimal route to link the existing railheads of the North Island Main Trunk Railway (NIMTR) across the centre of the island [27]. Some of it was motivated by how to define the best route for accessing the massive untapped stands of dense bush in the King Country, which were aptly described by Petersen [25: 74] as "a sawmiller's dream and a settler's nightmare".

The first and most urgent task was to undertake a triangulation survey of this huge unknown area, stretching from Kihikihi, south of Hamilton, to Mt Ruapehu. The work was led by Laurence Cussen (1843–1903), the Irish-born district surveyor for the whole Waikato Region, based in Hamilton.

Fig. 5.3 Ruapehu, the sacred mountain of the Ngati Tuwharetoa, with its snowy cap matching the white hair of the paramount chief of Tuwharetoa, Te Heuheu Mananui (see Fig. 4.3). *C.M. King*

Fig. 5.4 Te Heuheu Horonuku at the age of sixty, painted by Robert Atkinson in 1889. Contemporary photographs show that Horonuku had the same full head of grey hair as his father Te Heuheu Mananui, but Atkinson chose to show him as he might have looked when younger. Horonuku's decision to gift Tuwharetoa's sacred mountains to the government for a national park protected them from unscrupulous land developers. *Fletcher Trust Archive, image 62591*

By August 1884, Cussen and his staff had established 43 trig stations (Fig. 5.5) covering two million hectares of difficult country [32]. Inevitably, the relationship between surveying parties and the various iwi in the King Country was uneasy, and at times confrontational [13: 119–120].

Cussen's report to the Assistant Surveyor-General, accompanied by a map of the King Country [9], gives detailed accounts of the landscape and its potential for development, as well as describing encounters with Maori opposed to the work. Cussen climbed Ngauruhoe, Tongariro and Ruapehu (presumably with Te Heuheu's permission), and prepared the first detailed topographical maps of them in 1891.

That Cussen succeeded in this work suggests a considerable easing of Maori authority over the King Country which had prevented most European access since the end of the war. That does not mean that Cussen was unsympathetic to the Maori people of the area. He was in fact very interested in them, and he described their living conditions and took many photographs, which provide a valuable record of contemporary Maori life (Fig. 5.6). He noted with concern the evidence, from abandoned villages and gardens, of a greatly reduced Maori population—down to, he estimated, less than 4000 throughout the whole King Country. Only 3–4 decades earlier, the population of the much smaller Lake Taupo region alone had been around 1000–1600 [37].

Fig. 5.5 Laurence Cussen with telescope and survey gear at the Maraetua trig point, during the first survey of the King Country in 1883. *Hamilton City Library, image 412*

The decision to open the King Country to the NIMTR was a source of considerable anger to those Maori who disagreed with it. The surveyor Charles Hursthouse led the first survey of a possible route through Te Rohe Potae from the north. He started in March 1883 from the fortified town of Alexandra (now Pirongia) but got only about 15 miles before he and his surveying party were stopped and turned back.

After negotiations between Wahanui and the Maniapoto, and John Bryce, Minister of Native Affairs, Hursthouse started again, and this time got nearly to Te Kuiti. There, Hursthouse and Newsham, his European companion, were kidnapped by the dissident leader Te Mahuki, and held prisoner for almost two days. They were rescued by a party of Ngati Maniapoto led by Te Kooti—himself previously a prisoner of the Government, but recently pardoned—and returned to Alexandra with the help of Wahanui and Bryce [11].

Three months later, another surveyor, John Rochfort, made another attempt from the south, but access was not easy from that direction either. His report makes clear the dangerous work of pushing into unknown country occupied by armed and resentful warriors defending their ancestral homes [29].

Rochfort first started from Marton and pushed north, but was stopped at Karioi by armed resistance and threats, and he too was held prisoner for several days. He turned back to Wellington, and then tried again, with the same result when he got to Taumarunui. He had no choice but to go round, not through, the King Country and try again from the north—a diversion "by Tokaanu and … the west side of Taupo to Kikihihi, some 150 miles". From there, again with the help of Wahanui and Bryce, he finished the survey.

In 1885 the completion of Rochfort's exploration of the proposed central route for the NIMTR was celebrated by the planting of a totara sapling in Taumarunui. It grew into a magnificent tree, bearing for many years a plaque stating that it had been planted by chief Puia as a token of the promise of safe conduct given to Rochfort and his party of surveyors. Three other routes for the proposed railway were also investigated; two via Taranaki, and one via Hastings, but both were longer than the more direct central path from Marton to Te Awamutu.

For years after the NIMTR was finished, evidence of the surveyors' passage could be found in the bush, marking potential alternative pathways for the railway. Every so often there would be a small, lonely clearing marking their campsites along narrow but well graded tracks,

Fig. 5.6 Mr B. Cashel, a surveyor of Cussen's party, at Te Kumi, a Maori village near Te Kuiti, in 1883. Fencing was necessary to protect gardens from feral pigs. *Hamilton City Library, image 542*

still with a moss-grown pataka (storehouse on stilts) built by the surveyors to protect their provisions against ship rats [11: 117]. The same applied when later surveyors were working out prospective sawmill tram lines (Fig. 5.7).

The Select Committee charged with deciding the route met from 13 September to 9 October 1884, and heard from 84 witnesses. Its detailed report covers more than a hundred pages of small type plus a series of detailed maps. It accepted Rochfort's recommendation that the central route be chosen, even though it required excavating a tunnel at Poro-o-tarao, 53 chains long (1615 m) [27]. In October 1884 the Government agreed.

By 1885 the leaders of both Maori and Pakeha were reconciled enough for both to attend the ceremony marking the start of construction of the King Country section of the line. The first sod was turned on the south bank of the Puniu River on 15 April 1885 by Wahanui, Rewi Maniapoto and Premier Robert Stout.

The Poro-o-tarao tunnel was driven through from both ends at once, largely by manual labour. Within three years the two tunnelling teams met, with only half an inch deviation. The last brick was in place by April 1890, but the rails crawling south from Te Kuiti did not reach the northern tunnel portal until March 1895. So for nearly six years the finished tunnel lay idle, merely a convenient short cut for packhorses [8: 84].

As the railway line advanced, settlers and sawmillers followed, and gradually the sawmills developed into a principal provider of employment in the region for the next half century. Maori were strongly represented among the most skilled and highly valued axemen and drivers of heavy machinery in the bush and at the mills, as the later history of Pureora amply shows.

The fine stands of native forest (usually called 'the bush') in the central King Country had been recognised quite early, but it was not practical to exploit them until the completion of the NIMTR linking the standing timber with its potential markets. Laurence Cussen had reported from preliminary survey work in 1884, that:

> There is a great deal of valuable timber in the western Hauhungaroa, of which a good proportion would be accessible from the Te Awamutu and Marton railway line [9].

The value of the timber for financing the NIMTR was illustrated in a report to Parliament by Mr. Murray, Surveyor to the Lands Department, who had made a careful inspection of the vast forests in the Waimarino area. There is, he said,

Fig. 5.7 Survey party in the bush, selecting a route for a tramway for the Ellis and Burnand sawmill. Undated, probably early 1900s. *Hamilton City Library, image 11641*

> …an area of 91,000 acres of good milling timber…[for] about twenty miles along the line…and within six miles of the line on either side….the construction of the railway as proposed will greatly help the sawmilling industry, and render the forest timber an extremely valuable asset to the State. Mr Blow [Under-Secretary for Public Works] estimates that the value of the milling timber, at 1 s per 100 ft royalty, would pay for the construction of the line twice over [33].

The problem of accessing these forests was gradually removed as the construction gangs, approaching from both north and south, opened the prospects of markets in both directions. From Auckland, the NIMTR reached Te Awamutu in 1880, Te Kuiti in 1887, and passed through the Poro-o-tarao Tunnel in 1895, then to Waimiha early in 1901 and Ongarue by August 1901, leaving 14 miles to be completed to Taumarunui by 1903 [11]. The line from Wellington reached Marton in 1886, Taihape in 1904, and Waiouru in 1908 (Fig. 5.8).

It took five years (1903–1908) to resolve the most difficult engineering problems encountered on the route bridging the last gap from Taumaranui to Waiouru, including the construction of the Raurimu Spiral (enabling a climb of 550 m in 25 km) and several large viaducts. Finally, in 1908, the then Prime Minister Sir Joseph Ward officiated at a ceremony during which he drove the official last spike of the NIMTR (the title was changed from "Premier" (as held by Robert Stout) when New Zealand became a Dominion in 1907).

A modest monument, the 'Last Spike' Memorial, about 11 km north of Horopito on SH 4, commemorates this event. It is not easy to realise, looking at this unassuming concrete block, what catastrophic consequences were about to follow for the last and largest area of native forest remaining anywhere on the North Island.

The final completion of the railway from Wellington right through to Auckland enabled a huge expansion of sawmilling in the Central North Island. There was now a cheap means of getting access to extensive areas of bush, and then moving the sawn timber to distant markets, both to the north in Auckland and to the west and south in New Plymouth and Wellington. The completion of the line revolutionised travel, opened up the King Country to the Pakeha and, consequently, had enormous and long term ramifications for race relations, politics, land settlement, and trade [11, 16].

Rabbits

One of the unexpected consequences of the NIMTR was the arrival of a new and potentially devastating pest, the European rabbit.

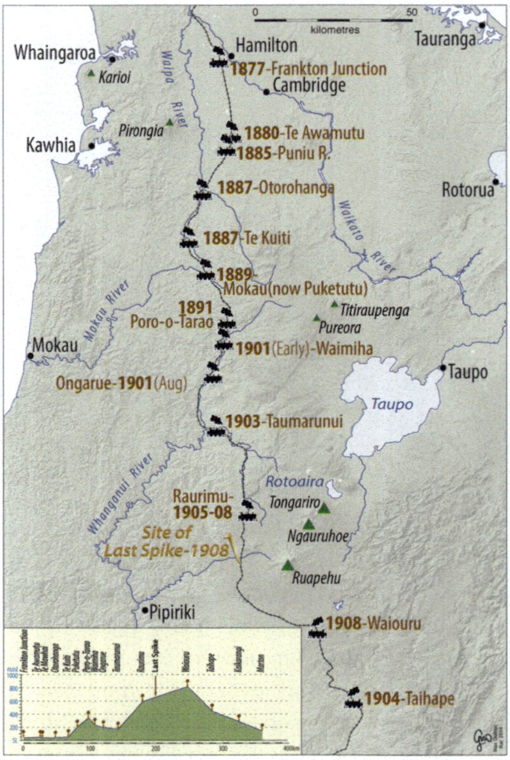

Fig. 5.8 Progress of construction of the North Island Main Trunk Railway. The actual line closely followed Rochfort's surveyed route, accepted by the Government in 1884. The separate advances of the northern and southern railheads, built in stages over 24 years, met at the Last Spike in 1908. The elevation profile illustrates the height of the central plateau, and the steep drop northwards that required construction of the famous Raurimu Spiral. *Drawn by Max Oulton from data in McKinnon (Ed)* (1997: *Plate 84*)

Rabbits are strictly herbivorous and strongly gregarious animals of open country and grassland. They did not appear in the King Country until well after forest clearing advanced close enough to provide them with suitable habitat nearby. The accelerated logging and clearing of all forests within easy reach of the railway line, and their conversion into rough pastures still dotted with tree stumps and fallen logs, turned the uninhabitable forests into a rabbit's idea of heaven. They followed the construction of the line from both directions, and were over-running virtually all open country available in the area by at least the late 1880s.

Burrowing by rabbits undermined the newly established pastures, and their selective grazing concentrated on the best of the introduced grasses, leaving only patches of short, soiled turf for the sheep. Their young are born in litters averaging five blind and naked kits at a time, and in the fields and mild climate of lowland North Island their breeding season could continue for most of the year [14]. Adult females can produce more than 40 young per season and live for 2–3 years. Hardly surprisingly, rabbits quickly became regarded by farmers as the worst animal pest in the area, at least until the arrival of the possums. In areas with the dry, easily burrowed soils that rabbits most favour, some farmers lost everything to the grey horde.

By 1882, the need to deal with the rabbit problem was seen as a matter of national urgency, requiring official Government support. The Government responded by sponsoring large-scale introductions of natural predators in the form of stoats, weasels and ferrets, the "enemies of the rabbit" that were assumed to keep rabbits from reaching such catastrophic numbers back "Home" [17]. These energetic predators were released only on pastures teeming with rabbits, where ferrets generally stayed. Stoats quickly moved into the forests, with catastrophic consequences, especially for kiwi (Chap. 13).

In the 1930s, farms in the Tapuiwahine Valley west of Pureora were "alive with rabbits". Children ran lines of 30 traps each and brought home rabbits to be boiled, baked and stewed, till their mothers never wanted to see another one [36]. Much of the farmland on both sides of PFP was "rabbit-sick" by 1940, despite the attentions of professional rabbiters armed with poisons, guns, dogs and ferrets. In 1942, one group of 11 trappers funded by a levy on farmers in the Maniapoto district accounted for 9284 rabbits. In due course, rabbits even invaded the newly planted tree nurseries at Pureora (Chap. 6).

Human attempts to "control" rabbits with guns, dogs, tame ferrets and traps only produced a perpetual and profitable harvest of carcases and skins, a powerful disincentive to trappers to work hard enough to put themselves out of business. Rabbits remained as numerous as ever, and provided a long-term supply of food for stoats and feral ferrets. So rabbits have played an indirect but significant role in the story of Pureora: the wider history of official efforts to deal with them is summarised by John Gibb and Morgan Williams [14].

The First Timber Mills

Sawmilling in the King Country began at Kawhia in the late 1840s, but further development was limited by King Tawhiao's edict after the Waikato War (1863–64) making the King Country off limits to Europeans. The Kingites (supporters of the King Movement) had retreated behind the aukati line, the northern limit of Te Rohe Potae, the King Country (Chap. 4).

With the advent of the railway, J.W. Ellis and his first partner Lewis established the first mill within the boundaries of the King Country at Otorohanga in 1890 [3]. This modest operation was the beginning of the Ellis & Burnand Company (E&B) which later became a dominant player in the local timber industry. In 1900 Ellis wrote to the Parliamentary Select committee which was considering the progress and probable benefits of the NIMTR to the timber industry. His support of the railway and his estimates of the value of the timber it would make accessible were of course expert, but not entirely disinterested.

The only other milling companies established before the turn of the 20th century were those based around Taumarunui. These mills had contracts to supply railway sleepers as the local construction work progressed [3], and by 1903 also a rail link to the Auckland market (Fig. 5.8).

Mill operators soon grasped the opportunities being offered as the construction of the NIMTR extended southwards. E&B established a big mill at Mangapehi (E&B No. 1), opened in 1904, which was at that time the largest sawmill in the northern King Country (Fig. 5.9). Business was so good that E&B added a second, smaller mill at Mangapehi (E&B No. 2) in 1908 [1: 16].

It could be argued that the invention of refrigeration was indirectly responsible for the development of sawmilling in the King Country in the 1890s [30]. The sudden prospects of huge profits from exporting refrigerated food to Britain stimulated a demand for kahikatea timber, because it was smooth and scentless, so was ideal for butter boxing. In response, cutting started in the Maori-owned Maraeroa C block (Fig. 5.2) in 1901 [22].

Sawmilling and land clearing progressed outwards from the railway line as the bush edge moved further away, until by 1907, NZFS data analysed by Somerville [31] showed 17 mills hugging the NIMTR through the King Country. E&B's Mangapehi operation, then the nearest milling settlement to the future Pureora Forest village, developed into a very large complex of mill buildings, railway sidings, timber yards and associated offices and houses [1: 107]. Most of the future output of timber from Pureora would be sent to market via Mangapehi.

Hauling the Logs to the Mills

In the early days of logging, one of the most difficult and labour-intensive parts of the operation was moving the logs after the trees were felled. Once the head (the canopy and branches) was cut off and the trunk cut into a log trimmed to length, the breaker-out attached strops or chains to the log using a set of dogs, heavy steel bars with a lug at one end and a curved hook at the other, which were hammered into the butt (one end). Each log, a deadweight of several tons, was then hauled by bullocks directly along the ground, in a process called skidding.

The work was so hard on the bullocks that they could not be worked for more than four hours a day, and they often suffered injuries [20: 19]. Drag on the logs, and damage to them, were also serious disadvantages of this early method. To minimise drag, skid roads were developed, whose surfaces were corduroyed with small logs laid crosswise.

A better method was to jack the log onto a strongly made wooden sled called a catamaran, using man-powered mechanical timber jacks. This lessened the ground resistance, more so if

Fig. 5.9 Ellis and Burnand's two timber mills and associated settlement at Mangapehi in 1916. The No. 1 mill built in 1903 (*left*) is wreathed in steam. The smaller No. 2 mill behind it was moved to that site from Tiroa in 1908. Later, two reinforced concrete drying kilns were built, roughly on the site occupied here by piles of sawn timber in the left foreground. The ruins of the kilns are now the only standing remains of this once extensive complex. *Alexander Turnbull Library, Wellington, New Zealand. ATL APG-0782-1/2-G*

the skid road was greased with a mixture of fat and kerosene. Even after this, the friction caused by the weight of the load would sometimes cause the runners of the catamaran to smoke.

Pulling heavy logs from the stump to a skid (loading point) became much easier after the introduction of steam haulers. A boiler, a steam engine and up to three winch drums were mounted on two foundation logs, which were tied back to a tree, a stump or a buried deadman (anchor point), to counteract the pull of the winch. Steam haulers could move large logs, but were limited by their need for supplies of fuel (normally mill slabs) and water, and their radius of use was dictated by the terrain and the length of the wire rope on the main winch drum. Haulers pulled logs to a central point from all directions, along tracks known as snig lines.

Steam haulers (Fig. 5.10) were typically crewed by 5 man teams: the hauler driver, one or more cross cutters, a breaker-out and a whistle-boy who signalled the hauler driver by a system of blasts on the hauler's steam whistle. The main haul rope was run out to where logs were lying, and a rope rigged alongside the winch rope circuit connected first to the hauler whistle and then onto a springy sapling to allow the signal rope to return after it was pulled. On some operations the boiler fire had to be attended over night if steam pressure was to be kept up for an early start the next day.

Getting from the skid to the mill was the next challenge. Tramlines were a huge advance in efficiency and they extended from sawmills into the forest, gradually lengthening as the cutting face retreated [22: 258]. Bullocks and horses

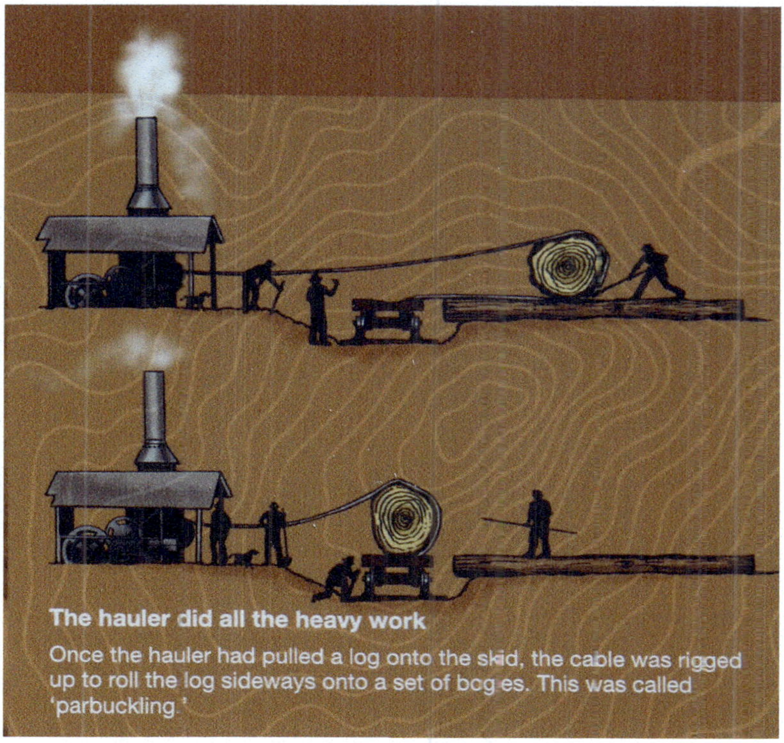

Fig. 5.10 The workings of a steam hauler are reconstructed for modern visitors in this interpretation sign along the Timber Trail (Chap. 14). Steam haulers had their own skids (*right*) for transferring logs onto tramway bogies (*centre*). *Crown Copyright, Department of Conservation Te Papa Atawhai (2013). Artwork by Alex Stone; sign produced by Snapper Graphics, Waiheke Island. Photographer: Greg Martin*

could pull enormous logs on bogies along these lines [22, 36]. Figure 7.6 shows the network of sawmills and logging tramways that grew up along the section of the NIMTR between Mangapehi and Kakahi [39: 8].

But as sawmills grew and tramways lengthened, more pulling power was required, and some progressive millers began to use steam engines along the tramways. Between 1871 and 1889, some 34 bush lokeys (steam locomotives running on steel rails) were in service around New Zealand [20: 14], but they were few compared with the need. All but a few of the 334 sawmills working in 1900 were still relying on horses.

Steam locomotives (Figs. 5.11 and 5.12) were better at hauling logs over steeper grades and the progressively longer distances from the mills. But only the larger logging companies could afford the cost of bringing logs many miles by tramway to their mill. Smaller operators were forced to shift their mills as the bush edge moved away from the railway, despite the associated social disruption to the milling villages.

The typical bush lokey in the early period was a small four-wheel direct-drive engine imported from England [36], weighing between five and ten tons, much like those used in railway and industrial service at the time. But in the new century, as sawmill capacity continued to increase, trams were increasingly extended into more rugged bush blocks over steep hills and round sharp curves, so creating demand for more powerful lokeys [20: 14].

In response a new type of lokey appeared, unique to bush tramways—the geared bogie locomotive. A boiler and engine were mounted on two or more bogies, and all the bogie wheels were driven by drive-shafts and gears. Four-wheel drive made them powerful and agile, able to negotiate uneven track and tight curves. At first, most bogie lokeys were Heisler or Climax engines imported from the USA, but by 1930 four New Zealand engineering companies had produced at least 98 bush lokeys, outselling the American competition right up until 1943, when geared bush trams were no longer needed [20].

Fig. 5.11 E&B's Climax 1650 bush lokey working on the Ongarue tramway in 1929, surrounded by tall forest. *Hamilton City Library, image 1613*

E&B continued their steam-powered Ongarue tramway operation (Figs. 5.11 and 5.12) until 1958, long after trucks began to replace steam engines for hauling logs to the mills from the 1940s onwards.

Settlers often preceded the mills into untouched country, taking up bush blocks in the expectation that mills would follow and would then pay royalties for timber harvested on private land [23]. Only the best or preferred timber was logged, and the rest was burnt and cleared.

Logging and burning were both used indiscriminately and wastefully. But small timber companies could not afford to, and perhaps saw no need to, invest in expensive regeneration planting [23]. Most of the lowland forest on the better soils west of the NIMTR was clearfelled and developed into farm land, but there was little farm development on the pumice lands on the eastern side of the line at this stage.

As the bush edge retreated, the cost of bringing huge logs many miles by tramway became prohibitive except for large mills like those of Ellis and Burnand. It was more economical to dismantle smaller mills, re-establish them near the bush edge and just take the sawn timber by tram to the railway. Hence the small mills were relatively short-lived, and so were the settlements housing the timber workers and their families.

Fig. 5.12 E&B's Climax 1650 lokey cautiously pulling a rake of logs downhill across the Mangatukutuku viaduct towards the mill at Ongarue in 1955. The complex structure of the wooden bridge and the depth of the drop below it are well shown. The engine is running backwards so that the driver can keep a close eye on the heavily laden bogies. This part of the Ongarue tramway is now incorporated into the Timber Trail, and a new suspension bridge replaces the viaduct, demolished in 1958. *Peter Mellor*

Logging Along the Margins of the Future Pureora Forest Park

Logging of the forests on the western flanks of the Hauhungaroas, and in what later became the Pureora Forest Park, started from Taumarunui, a Maori village at the confluence of the Ongarue and Whanganui Rivers. The slopes and valleys east of the village were renowned for the size and quality of their totara trees, a resource greatly valued for building canoes. Taumarunui was an important centre because it was the gateway to the central King Country from the south, especially after the 1890s when W. Hatrick established a steamboat service connecting Taumarunui with Whanganui. Ellis and Burnand opened a large sawmill at nearby Manunui in 1901, and Taringamutu Totara Sawmills Co. Ltd operated several mills up the Taringamotu valley from 1907 to 1956 [2].

The pace of sawmilling accelerated after 1903 when the section of the NIMTR linking Taumarunui north to Te Kuiti was opened, and again after the post 1914–18 war restart of E&B's big mill in Ongarue in 1921. The mill was serviced by the Ongarue tramway (Fig. 7.6), which, over the 54 year history of this massive enterprise, eventually extended some 45 km north-east from the mill to the precipitous slopes above the Maramataha River, as described by McMillan and Walker [22: 257–332] and Anderson [1]. The old tramway now forms the southern section of the largest new development in Pureora Forest Park, the Timber Trail (Chap. 14).

By 1908 the Government had acquired large areas of Maori land in the King Country for settlement (Fig. 5.2). Shortly after the arrival of the main trunk line in Waimiha, and the establishment of the railway settlement there, much of this Crown Land was made available to alleviate unemployment caused by the depressed state of the economy.

When the first settlers arrived at Waimiha, scrub and fern covered the easy country and there were fine stands of unlogged forest on the rest. Within twenty years, 28 sawmills were established around Waimiha and five major tramway networks enabled the transport of logs to these mills, mainly from the Piropiro and Okauaka catchments [31, 36].

In 1923 the Hayward Timber Company established a mill in the Piropiro area. Within a short time, Hayward's had built a camp with a school and over 10 km of tramway into the Piropiro bush, and two more mills at Piropiro (where the DoC campsite is now located, Chap. 14). The three Hayward's mills, and the mills operated by Endeans and other companies active in the area around then, are significant to this story, because all but one of these logging operations worked areas which are now part of PFP and are traversed by the Timber Trail (Chap. 14).

By 1936 the Hayward Timber Company was in financial difficulties, and sold its Piropiro mills, tramway network and logging leases to E&B. The mills then became known as E&B No. 1 (1936–1944), No. 2 (1936–1937) and No 3 (1936–1945). The No. 2 mill was closed and dismantled in 1937 because a beetle was attacking the sawn timber stacked there, although half a mile down the road at No 3 mill there was no damage [36].

In 1945 E&B's cutting rights at Piropiro expired, and the No 3 mill was dismantled and shifted to Maraeroa. The area where the No 3 mill had stood was prepared and planted in pine trees in 1985, which were logged in the summers of 2010–11 and 2011–12.

Tiroa and Maraeroa C

In the Rangitoto-Tuhua RT36 block at Tiroa, east of Mangapehi, there was an area of 3600 acres (1457 ha) of Maori-owned bush. The timber cutting rights over this area were vested in Ellis and Burnand in June 1904, and continued until the last Mangapehi mill closed in 1968 [1: 11].

To access the Tiroa block, E&B constructed a horse-drawn tramway (the Mangapehi tram, Figs. 7.3 and 7.4), and progressively extended it as the areas closer to the mill were worked out. Construction offered no great difficulties because the line traversed fairly easy pumice terrain. In January 1905, the wooden rails were replaced with steel tracks capable of carrying a Climax lokey. By the time E&B was ready to phase out tramways in favour of trucks in the 1950s, the Mangapehi tram line and its various branches totalled 58 km.

As the milling of the Tiroa timber proceeded, in December 1912 the company secured rights to the adjacent Maraeroa C block. This large area (13,727 acres, 5555 ha) had escaped being purchased by the Crown in the late 1890s, and had instead been vested in the Waikato-Maniapoto Land Board under part 14 of the Native Land Act 1909. The Maraeroa and adjacent Pouakani blocks included several sawmilling areas important to this story, as shown in Fig. 5.2.

E&B reached agreement with the Maori owners in March 1913, securing cutting rights for 25 years for the same royalty as before from the Land Board, a government agency, which in turn passed some of the payment to the Maori owners [1: 64–65]. Harry Burnand's original estimate, that there was 10,000 acres (4047 ha) of milling bush on the block, including 65 million log feet of rimu, 27.5 million feet of matai and 7.5 million feet of totara, turned out to be very conservative, since by 1941 there was still an estimated 66 million log feet to cut from the remaining 44 % of the area still uncut.

E&B extended the Mangapehi tramway into the Maraeroa C block from its northwestern corner (Fig. 7.3) from 1918, and worked southeast-wards from Pukemako Camp (built in 1917 to house the work force), 13 tram miles (20.8 km) to the mill. Within a few years branch lines had been laid throughout the block. Access to the camp was by tram only until the late 1930s, when the Mapara Timber Co. built a rough road to their mill at Barryville [1: 64].

In 1945 it became uneconomic to tram the logs to Mangapehi, so E&B erected a new mill at

Fig. 5.13 Ellis & Burnand's Maraeroa mill, with stacked timber in yard, 1965. *Fletcher Trust FT 6259P/147*

Maraeroa (Fig. 5.13). The mill incorporated most of the machinery from the E&B No 3 mill at Piropiro, which had operated from 1936 until 1945 when E&B's cutting rights at Piropiro expired.

The sawn timber was transported on the tram to the Mangapehi mill and its associated railway station (Fig. 5.9). By the 1950s, building roads was a much more economical proposition than further extending tramways, so road construction spread into all the bush areas scheduled for logging. The Mangapehi tramway was closed in 1953 after Flemings' logging trucks were contracted to convey the logs to the mill.

Over the years as milling progressed deeper into Maraeroa C, the old Pukemako Camp buildings became uneconomic to repair. So E&B decided to build a new settlement for the timber workers 2.5 miles (4 km) away, nearer to the Maraeroa mill. In 1950, E&B closed the Pukemako Camp and shifted the inhabitants to Maraeroa. The 12 houses and cookhouse were moved to Maraeroa for £2500.

As the Pureora project got started in 1945 (Chap. 7), NZFS began supplementing the Maraeroa mill's supplies from State Forest 96 (Pureora Forest). But by September 1965, E&B found there was only an estimated 30 million log feet left to mill on Maraeroa C, which would provide only two years of cutting at the current rate, so they urgently sought more logs from NZFS to keep the Mangapehi and Maraeroa mills going. These mills employed 96 men at that time, of whom 52 were Maori, with 251 dependents in total. The people lived in Mangapehi, Benneydale and at Maraeroa. Another 24 men were employed on contract by Hohneck Contractors and Fleming Brothers, and these people had 102 dependents [1: 159].

But NZFS refused to supply E&B with any more logs from State forests. E&B had no choice but to start winding down their operations, so the end of E&B's indigenous logging was already in sight long before the environmental movement took up the cause. The first mill to close, in June 1967, was the original big (No. 1) steam driven mill at Mangapehi (rebuilt after a fire which caused damage of £15,000) [10]. The Maraeroa mill followed in December 1967.

The smaller No. 2 Mangapehi mill (rebuilt with electrically-driven saws after another fire in December 1947) lasted another 12 months, until the last log went through it on 19 December 1968. Nothing is left of E&B's formerly huge operation at Mangapehi now except a few piles of rusting iron, some concrete foundations, and one of the reinforced concrete timber drying kilns, still with its rusting circulation fans hanging silent in the ceiling.

Over the 68-year life of the milling settlement in Mangapehi, 900 million board feet of timber (defined in Box 6.1) was produced. At its peak in around 1952, the town had been home to about

600 people [18]. The closure of the mills had a profound effect on the whole area, just as at countless other short-lived sawmilling communities all over New Zealand as the bush nearest to each was exhausted. With no work on offer, people moved away, houses were removed or abandoned, and the area slid into depression. Mangapehi has never recovered [1: 159], and neither have two other formerly large and thriving timber/railway communities further south along the NIMTR, Waimiha [36] and Ongarue [22].

Much the same happened on a smaller scale when the 29 men laid off from the Maraeroa mill were forced to leave the district and find employment elsewhere [1: 159]; and again in the Rangitoto area when Smyth Bros and Boryer mill at Ngaroma closed in 1975 [15]. None of these broken sawmilling communities ever got the publicity that surrounded Pureora and Barryville a few years later.

After the logging of Maraeroa C was completed, the land was incorporated and converted to pine plantations. There were 702 Maori owners of the block at the end of 2007. Carter Holt Harvey has a 99 year lease on the block, and the first crop of *Pinus radiata* timber has been clear-felled, milled and replanted (Fig. 5.14) [1: 85]. The old outline of the Maraeroa C block is still visible on Google Earth, and zooming in shows the orderly ranks of pine trees, the logging tracks and the skids.

Likewise, the cleared land on the Rangitoto-Tuhua RT36 blocks was partitioned into farms and taken over by the Department of Maori Affairs. Google Earth now shows rolling green pastures on the hills that were once covered with native forest on both sides of SH30.

Fig. 5.14 The first pine plantations on the Maraeroa C block (which replaced the clearfelled native forest) were themselves logged and cleared in about 1990, ready for the second rotation. Looking SW from Pureora, across Maraeroa C, the western edge of PFP and the Taumarunui district, towards Taranaki's snow-covered peak (*top left*). *Crown Copyright, Department of Conservation Te Papa Atawhai. Photographer unknown. DOC image 10067780*

A3B2

The Mapara Timber Co (owned by the Knight Bros) had obtained access to a block of timber of 1981 acres (802 ha) known as Maraeroa A3B2 (Fig. 5.2), some time before 1939 [38: 18]. But road access from the block to the railway was virtually non-existent. The Mapara Timber Co formed a rough wagon track to get the logs from the block to Benneydale and thence to their mill at Poro-o-Tarao (just south of Mangapehi), which is marked on the maps shown by Anderson [1: 24–25]. Unfortunately, the mill burned down in 1938, along with Mrs Knight's house next door [5]. The bank called in their overdraft of £5000, and the Mapara Timber Company was wound up [4].

Marton Sash and Door Co Ltd had had a mill and tramway near the NIMTR at Waione Siding (Fig. 7.6) since 1927, but it too burned down in 1939. They put in a successful tender of £20,000 for the lease on Knight Estate's cutting rights on the A3B2 block, for a royalty of two shillings per 100 ft of logs, and set up a new mill there in 1941 (Fig. 5.15). They called it their Mangapehi mill (NZFS register number 82: see Box 7.2), but it was 18 miles east of Mangapehi, at what later became the site of Barryville [1: 27]. They built a tramline to bring logs to the mill from the bush, but it was never connected to the NIMTR.

The road from Te Kuiti to Benneydale had been built in 1939, but was still only a rough pumice track. Marton Sash and Door made considerable improvements to it, with NZFS help in later years, and from 1941 they carted sawn timber to Mangapehi Station until their Barryville mill closed in 1955. Despite the access difficulties, Wilson comments that this was one of their most profitable mills [38: 19].

Nothing now remains of any of these old sawmills, so it is hard for people now to visualise the vibrant life that once revolved around them. The nearest picture we can get is provided by one significant historic relic of the old logging days, Endean's Sawmill on Ongarue Stream Road, near Waimiha. It was once a large, busy industrial site (Fig. 5.16); now it is New Zealand's equivalent of an archeological ruin, the most complete set of remains of a native-timber sawmill settlement in New Zealand, established on that site in 1928 [26]

The mill stands silent at the side of the road exactly as it was left when it was last turned off in 1996, complete with parked vehicles, piles of sawn timber and heaps of rusting machinery and tram bogies. Even the conical furnace where the sawdust was burned still stands, although battered and out of shape.

Buildings, equipment, manager's office and worker's cottages remain just as they were left—a stark reminder that all the old bush mills were not just places to work, they were also small communities of people who lived, worked and played on the site.

Some of the stacks of processed native timber, all weathered to silver, are still as neatly piled as the day they were made; others are disordered. Sheep wander peacefully among the ruins, maintaining a neat sward of living grass around the evidence of the busy industry of the past.

Coal Rides the Timber Tramway

The Mangapehi coal seam was discovered relatively late, in 1931–32 [12: 18]. The seam was 3.7–4.5 m thick, and provided the only coal in the wider Pureora area, so was a valuable resource for local steam engines and haulers. Coal was originally extracted through the Mangapehi Mine (1934–1962), and then between 1978–98 from the adjoining Benneydale Mine. The two mines are separated by the Benneydale fault and about 50 m of unworked coal.

The township of Benneydale lies about 2 km west of the mine (and 22 km west of Pureora). Benneydale was built to house coal mine workers, so did not exist until 1941. The name is a combination of the surnames of an Under-Secretary of Mines, C.H. Benney [19], and the first mine manager, R.T.H. Dale [1: 136]. Benneydale was important to the residents of Pureora village from 1945 onwards, because it was the nearest place where they could find alternative

Fig. 5.15 The first mill at Barryville in 1953, built by Marton Sash & Door Co. in 1941. Tramrails lead into the mill from the bush in block A3B2. The truck parked beside the water tower has an MSD sign on the door. This mill closed in 1955, but the village was unaffected as Carter's (Morningside) had already (in 1949) built another mill alongside it. *Auckland War Memorial Museum – Tāmaki Paenga Hira. PH-NEG-H549*

Fig. 5.16 Endean's mill near Waimiha, built in 1928 and finally closed in 1996 but never dismantled, is the only surviving nearly complete relic of the King Country timber industry. *Crown Copyright, Department of Conservation Te Papa Atawhai* (1986). *Photographer: J. Gaukrodger*

housing and shops, plus a local doctor, and a rival rugby club (Chap. 8).

The Mangapehi mine workings were deeper than those at the Benneydale mine. One million tons of coal were produced from the Mangapehi seam between 1934 and 1998. In its peak years the coal industry employed 130 people [12: 3].

The mine owners signed an agreement in November 1936 with E&B to use their Mangapehi tram to haul coal to the Mangapehi railway station. A siding had already been established at the 4 mile mark (6.4 km) on the tramway. Later E&B was contracted to take NZR wagons to the mine siding and return them loaded to the station [1: 136]. The agreement ended in 1952, when it became cheaper to shift both logs and coal by truck, and E&B uplifted the tramway.

After an explosion and fire in February 1962, the Mangapehi mine closed and was allowed to flood. The mine and its associated land holding was handed to the Department of Lands and Survey in 1964 for disposal. All except about 235 ha was sold to private farmers.

The coal remained undisturbed after the Mangapehi mine closed, but in 1978 a new mine, the Benneydale mine, was opened by J.&T. Hughes, private coal miners, working under two Crown Mineral Licenses [12: 18]. Over the 20 years of its working life, this mine produced around 200,000 tons of coal. But over time, the coal became hard to sell, because it had a relatively high sulphur content [35]. The mine was closed in early 1998, and allowed to flood.

Within a year the outflow of contaminated water from the main portal of the Benneydale mine into the Mangapehi Stream required a seal on the mine portal and the construction of a mine water treatment system to clean up the discharge water, which continues today [28].

References

1. Anderson, K. (2008). Maoriland sawmillers. Ellis & Burnard Ltd, Sawmillers and timber merchants of Mangapehi. Manurewa: Ken Anderson Family.
2. Anderson, K., & Cooke, R. (2014). "Tringamoot": The story of Taringamutu Totara Sawmills Ltd in the Taringamotu Valley. Taumarunui: Taumarunui and Districts Historical Society Inc.
3. Anon (1953). Ellis and Burnand: Half a century of progress in the timber industry in New Zealand 1903–1953. Hamilton: Ellis & Burnand Ltd.
4. Auckland Star, 23 September 1939.
5. Auckland Star, 10 December 1928.
6. Burdon, R. M. (1948). The life and times of Sir Julius Vogel. Christchurch: The Caxton Press.
7. Cowan, J. (1913). Tales of Tongariro: How the fire-peaks came to be a national park. Otago Daily Times, 22 July, p 26.
8. Craig, D. (1962). South of the Aukati line: A history of the King Country. Te Kuiti: King Country Chronicle Ltd.
9. Cussen, L. (1884). Mr Lawrence Cussen to the Assistant Surveyor-General. Appendices to the Journals of the House of Representatives, 1884 II C1, 29–32.
10. Evening Post, 14 July 1945.
11. Fletcher, R. S. (1978). Single track: The construction of the North Island main trunk railway. Auckland: William Collins Publishers.
12. Fowke, N. (1997). Geological assessment of Mangapehi Coalfield. Wellington: Coal Corporation of N.Z.
13. George, M. (1990). Ohakune: Opening to a new World. Ohakune: Kapai Enterprises Ltd.
14. Gibb, J. A., & Williams, J. M. (1994). The rabbit in New Zealand. In H. V. Thompson & C. M. King (Eds.), The European rabbit: The history and biology of a successful coloniser (p. 158–204). Oxford: Oxford University Press.
15. Hunwick, E. (1990). Logging of the northern Rangitoto Bush. Footprints of History, 4, 73–79.
16. Johnson, C., Larsen, R., & Ramshaw, K. (2008). Main trunk: Portrait of a Railway. Wellington: Grantham House Publishing.
17. King, C. M. (1984). Immigrant killers: Introduced predators and the conservation of birds in New Zealand. Auckland: Oxford University Press.
18. King Country Chronicle, 12 April 1977.
19. King Country Chronicle, 3 October 1944.
20. Mahoney, P. (1998). The era of the Bush Tram in New Zealand. Wellington: IPL Books.
21. McKinnon, M. (Ed.) (1997). Bateman's New Zealand historical atlas. Wellington: David Bateman.
22. McMillan, L., & Walker, A. (2005). Ongarue: A place of the heart. Ongarue: Ongarue School Centenary Committee.
23. New Zealand Forest Service (1983). Sawmilling in the King Country: An historical review. Wellington: New Zealand Forest Service.
24. Noonan, R. (1975). By design: A brief history of the public works department, Ministry of Works 1870–1970. Wellington: Ministry of Works and Development.
25. Petersen, G. C. (1965). Pioneering the New Zealand bush. In R. F. Watters (Ed.), Land and society in New Zealand: Essays in historical geography (p. 56–79). Wellington: A H & A W Reed.

26. Pollock, K. (2011). King Country places—Benneydale. In (Ed.), Te Ara—the encyclopedia of New Zealand. http://www.TeAra.govt.nz/.
27. Richardson, E. (1884). Report of the North Island Main Trunk Railway Committee. Appendices to the Journals of the House of Representatives, 1884 II I 6, 1–105.
28. Ritchie, N. A. (2012). Report on historic values, former Benneydale coal mine. Hamilton: Department of Conservation.
29. Rochfort, J. (1884). Reports on the main trunk line, Auckland to Wellington: Central Route. Appendices to the Journals of the House of Representatives, 1884 II I6, 87–94.
30. Smith, M. W. M. (1953). Some aspects of family and community life in a New Zealand saw milling village. Diploma of Social Science, Victoria University College.
31. Somerville, J. C. (1965). The timber industry of the King Country. MSc thesis, University of Canterbury.
32. Stanley, R. D. (2013). Cussen, Laurence. In (Ed.), Dictionary of New Zealand biography. URL: http://www.TeAra.govt.nz/en/biographies/2c43/cussen-laurence. Te Ara—the encyclopaedia of New Zealand, updated 7 June 2013.
33. Stevens, J. (1900). Report of the North Island Main Trunk Railway Committee. Appendices to the Journals of the House of Representatives, 1900 I-13, 1–16.
34. Waitangi Tribunal (1993). The Pouakani report. Wellington: Waitangi Tribunal.
35. Waitomo News, 5 March 1998.
36. Walker, A., & Cooke, R. (2003). Waimiha: People of character. Waimiha: Waimiha Reunion 2001 Committee.
37. Walton, A. (1986). The population of the Lake Taupo Region, New Zealand, 1839–1859. New Zealand Journal of Archaeology, 8, 73–88.
38. Wilson, R. A. (1957). The first fifty years of the Marton Sash, Door and Timber Company Ltd. Palmerston North: Marton Sash, Door and Timber Co. Ltd.
39. Yonge, J. (1993). New Zealand railway and tramway atlas, 4th edition. Exeter, UK: Quail Map Company.

Management of Native Forests in the Central North Island, 1919–1977

6

C.M. King, A.E. Beveridge and M.C. Smale

Abstract

This chapter summarises the earliest official (European) attitudes to exploitation of native forests, and the foundation of the New Zealand Forest Service (NZFS) in 1919. The short-term and uncertain nature of early timber mills greatly influenced the development of native forest policy. Further changes in official attitudes to logging of native forest began after the implications of the National Forest Survey of 1946–55 became clear. Analysis of this huge database reinforced earlier predictions that the native timber resource would be exhausted sooner than anyone had expected, and led to a series of changes in management strategy, new methods of inventory and assessment, studies of regeneration rates, trials in selective logging from 1959 onwards, and eventually to new management proposals for native forest including the cessation of clearfelling from 1977. We also describe the planned transition to fast-growing exotic forests, the only kind that would be able to meet future demand. The early development of this forest conversion policy aiming to supply the sawmills with exotic logs in the foreseeable future required a lot of experimental study by scientists of NZFS on replanting clearfelled areas in exotics, and some of this was done at Pureora.

C.M. King (✉)
School of Science, University of Waikato, Private Bag 3105, Hamilton 3240, New Zealand
e-mail: cmking@waikato.ac.nz

A.E. Beveridge
19/10 Gerard Way, Auckland 1072, New Zealand

M.C. Smale
Landcare Research NZ Ltd, Private Bag 3127, Hamilton 3240, New Zealand
e-mail: smalem@landcareresearch.co.nz

C.M. King et al. (eds.), *The Drama of Conservation*,
DOI 10.1007/978-3-319-18410-4_6

Keywords
New Zealand Forest Service · Forest policy 1919–77 · National Forest Survey · Selection logging · 1977 end of clearfelling · Transition to exotic plantations · Pureora Forest nurseries · Traditional log measure · Conversion factors

The huge forests that covered the central North Island at the end of the 19th century looked inexhaustible. European settlers needed to develop the land into productive farms, so no-one objected to extensive logging and clearing of the native bush. Moreover, the official policy of clearfelling was supported by professional foresters.

The forests teemed with pigeons, kaka and other bird species (Chap. 2), but little official notice was taken of them. Hence there was almost no public or government interest in retaining blocks of forest for future purposes that might or might not include wood production—especially as, until large areas of exotic forest reached maturity, the welfare of many businesses and communities depended on avoiding a decline in wood production from native forests.

Licences of various kinds intended to regulate the harvest of native timber appeared in the early days of the British colonial administration, and a Forests Act 1874 provided for the first designated reserves and a department to manage them [25]. The infant department did not last long, and neither did an unsuccessful attempt to revive it a decade later. For most of the 19th and well into the 20th century, the usual practice was simply to sweep away the native forests for farm development, with or without first extracting prime logs [21]. In the absence of any legal regulations controlling the rate and manner of land clearance, vast areas of the most accessible lowland forests simply disappeared.

A Royal Commission on Forestry was set up in 1913 to examine the extent of existing native forest, to decide how much of it must be retained to protect soils water and scenic values, and how much could be released for sawmilling and cleared for settlement [26]. The Commissioners produced the pessimistic conclusion that current practice could not continue. They recommended that the national forest resource should be managed by trained foresters, and the few existing plantations of exotic timber species should be greatly expanded. The Government accepted their report, although a proper forestry organisation with professional staff had to wait until after the end of the First World War.

Origin of the New Zealand Forest Service

On 5 October 1917, Prime Minister William Massey announced his intention to set up a special department of forestry. Initially part of the Lands Department, the State Forest Service was officially established on 1 September 1919, under L. MacIntosh Ellis [1]. The name was changed to the New Zealand Forest Service (NZFS) in 1949 (Fig. 6.1).

Fig. 6.1 The logo of the New Zealand Forest Service, 1949 to 1987. *Max Oulton*

Ellis had firm ideas about what was needed. He established a Head Office in Wellington and seven regional conservancies for administration, and recruited a corps of professionally trained and experienced staff. These were not yet available in New Zealand, so the first new senior salaried foresters were mostly graduates from Edinburgh.

European managers trained in forestry science appreciated the grandeur of the emergent podocarps, although they did not see the forest either as a sacred food basket in the way that the Maori had done, or as the romantic tragedy envisaged by many European explorers. Rather, the main cultural assumption underlying the descriptions and classification systems used by the early NZFS foresters was that the forest was a resource to be managed for the production of timber and the protection of soil and water. That was at least an advance on the prevailing attitude of the previous 100 years, which regarded mature native forest "both positively as a source of timber and firewood and negatively as an impediment to agriculture" [11: 374].

Even though the New Zealand environment was quite different from anywhere else in the world, and home to irreplaceable endemic species dependent on it, the universal assumptions of formal forestry training based on overseas experience were applied here as well. As David Young [35: 180] put it,

> The Forest Service brotherhood of highly trained scientists and skilled foresters were united in a belief, reinforced for an elite by postgraduate training at Aberdeen, Oxford and Canberra, that forests were there to be managed..... Management meant some form of logging.

Of course, the role of protection forests on steep mountain land was clearly recognised, and policies were developed to protect water supplies and reduce soil erosion. The Forests Act of 1921 gave statutory protection to large areas of state forests in mountain lands, whilst public and political attitudes strongly favoured replacement of lowland native forests with grassland for livestock. It was a long time before endemic animal species got the same attention.

The first National Forest Inventory was conducted in 1921–23, and estimated that there was about 4.6 million acres of standing native forest, but only a small proportion of it carried merchantable timber [1: 6]. The total stock was between 35,000 million and 60,000 million board feet (bd ft) of timber (82,590,802–141,584,232 m^3: for metric conversions, see Box 6.1), comprising 63 % softwoods (southern conifers, see Box 2.1), almost half of it rimu, and 37 % hardwoods (broadleaved trees), nearly all southern beech [25: 185]. But Ellis reckoned that demand for sawn timber was likely to rise to 1000 million bd ft a year "within a generation".

Box 6.1. Log measure

From [20] and online reference tables

To convert	To	Multiply by
1 acre	Hectares, ha	0.4047
Chains, surveyors	Metres, m	20.1
1 board foot (=super foot)	Cubic metres, m^3	0.00236
1 cubic foot	Cubic metre	0.0283
Cubic feet/acre	Cubic metres/ha	0.069
Miles per gallon	Kilometres/litre	0.354

The common large-scale measure of standing forest traditionally used in NZ was millions of board feet (1 million bf = 2360 m^3), or cubic feet/acre (1 acre foot = 1233.6 m^3 on 0.4 ha). The terms "board measure" refers to units of board feet.

1 board foot = 1 superfoot: 12 × 12 × 1 inches = 0.083 cubic feet
1 cubic foot = 12 super feet
1 cubic metre = 424 board feet, 35.3 cubic feet
Cutting rights: a fixed term contract allowing a sawmilling company to log a specified area

Stumpage rate: sale price for a block of standing timber, paid as a royalty to the owners
SA volume: volume of merchantable timber available on a given sawmill area.

An average single family home of about 2400 sq ft (223 m^2) requires about 14,000–16,000 bd ft (33–38 m^3) just for the framing, and about 14,000 sq ft (1300 m^2) of flat panelling.

Why such a large demand from a relatively small population? Mainly because timber was plentiful, the risk of earthquakes has always made building houses in timber a safe option, and New Zealand has no long tradition of brick or stone architecture, except in places where it is a speciality, such as in Oamaru. Hence, a New Zealand-trained builder is usually a carpenter rather than bricklayer. So it was clear from the beginning that exploitation of the national timber resource had to be efficient and carefully rationed.

Ellis proposed a master policy for the future development of forestry in New Zealand. He recommended a combination of far-sighted management proposals to conserve native timber supplies. The hitherto unlimited freedom of independent sawmillers, who had become used to operating under conditions of intense competition and loose official control without consideration for the future, would have to be confronted [1: 6].

More than that, if New Zealand was to remain self-sufficient in wood supplies, another source of timber had to be created. Ellis believed that the only way to ensure continuity of supply and a stable and competent management was to eke out the native timber stocks while at the same time transferring production to exotic forests. The idea of *managing* native forests for a sustained yield of slow-growing native timber was generally considered unrealistic. This attitude, commented Malcolm Conway, probably influenced forest policy for another 40 years [21: 4].

Ellis estimated that if the State Forest Service increased its area of exotic plantations from the 13,000 acres (5260 ha) available in 1925 to 300,000 acres (121,405 ha) by 1935, they would yield 450 million bd ft of timber (1,061,880 m^3) by 1965, by which time the yield from native forests would be down to 50 million bd ft (117,987 m^3). Ellis estimated that alternative suppliers and importations could produce only another 200 million bd ft (471,947 m^3), making a total of only 700 million (1,651,813 m^3) of the 1000 million bd ft (2,359,737 m^3) he expected would be needed. Ellis's projections were remarkably accurate: the average sawn timber output in 1964–65 was 696 million bd ft (1,642,377 m^3) [1: 14]. One can see why later Directors of NZFS were so committed to establishing exotic plantations.

In 1939, as New Zealand again sent men overseas to a second world war, the new Director of the State Forest Service, Alex Entrican, recognised that forest policy must in future begin to consider not merely production versus protection, but additional, multiple uses, not only concerned with production and protection—the first time any such idea had been broached, even if not in the modern form [25: 204]. But operational staff with relevant skills and experience were hard to get. The initial task of the Timber Production Advisory Committee established in 1943 was to overcome the labour shortage and to arrange release of volunteers from the war in the Pacific. Further development of the idea had to wait until after the war [25: 272], but Entrican never stopped thinking about it and searching for new ideas—once, with a unexpected consequence.

> [Alex Entrican] made frequent overseas trips to attend forestry conferences - always on the lookout for new, improved techniques and ideas. At one such conference in the United States a local forester approached Entrican and introduced himself. "And who are you?" asked the American. Entrican reputedly drew himself up to his full imposing height and declared: "I am the Director of the State Forest Service". "Yeah – yeah", said the American, "but which State?" Legend has it that immediately upon his return home Entrican had the title State Forest Service changed to New Zealand Forest Service [8: 112].

Fire was a constant hazard in the early days, and bush burns often raged out of control, destroying not only valuable stands of prime timber but mills as well. Before 1918, 42 mills were working on Crown, Maori and settlers' lands along the railway north of Taihape. Then a massive conflagration in March 1918 destroyed at least 12 sawmills, 120 houses plus haystacks, woolsheds, telephone lines, thousands of stock, and huge areas of forest around Raetihi, Ohakune and Rangataua [2]. Part of the reason for establishing a State Forest Service was to control indiscriminate firing [22].

Policy Developments 1946–1977

The National Forest Survey conducted in 1946–55 [14] provided masses of new ecological data on New Zealand's native forests, plus field experience and inspiration to a generation of foresters, trainees and students [16]. Considering the extent of the area covered by the survey, it was extraordinarily intensive, as it was based on one-acre sample plots (5 × 2 chains, 0.4047 ha) spaced 20 or 40 chains (402–804 m) apart on E-W lines one mile (1.6 km) apart. Throughout the ten years it ran, the survey was expertly co-ordinated by the State (later NZ) Forest Service.

The survey reached the central North Island in 1947, and Peter McKelvey [15] used its results to compile the first comprehensive ecological account of the west Taupo forests, including the present Pureora Forest Park. The composition and patterns of the forests were described, accompanied by detailed maps showing the forest canopy composition as it was in about 1952. Subsequent logging and clearfelling give these maps considerable historical value. An example of the data collected (lists of the five main forest species at each sample point as they then stood) is shown in Box 13.6, and a map of the survey lines as they passed through what later became the Pikiariki Ecological Area in Fig. 13.17.

In his 1952 report to Parliament, Entrican stated that the early results from the National Forest Survey showed the volume of native softwoods (podocarps) remaining was much less than previously believed. At the present cutting rate, he predicted, these resources will be exhausted in less than 20 years, and in less than 11 years in the North Island [16: 71].

The enthusiasm engendered by the National Forest Survey led to the establishment of a new Indigenous Silviculture (renamed Indigenous Forest Management from 1978) group at the Forest Research Institute (FRI) at Rotorua. Roger Cameron was appointed in 1956 to run it, and Tony Beveridge was a research forester stationed at Pureora from 1958 to 1960 [5].

The first professional appointments within NZFS of staff concerned with the possibilities of regeneration and management of native forest worked in Roger Cameron's group. The objective was to find ways of achieving a sustained yield of timber, so the group's members were still working firmly under the NZFS policy of managing forests for continued production for economic value rather than protection of native fauna, but they were certainly future-oriented and protective in ways that forest management had not been before. Much of their work concerned developing selection logging techniques and assessing their impact on the forest community, to a degree which varied with NZFS policy changes down the years.

Foresters had traditionally preferred to keep people out of production areas, but in 1954, in response to public expressions of need for legitimate opportunities for recreational uses of state forests, NZFS developed a new working plan for Tararua State Forest, a 280,000 acre (113,312 ha) protection forest north of Wellington. This plan allowed freedom of entry and recreational use to be incorporated into the operation of forests which were also managed for other objectives, including timber production where appropriate. This idea evolved into the State Forest Park concept, and eventually into true multiple-use management (Chaps. 11 and 14).

Beveridge's Review

Tony Beveridge (Fig. 6.2) recognised four periods of development in NZFS thinking about native forests [5]. The first, 1956–61, was the era of preliminary ecological forest surveys. Exploratory work by Beveridge and others, including David Preest and Harry Bunn, considered improvements to management of silviculture; seed production and dispersal and the effects of rodents on it; the growth rate of major tree species; manipulation of the canopy to release natural regeneration or make gaps for nursery- raised seedlings; planting out of seedlings and assessment of their performance; and two trials in Pureora and Whirinaki to measure the effects of logging on changes in forest structure. More than 50 permanent transects were established through forests of various types and logged to different intensities in order to calculate volume increments and changes in structure.

During the second period, 1962–71, policy changes reduced FRI's native forest research team down to two, and exploitative logging and conversion to exotics proceeded apace. The South Island Beech Scheme, a very controversial plan to log large areas of beech forest in Westland, was being developed, providing a huge distraction for forest planners. Three staff (Dudley Franklin, Roger Cameron and David Preest) had to leave their study plots to be managed by a reduced team, and at risk of being surrounded by exotic forest.

Few new silvicultural studies were started and little was done on representative reserves because the focus was still on managing native forest for timber production or for conversion. Cutover sites in tawa-dominant forest cleared for planting of exotics were shown to be inhospitable for new pine planting because of weeds, high costs and too much *Armillaria* root rot.

Fig. 6.2 Forest Research Institute scientist A. E. Beveridge revisiting Pureora forest in 2009. *M.C. Smale*

The mid-1960s was the beginning of a period of rapid change in New Zealand society. After the Second World War there had been pressure to develop New Zealand's infrastructure—to build hydroelectric dams, roads and houses. But environmental damage, once seen as the inevitable consequence of development, was now being challenged by growing environmental activism (Chap. 9).

At the first Forestry Development Conference in 1969 (the 50th jubilee year of NZFS), the emphasis was still on economic growth [21: 5]. The main objective of the conference was to recommend policies which would ensure that New Zealand would make the best use of its capacities to grow wood and to develop sustainable industries based on the forests. The social and environmental aspects of forestry and forest industries were also recognised, and particularly the creation and preservation of scenic and recreational facilities and protection against erosion, but these were not dominant themes.

So by the start of Beveridge's third period, 1972–79, the gathering strength of the environmental movement was creating a lot of work for the Indigenous Forest Management group, especially a demand for advisory work on the concept and design of ecological areas. John Herbert conducted surveys of regeneration in Tihoi SF and in the Hauhungaroas; two more logging trials were established and growth rates assessed.

After digesting the results of the 1974 Second Forestry Development Conference, NZFS eventually came up with a new, multiple-use zoning policy for managing native forests. It was announced in 1975 by the Director-General of Forests, Malcolm Conway, and published in 1977. For the first time, NZFS officially abandoned clearfelling in favour of selection logging and the reservation of large forest areas.

The 1977 change in policy was surprise enough to NZFS staff, but after 1979, the game changed completely, and the fourth and last of Beveridge's phases played out against a completely different administrative landscape [5]. Five new scientists or foresters were appointed, most with ecological or botanical skills and some with a good knowledge of the new quantitative techniques that were now emerging as essential research tools.

Pureora Forest Park was gazetted in 1978, and this new generation of foresters worked largely in the Ecological Areas, with special attention to developing a philosophy of management for New Zealand's now extensive network of reserves in State Forests and other protected areas. Mark Smale, George Pardy, and David Bergin analysed silvicultural trials, regeneration assessments, and succession and performance of planted native species, and re-measured some of the transects established by ecological surveys of the late 1950s.

Selection Logging

The National Forest Survey had already produced the alarming conclusion that the cutting of native forest could not continue at current rates. The consequences of this revelation varied around the country, but one measure taken in the west Taupo forests was to set aside the Tihoi and Waihaha area as a long-term timber reserve, while logging was continued in Pureora State Forest. NZFS also began trials of selection logging in the Waipapa section of the Pureora SF, as a further measure to ensure long-term protection of forest cover for soil and water conservation purposes as well as future timber supply.

The first trial of selection logging at Pureora started in 1961, in podocarp/tawa forest in part of the Pouakani block included in the present Waipapa Ecological Area. In February 1974 the Director-General of Forests A. P. Thompson, NZFS Conservator G. M. O'Neill, and advisers Harry Bunn and Tony Beveridge visited Pureora to evaluate the trials and consider the remaining resources.

Two more trials were established in 1975 and 1976 in podocarp forest at Tihoi (Fig. 6.3), and a fourth at Whirinaki in 1979. The aim of these trials was to provide data to support an alternative to destructive logging, which could leave the old podocarp element in a relatively stable condition while enabling limited logging and natural

Fig. 6.3 An aerial view of an experimental trial of selection logging in Tihoi Forest, 1981. The plot is accessed from Swamp Road, off Link Road (foreground). The grey patch at centre is a small wetland. *Photographer unknown, copyright assumed SCION 8043209*

regeneration to proceed together. The long-term results did not quite support that expectation.

Preservation and timber extraction made uncomfortable bedfellows, and cartoonists gleefully lampooned the idea of selection logging. Burton Silver's famous Bogor strip published regularly in *The Listener* envisaged several pointed questions for the (mythical) annual woodsmen's examinations. The best-known one is shown in Fig. 6.4. Others asked Bogor to define the ecological consequences of clearfelling (he replies "If you take all the trees away, the birds fall down"), or to fill in the missing words in a statement supposedly quoted from the Director of Forests, who favoured selective (a) ——, though there is a case to be made for (b) ——; (Bogor

Fig. 6.4 One of a series of *Bogor* cartoons by Burton Silver, first published in *The Listener* in 1982 during the anti-selection logging controversy of the 1980s. *Reprinted with permission from* The Best of Bogor, *Silverculture Press, Wellington, 1994*

suggested "(a) Morality, and (b) Consciences") [27]. The reality was, of course, much more complicated, as explained in a full account of the selection logging trials done by NZFS since 1961 [3], and summarised in Box 6.2.

Box 6.2 The effects of selection logging

John Herbert, Tony Beveridge and Mark Smale have supplied the detailed arguments central to the debate about the implications of selection logging for management of the forests and their subsequent recovery [9, 28, 29]. The first trial was established in 1961 in podocarp/tawa forest of Pureora Forest (in part of the Pouakani block included in the present Waipapa Ecological Area). Then further attempts at selection logging of rimu-dominant dense podocarp forest were tried in the Tihoi Block of the present PFP in 1975.

Measures to reduce logging damage were tried in blocks in which 30 and 55 % of the merchantable volume were removed by tractor. These fixed proportions were reasonable in theory but hard to achieve in practice, because the forest was so complex [3]. The objective was to remove small groups of trees by directional felling, leaving other groups intact. Damage to the residual trees was assessed in terms of undercutting of root systems, compaction of root plates by tractor, debris left on root plates, debarking of tree stems and butts, and crown damage.

Severance of lateral roots by undercutting (8 % of trees) was considered to be the most serious type of damage, likely to cause instability of the surviving large podocarps, most of which were aged between 400 and 600 years. Nearly a year after logging, a windfall assessment showed that 39 trees had fallen over in 36 ha (6 in the unlogged control) although 74 % of these windfalls had been classed as 'culls' owing to internal or external butt and stem rots. Subdominant miro were prone to uprooting. In the logged blocks, logging machinery had disturbed 74 % of the ground to various degrees.

When John Herbert returned to Tihoi to check the results, he estimated that, in the 12 ha unlogged control block, the estimated gross stand increment was 1.79 m^3/ha/year (including culls), less than the total measured natural losses over 3 years (1975–78) of 2.48 m^3/ha/year, giving a net loss of 0.69 m^3/ha/year [10].

Logging resulted in an increase in this natural mortality, producing an estimated net loss after selection logging of 3.44 m^3/ha/year in the 30 % logged block and 7.89 m^3/ha/year in the 55 % logged block. If only merchantable trees are considered, the net annual increment for the control block was +0.17 m^3/ha/year, but −1.47 m^3/ha/year in the 30 % block and −1.71 m^3/ha/year in the 50 % block.

If this accelerated short-term rate of loss were maintained in selectively logged forests, Herbert warned, the value of the remaining forest would be seriously diminished. Forest ecologists point out that the merchantable trees are only part of the forest community. 'Cull' podocarps with stem or butt rots comprised 18 % of all trees in the original stand of the control block, but 40 % of windfalls.

By 1998, a new review [23] confirmed a general net volume decrement in both logged and unlogged forest at Pureora. Even where large podocarps aged mainly from 400–700 years survived, many were senescent and at risk of windfall or standing death. Recovery over the 24 years since harvesting of podocarp/tawa forest had produced thickets of putaputaweta and small leaved coprosma over dense ferns on compacted sites, often with rimu saplings.

Less disturbed sites in the 1961 Pureora trial area were originally colonised by wineberry and fuchsia, and slash by wineberry, fuchsia, pate, fivefinger, kamahi and ferns. Most wineberry had collapsed after 15–20 years, and when possums arrived, nearly a decade after logging ceased, fuchsia and other broad-leaved species mostly succumbed to browsing. Rimu and tawa saplings and matai seedlings were now frequent on these sites, with tree ferns and ground ferns.

Smale and Beveridge [29] made the most recent re-measurement of the long-term (43 year) impact of group selection harvesting in the two 15 ha harvested blocks. They found that, although selection harvesting obviously reduced the overall density of mature trees, it had not substantially altered the canopy composition or the population structure of conifers. It did not adversely affect the stability of the residual forest. Mortality rates of merchantable conifer trees were similar in the 15 ha control (0.7 %/year) and of the residual conifers in the two harvested blocks (0.6 %/year, 0.8 %/year). Tree mortality in harvested blocks was not at that time considered to be related to harvesting disturbance. Productivity was somewhat reduced by selection harvesting (0.3, 0.2 m^3/ha/year in harvested blocks; 0.4 m^3/ha/year in the control) because of reduced tree densities. Net merchantable volume increment was still negative in conifer species in all blocks, ranging from −0.4 to −0.6 m^3/ha/year. A later more detailed analysis of all residual trees—merchantable and cull—found that despite harvesting to the highest standards, there had been a brief period of elevated mortality after it [30].

For a start, logging trials in complex forest ecosystems have to be done on large areas to be realistic, but it was often next to impossible to find suitable, large enough blocks to compare, and to ensure that they were similar before any logging began. Then, extensive roading was necessary to get tractor access to the trees chosen for felling, so even in the Tihoi block committed to the lowest level of harvest (30 %), nearly 60 % of the ground surface was damaged by dumping of debris and churning by tractor treads. Even very large trees have a superficial mat of "feeding roots" in the surface humus, and many large roots less than 60 cm below it, so damage to these sensitive roots by heavy machinery greatly contributed to post-logging mortality and wind-throw [18: 87].

Three of the four trial areas were within Pureora Forest Park and were logged in 1961, 1975 and 1976. The low density podocarp trial area (1976) was part of several hundred hectares of partially-logged forest in the same block, and contains exceptionally abundant and well-advanced regeneration of podocarps to pole size, sufficient to replace the present large, scattered podocarps. Prolific wineberry germinates in the summer after logging in these forests, but starts to die back after 10–15 years, which is also the time taken for smaller logging debris to disintegrate. Epiphytic growth on head logs (or on fallen stems) and large branches include kamahi and broadleaf (future podocarp nurses) and podocarp seedlings, a few of which become terrestrial and continue growing.

An early review of the effects of selection logging trials was done as part of the work of the Forest Bird Research Group during the 1978–81 logging moratorium (Chap. 10) [13]. The team sampled and described the forest types, and measured the forest structure and density of vegetation and species abundance. The conclusion that seemed most reasonable at that time was that selection logging, at the levels used and in the study blocks they surveyed, had little impact on forest structure.

This optimistic assessment supported the official view, expressed by the then current OC at Pureora, Jack Walker. His 1979/80 Annual Report records this comment:

> I am of the opinion that selection logging is a legitimate form of management, which, if carried out with care and sensitivity, will not only produce some very valuable and sought after timber, but also help to perpetuate the forest itself. From my observations over the past 6 years the forests of the Pureora region appear to be undergoing a change

> in that Tawa is succeeding Podocarp in most cases. The reaction of light wells where podocarps can thrive, combined with fertilizing and releasing, would appear to be the only way in which the mixed podocarp-hardwood forest as we know them today, will be available to future generations [34].

Other interested parties strongly disagreed. The fiercest criticism of the policy of selection logging and of its effects on a virgin forest came from The Whirinaki Forest Promotion Trust, a group of prominent senior conservationists led by John Morton, formed to argue against the continuation of selection logging in Whirinaki [18]. They rejected the assumption cited by Jack Walker, that the podocarp forests cannot sustain themselves without the local regeneration stimulated by selection logging. They pointed out that although regeneration is visible in many places in Whirinaki, the age-structure of an ancient podocarp forest is always unbalanced. Standard arguments about managing such forests for sustained yield often invoke the example of managed fisheries, in which the harvest of an important species such as snapper can be monitored to ensure that it extracts only the equivalent of the annual increment. Over- or under-harvesting in one year can be adjusted by changing the following quotas, and the population can be brought back to balance within a few years.

But an old-growth podocarp stand is not like a fishery, because the standing crop is an immensely complex association of many different species, with an extremely slow rate of recovery after harvesting. The time scale for fisheries-style management of a podocarp forest would need to be measured in centuries, far too long for any stable human oversight, and the losses and damage recorded cannot be offset by the increment of the remaining trees (Box 6.2). Furthermore, the background assumption made by Walker and many other foresters, that tawa will inevitably replace the podocarps, did not survive the later detailed analyses of forest structure in PFP by forest ecologist John Leathwick (Chap. 2).

The arguments for and against selection logging stimulated a lot of independent research attempting to predict the effects of selection logging on native birds and invertebrates. One study by Colin O'Donnell and Peter Dilks [23] was based on detailed knowledge of the habitat requirements of forest birds, with a view to estimating the amount of preferred habitat that would remain to them after logging. Their model showed that the effects on wildlife of selection logging would be severe, even when only a small proportion of trees were extracted. Very large and old trees, including those that are senescent, unstable, standing dead or vulnerable to windthrow, are the ones most likely to be targeted for removal, yet these support the most invertebrate larvae, fungi, heavy fruit crops and nesting sites—all critically important resources for forest birds. The removal of those particular trees would affect the native fauna out of all proportion to their numbers, even if tractor damage to the forest floor was avoided by the use of helicopters for log extraction.

Many years of re-measurements of the trials at Pureora produced long-term results that contradicted the earlier ones. They tended to support the Whirinaki Trust's argument, and future policies were shifting. By 1998 it was clear that in the forest types chosen for the trials, dominated by large podocarps aged mainly from 400–700 years, the remaining trees were liable to windfall or standing death [28]. Miro had usually been left because its timber was not saleable, and because its fruit is an important food for pigeons, but proportionately more miro died from windfall than any other species—"Hard luck, pigeons", commented Morton et al. [18: 97]. In the 1961 Pureora trial area, the net timber volume decreased in both logged and unlogged forest. Broad-leaved species (wineberry, fuchsia, pate, fivefinger, kamahi and ferns) quickly colonised areas of slash, but these were browsed out when possums arrived nearly a decade later, leaving only tree ferns and ground ferns

In the 1975 and 1976 trial areas, thickets of putaputaweta and small-leaved coprosma over dense ferns were established on compacted sites, often nursing rimu saplings. Less disturbed sites were colonised by wineberry and fuchsia, but again, most wineberry collapsed after 15–20 years, and fuchsia succumbed to browsing by invading possums. Rimu and tawa saplings and matai seedlings now frequently replace them on these sites.

The most recent reassessment, 43 years after the trials began, concluded that, although regeneration of major canopy species was well advanced within a decade of harvesting, much more time would be needed for the forest as a whole to recover fully [29].

The 1977 Management Policy: The End of Clearfelling

In 1971 the Government published a document entitled *Utilisation of South Island Beech Forests*. It proposed large-scale milling of South Island lowland beech forest to provide timber for one or more pulp mills. Half of the milled area was to be replanted with exotic *Pinus radiata*, while part of the remainder was to be selectively logged and replanted with eucalypts.

The Nature Conservation Council and representatives of the Royal Forest and Bird Protection Society (RFBPS) inspected the Westland beech forests involved in the proposals, and considered their response. Eventually it was decided that, although there was no objection to some beech forests being used for sustainable management, the plans for clearfelling and burning of others for the establishment of exotic forest were regarded as unacceptable. RFBPS decided to organise a Parliamentary petition to effect changes.

NZFS's proposal also galvanised many other more militant groups, including the Beech Forest Action Committee, which later became the Native Forests Action Council (NFAC), and later still, the Maruia Society. It produced the Maruia Declaration, signed on the banks of the Maruia River—one of the first areas in Westland designated for clear-felling—on 4 July 1975. This was circulated as a public petition before being submitted to the government in 1977 with 341,159 signatures [25: 420]. It demanded legal recognition of native forests and an end to their logging. The Maruia Declaration seemed radical at the time, but in fact almost all of its demands were met within the next 30 years.

Discussions of the proposal for landscape-scale logging of beech forest had raised important issues about the adequacy of existing reserves. Although there were large areas of beech in mountainous national parks, there was little remaining lowland forest, and most of it was earmarked by the Forest Service for milling on the grounds that logging must continue to maintain timber supplies and full employment.

Subsequent opposition from an increasing list of environmental groups escalated from specific concerns about the West Coast beech scheme into an assault on the basic tenets of the Forest Service in general. NZFS, which had since its inception regarded itself as a conservationist department concerned with wise use and protection of forest and land—in contrast with other government bodies which promoted exploitative and consumptive use [25]—found it difficult to understand why there was such persistent opposition to its plans, which it perceived as being for the good of the nation.

A comment in the 1973/74 annual report of Pureora State Forest shows how far official thinking was shifting during this period, long before the protests began.

> "Conservation" and "environment" are two words that now occupy a very important place in Pureora's vocabulary. Virtually every action concerned with logging or land conversion is now examined with the environment in mind [34].

Over the next few years, things changed remarkably quickly. The Second Forestry Development Conference of 1974 brought together a wide range of people interested in the many possible uses of forested land, and enabled them to contribute to the discussions. It was preceded by extensive surveys and preparation of background papers on existing resources and policies, and followed by 33 written submissions, which were received and formally considered. NZFS eventually came up with a new multiple-use zoning policy for managing native forests, specifically excluding clearfelling and promising protection for large forest reserves.

Parliamentary elections in 1975, and their aftermath, caused so many distractions that it was another two years before any government could enact this change in policy, but the intention was picked up by the candidates during the campaign. The National (conservative) Party election candidate for Taupo, Ray La Varis, promised that his party would not allow any milling in the Waihaha forest. So the electoral issue in 1975 was not so much that no change was in sight, but that it was a long way away and the need to bring it forward was getting very urgent.

The election was won by the National Party, and in 1976 it amended the Forests Act 1949 to give people greater statutory opportunity to offer comment and advice on management proposals. The final version of the new policy, covering all remaining native forests in the country was published in March 1977 [21].

In his official capacity as Minister of Forests granting Government approval to the new 1977 NZFS policy for native forest management, Venn Young wrote that

> New Zealand's forests, native and exotic, support a thriving and strongly developed industry. In the early days of European settlement our native bush was a barrier to land occupation and settlement. Later it was regarded as a resource to be exploited for the production of timber for domestic use and sale overseas [21: 3]

Young might have put "Currently" instead of "Later", because there were still many people who thought that way. But history shows that the 1977 document was simply the latest instalment in a long series of policy changes in favour of preservation of forests and wildlife by an organisation that started off with very different assumptions. The changes announced in 1977 were not driven by professional foresters, but they accelerated a critical process which undermined NZFS's traditional management plans from 1978 onwards (Chap. 9). After 1984, even more radical, politically motivated changes were imposed on forest managers, against the bitter opposition and muzzled objections from their own trained staff (Chap. 11).

One might not expect such a slim brown booklet of only 15 pages plus four pages of colour photographs to have such a significant effect on an important national industry, but so it was. The policy not only stopped clearfelling completely, but also reduced the timber available for milling by about two-thirds, and allowed for extensive ecological reserves. In other words, usage of the remaining areas of native forests would be divided in the ratio of about 2:1 between preservation and low-level timber production.

NZFS envisaged a management regime under which logging would be prohibited in areas to be set aside as ecological reserves, and in protected forest which was to be preserved for soil and water conservation purposes. It provided for only limited further timber production from some North Island podocarp forests, including some of the west Taupo forests zoned as 'periodic-yield areas' or 'partial logging areas'.

The extent of this intended self-restraint was an astonishing contrast to the 1960s when cut-over forests were routinely either planted in exotics or transformed into farm land in the climate of development that prevailed then [32]. NZFS management hoped and expected that it would be seen as radical enough to disarm the growing environmental lobby.

Local staff realized that implementation of the new policy would have some dramatic effects on Pureora Forest, its staff and their work [34]. Tony Beveridge, an experienced FRI scientist familiar with Pureora, commented that

> A vital decade for decisions that will have great influence on the future nature and extent of our native forests and the ways it can be used and enjoyed by man began in 1975 with the introduction of the revised policy for our native state forests....it is generally accepted that "multiple-use forestry" is now a reality and that wood production is only one value, and an increasingly minor one, in many native State forests [4].

He was right, although no-one realized at the time that the protracted transition to the new policy was only just beginning, and would be far more painful than anyone in NZFS expected.

By the time Beveridge's paper was published, the tide of events was taking the key decisions out of the hands of those who had previously held all the cards.

Part of the urgency of introducing the revised policy to Pureora arose from the clear recognition that hitherto legally committed timber supplies were not going to be able to meet two existing 15-year contracts signed in 1968 and 1970 (Chap. 7). The total resource available in PFP at 1 November 1977 was 142,000 m^3, which, at the then current cutting rate of 43,000 m^3 a year, would be exhausted by December 1981 [17].

Negotiations with the affected sawmilling companies holding current contracts started in May 1977, aiming to find suitable alternatives [17]. The timber companies were staunch at first, because they were convinced that their contracts were inviolable, and they did not believe that the rising public criticism of NZFS policy could ever change that view.

But it was already too late. The 1977 policy did not have time to prove itself before the end of that year, by which time events were moving out of NZFS's control. On 20 December 1977, NFAC requested a total halt to logging in the Pikiariki area of PFP, which was refused. In January 1978, the sky fell in (Chap. 9).

Growing Pains: The Transition to Exotic Timber

In the 1950s, New Zealand was still constructing almost exclusively timber buildings in most parts of the country, but supplies of the slow-growing native timbers were rapidly running out. Ellis' prediction of 1920, that only 50 million board feet of native timber (117,986 m^3) would be left by 1965 [25: 278], was proving to be remarkably near the mark. By the 1950s only nine native species were still being used to any extent, i.e., kauri, rimu, matai, totara, kahikatea, miro, tawa, red and silver beech. The inevitable end of milling of native timber began to appear uncomfortably close [22].

There was no short-term alternative but to reduce the harvest of North Island native timber substantially. Needless to say, politicians, sawmillers and consumers admonished the Forest Service as proposing strategies damaging to the national economy, but on 10 July 1952 Cabinet authorised NZFS to reduce native timber sales "as far as practicable" [25: 279]. At the same time, since regeneration of native forest takes about five times as long as one growth-to-harvest cycle of a suitable exotic species, it had long been equally obvious that only efficiently managed exotic plantations would ever be able to meet the relentless demand for wood and to support wood-processing industries of great value to the national economy [21].

As early as 1890, the Forest Branch of the Lands Department had established exotic plantations near Whangarei, and from 1897 they began experimental tree planting at Rotorua (hence the redwood forest there), and in the barren pumice on the Kaingaroa Plains northeast of Taupo. By 1909 the rapid growth of *Pinus radiata* trees there was recognised as making pine plantations a potentially attractive commercial proposition, so that a 1923 estimate by the State Forest Service reckoned that the future return from pines would be between £250 and £500 per acre.

In the late 1920s, New Zealand was struck by the global economic depression and widespread unemployment. Use of the unemployed for afforestation was seen by government as constructive relief work, so an increase in planted areas during the period 1929–1931 became an unplanned benefit of the 1930s economic meltdown. Large areas of open fern, scrub, tussock and heath lands of the central North Island—mostly outside PFP—were planted in exotic forest by the Government and by private companies.

From then on, NZFS had to devote substantial resources to the establishment and management of exotic forests. They knew the transition to dependence on them would need to be completed by the 1980s, when they expected the available native forests would be exhausted and the remainder largely protected.

By 1949, as the dwindling supplies of native timber in the state forests of the west Taupo ranges began to raise apprehensions, the newly renamed

New Zealand Forest Service extended the official policy of planting exotic species on logged-over areas of native forests to the Pureora area. Experimental planting of exotic species in the King Country began in Hurakia State Forest (now part of PFP) in 1937. Although the first batch of Douglas fir was significantly damaged by frost, the policy of exotic tree experimentation was continued, because forest managers expected it to ease anticipated pressures on native forests and to create ongoing employment in the rural areas.

In April 1951, the Director-General of NZFS proposed to discuss with the Minister of Forests the possibility of establishing Pureora SF as the hub of a sustained-yield project based on complete clearing of logged native stands and conversion to future plantings of exotics, except along stream banks that were to be set aside as Scenic Reserves [34].

Production of sawn timber and wood-pulp from the exotic forests planted during the 1930s Depression was expanding rapidly, and so was the market for wood products. To meet the escalating demand, the Pureora Working Circle Plan drawn up in 1952 [19] reckoned that some 15,484 acres (6265 ha) within its area could eventually be stocked (at between 25 and 80 %, depending on the soil type) with exotic plantations (Chap. 7).

The first step was to establish a local nursery where experimental plantings could check the suitability of exotic trees to the climate and soil conditions of Pureora Forest, and then to produce large numbers of hardy seedlings ready for planting out in cleared areas. The plan depended on a silviculture gang under a good foreman, and it had to get going well before weeds and second growth took over. The Pureora Forest village community was planned and equipped to meet all these requirements.

The Pureora Nurseries

Two local nurseries were established near Pureora Forest village—the first in 1950 opposite the access to the Totara track, and a larger one in 1953 (Fig. 6.5). Young trees were brought in and acclimatised ("lined out", planted in the nursery for hardening) for one or two years, and then transplanted out into cutover bush (Fig. 6.6). Planting began with Douglas fir, the preferred tree at the time [7].

Other species besides Douglas fir that were established in this way included western red cedar, Japanese cedar, and macrocarpa. To begin with, the idea was to "enrich" cutover stands by planting exotic seedlings in cleared areas, but this did not work well, so the policy switched to burning of cutover forest remains followed by total conversion to plantations.

In 1951 the nursery was being run by Mrs Barrett (forewoman) and Miss Hona, and lining out was going well, although more girls were needed. From October that year, nursery staff also began to raise *Pinus radiata* from seed. Colin Sutherland was the officer in charge of the nursery, and Harry Bunn, a new graduate from

Fig. 6.5 The Pureora forest nursery. *Crown Copyright Department of Conservation Te Papa Atawhai (1950) Photographer: Eric Johnstone*

Fig. 6.6 Interplanting cutover native forest with Douglas fir at Pureora, illustrating Harry Bunn's diamond-shaped group method, their centres spaced as marked by poles. *Crown Copyright, Department of Conservation Te Papa Atawhai (1956). Photographer unknown*

the Australian School of Forestry, was appointed in February 1953 to solve the technical problems. Harry worked closely with FRI scientists and, having been trained in Australia, was an expert on eucalypts. He watched closely what was being done, and was soon asking whether there might be better ways [12: 13].

By the end of 1954, 97,600 trees had been lined out. Included in that total were: 67,650 Douglas fir; 9250 Japanese larch; 15,000 Monterey (radiata) pine; and 5700 mountain ash (*Eucalyptus delegatensis*). Initial progress was very encouraging, and it was believed that the excellent progress of the first Douglas fir plantings had proved the suitability of Pureora for establishing exotic plantations, although it was recognised that these trees would take 40–50 years to mature.

The nursery programme started well, but over the long term, several serious problems appeared. First, the young trees struggled to cope with the cold, wet climate of the high country (Box 1.2). The field set aside for the new nursery was a frost hollow, which meant that only certain frost-hardy species could be used. Pureora's frosts were deadly to non-acclimatised seedlings of most species except, possibly, Douglas fir. "We were still making the same mistakes even in the 1980s", commented John Gaukrodger much later; "on one occasion we planted 20,000 rimu seedlings from Sweetwater and lost the lot. Eucalypt seedlings planted in September at 40 cm tall might have shot up to a metre high by Christmas and then got wiped out completely by a 5° frost in January".

Second, it had been clear for some time that not enough local labour was available to deal with the planned exotic establishment programme, originally expected (in the Pureora Working Circle plan) to cover 6265 ha (Chap. 7). By September 1955, the shortage of labour put the planting programme in arrears, so immigrants were bought in, including Gordon Gillespie from UK who later became officer in charge of exotic planting.

Third, Harry Bunn found that only 4048 ha were suitable for planting, i.e., less than two-thirds of the cutover land could be replaced with exotics as planned. Doubts began to arise about Douglas fir, because it is susceptible to the needle-cast fungus *Phaeocrytopus gaeumannii*, and takes longer to harvest (Box 6.3) than shorter-rotation species like *Pinus radiata* [7]. Then came the first appearance in 1965 of *Dothistroma pini*, a damaging needle blight disease of pines, and by 1975, large-scale mortality due to *Armillaria mellea* root rot fungus that could cause losses of up to 40 % of young pine trees planted on cutover native forest.

There were continuing difficulties with finding the best species to plant in the area to compare with the early favourites, Douglas fir and larch. However, despite all the dangers of relying on a monoculture known to be vulnerable to diseases, radiata pine has continued to dominate new plantings to the present day.

Fourth, there were the pests. Rabbits dig burrows or chew the bark of young trees, and were such a nuisance in the nursery at Pureora village in the 1950s that nursery staff had to lay poison for them.

Much more annoying, and harder to deal with, were the horses. When the open ground in front of the village (Figs. 1.8 and 14.3) was let out for farming and fenced off during the 1956/57 year, the semi-feral horses loosely attached to the village families had to find other grazing areas. They became a considerable nuisance around the village, breaking fences to get at vegetable gardens and rubbing their haunches against the houses. One report complained about the danger to children in the village from uncontrolled horses, including a stallion chasing five mares around the houses [34]. Worse, the horses found the freshly turned earth of the nursery ideal for rolling in. Every roll damaged a few more planted seedlings, but nothing was done about them until Colin Sutherland stormed into the OC Kitch Pedder's office one Friday afternoon —"and he was *mad*!!", remembered Pedder—because he had just planted out an area and the horses had come in and stirred it all up.

Official protests and threats, and a rodeo-style round-up made no long-term difference, so inevitably, several of the offending animals were shot over the next couple of years. Harry Bunn reckoned that it was all done according to the right procedures—advertising first, and so on—and Colin deliberately shot only the oldest ones. On tape, Pedder denied doing it himself, but the next OC Ivan Frost did not mind admitting it was he who gave Gordon Gillespie a box of ammunition and told him to go for them. He took out eight or ten himself, but commented that horses are very hard to kill, so it was the worst thing he ever had to do [33].

But it did not help much. The 1958/59 annual report complains that the revised working plan for silviculture was still two years overdue. Only 50 % on average of cutover land a year was being planted because of insufficient nursery stock. Problems in the nursery continued with old roots in the soil, frost scorching, poor drainage, problems with irrigation, and wild horses still damaging plantings despite a prolonged effort to get rid of them. *Armillaria* continued to damage planted-out radiata pines.

By 1963 the horses had gone, but possums were moving in and the nursery was still suffering from bad seed, late frosts and *Armillaria*. The staff knew the urgency of the problem, and went on trying.

National officials were still confident, despite local problems. The then DG of Forests, A. L. Poole, told a meeting of the Dominion Sawmillers' Federation that

> We are all familiar with the changes occurring during the transition from milling based primarily on native logs to our present milling based primarily on exotic logs.....The native resource is but a shadow of its former self and the end of that in the North Island can be forecast.......The Forest Service has the greatest trouble in exerting control over the cutting.....[to provide] the breathing space so urgently needed to give time to plant exotic forests.....now the gap between the foreseeable end of the native timbers and the coming in of the exotics has been almost bridged [24].

Poole predicted that the volume of sawn exotic timber produced in the Auckland Conservancy alone would rise from a mere nine million cubic feet (21,237 m^3) in 1966 through 19 million (44,835 m^3) in 1980, to reach 42 million (99,108 m^3) by the year 2000, by which time the production of sawn native timber would have dropped to 8 million cubic feet (18,877 m^3). Poole's forecasts were based on 1967 estimates valid at the time; he had no way of knowing how events at Pureora would affect them within the next decade.

Under the new native forest policy published in 1977, NZFS restricted the west Taupo area to be converted to exotics in future. The total existing area of plantations totalled 5895 ha in

Box 6.3. Areas of exotic forest plantings (ha) around Pureora as at 1978, by species and period planted [7]

Sub-region	Species	Before 1950	1950–54	1955–59	1960–64	1965–69	1970–74	After 1975	Total
Pureora	Douglas fir	10	40	160	340	790	760		2100
	Radiata pine				40		380	1180	1600
	Other	5	12	46	149	150	38		400
Hurakia	Other	200							200
Tihoi	Douglas fir		5	45	36	235	441	198	960
	Radiata pine			2			88		90
	Eucalypts							135	135
	Other			10	3		2		15
Waituhi	Radiata pine						395		395
Total		215	57	263	568	1175	2104	1513	5895

Projected sawlog yield from radiata pine: 550 m^3/ha after 25 years; from Douglas fir, the same but after 45–55 years.

Under NZFS plans of the time, the sawlog volume from exotic plantations from 2000 onwards was expected to total 171,000 m^3 per year.

March 1978 (Box 6.3), and a further 1100 ha were to be planted by the end of the year. But only about 400 ha of land suitable for planting [7] remained.

The nursery was closed in 1966/67, meaning that future planting requirements would have to be met from Rotorua or Kaingaroa. The plantings at Piropiro were successful with stock grown at the Cambridge nursery, but elsewhere, NZFS continued to ignore previous experience, that often made planting imported stock a lot of wasted effort.

Large-scale planting of exotics ceased in Waituhi in 1973, in Tihoi in 1977, and was expected to end at Pureora in 1978. Some areas within PFP were still being planted with exotic species under a special arrangement to provide employment up to 1985 (Chap. 10), but NZFS ran out of plantable land before that. Further conversions on DOC land were prohibited under the Conservation Act of 1987. This Act provided the park with its long-awaited formal legal protection, whilst at the same time allowing for some cultural harvest of totara for carving [6].

Coping with the Transition to Exotics

The men who had spent their lives handling native timber in the bush and at the mills developed a real feel for it. They loved a beautiful, clean matai or rimu log, and took great pride in cutting it correctly. Sometimes, remembered Rusty Russell, a long-time Pureora resident who had never known any other trade (Chap. 8), you might run your hand along a totara log on the skids and say “Man, great stuff!”.

The old-time loggers regarded radiata pine as fit only for firewood. “My father”, said Rusty, “used to say, when driving past a plantation, ‘I dont know what they are going to do with all this stuff, they are only bloody weeds…. There’ll be enough timber in the King Country to see us both out’, he said, but he was wrong” [33].

Magnus Russell was not alone in his opinion. Even as late as 1965, Somerville [31: 75] concluded his analysis by stating that: the “reafforestation scheme has meant that the mills of Ranginui Timber Co. and C. & A. Odlin Ltd

Fig. 6.7 Thinning a pine plantation in Compartment 124 (on the corner of Pikiariki and Barryville Roads). Les O'Leary is gathering stems of 7–8 year old pines cut for posts; the rest matured and were logged in 2010. *Crown Copyright, Department of Conservation Te Papa Atawhai (1983). Photographer: John Mason*

have an indefinite life expectancy. The native forest in this area will continue to produce logs for another 25–30 years, and by that time the first and second thinnings (Fig. 6.7) of the exotic timber will have taken place and the mills will be able to carry on cutting *Pinus*. Thus for these two mills the future appears secure".

Harry Bunn's innovations, such as a diamond pattern for interplanting within cleared patches, earned the predictable blunt reactions from traditional bushmen such as logging officer Jack Fyffe [33]. Nevertheless, Harry did the job for more than three years, of which he remembered mostly the perpetual problems of shortage of skilled labour and the constant battle to get appropriate stock for planting.

Harry's conclusion, after many years of service with NZFS, was that "at Pureora they went into things on a management scale without research to establish how feasible it was. Research has to look well ahead of management. By the time you've got an answer, management is no longer interested" [33]. But, hard though it was for some, the transition to exotic timber was inevitable, and only those foresters and mill managers willing to adapt to producing and handling it survived past the 1980s.

References

1. Allsop, F. (1973). The first fifty years of New Zealand's forest service. New Zealand Forest Service Information Series, 59, 1–123.
2. Auckland Herald, 20 March 1918.
3. Beveridge, A. E., & Herbert, J. W. (1978). Selective logging trials and their implications for management of the West Taupo forests [*Transcript of the Taupo seminar*]. Wellington: New Zealand Forest Service.
4. Beveridge, A. E. (1979). Management of the central North Island indigenous forests (Sanderson Memorial address). Forest and Bird, 214, 25–33.
5. Beveridge, A. E. (1985). Ecological research of the Indigenous Forest Management Group at the Forest Research Institute, Rotorua. In B. Veale & J. G. Innes (Ed.), Ecological research in the central North Island volcanic plateau region: Proceedings of a NZ forest workshop (p. 18–26). Rotorua: Forest Research Institute.
6. Burns, R. (1998). Pureora Conservation Park strategic plan. Hamilton: Department of Conservation, Waikato Conservancy.
7. Field, D. A., & Robinson, E. D. (1978). Production forestry potential in the state forests of the Hauhungaroa-Rangitoto ranges [*Transcript of the Taupo seminar*]. Wellington: New Zealand Forest Service.
8. Harris, D. (Ed.) (2009). Timber cruising and other forestry stories. Wellington: FirstEditionPublishers.com.
9. Herbert, J. W. (1980). Structure and growth of dense podocarp forest at Tihoi, central North Island and the

impact of selective logging. New Zealand Journal of Forestry, 25, 44–57.
10. Herbert, J. W., & Beveridge, A. E. (1977). A selective logging trial in dense podocarp forest in the central North Island. New Zealand Journal of Forestry, 22, 81–100.
11. King, M. (2007). The Penguin history of New Zealand illustrated. Auckland: Penguin Books.
12. Klitscher, K. (Ed.) (1997). Characters of FRI. Rotorua: New Zealand Forest Research Institute Ltd.
13. Leathwick, J. R. (1981). The vegetation of kokako and general bird study areas in some central North Island indigenous forests. Wellington: New Zealand Wildlife Service, Forest Bird Research Group.
14. Masters, S. E., Holloway, J. T., & McKelvey, P. J. (1957). The national forest survey of New Zealand, 1955. Wellington: New Zealand Forest Service.
15. McKelvey, P. J. (1963). The synecology of the West Taupo indigenous forest. New Zealand Forest Service Bulletin, 14, 1–126
16. Millen, J. (1999). Through trackless bush: The story of the New Zealand national forest survey 1946–55. Rotorua: P. F Olsen & Co.
17. Molloy, G. J., Rockell, J. D., & Ure, J. (1978). Introduction to the State Forests of the Rangitoto and Hauhungaroa ranges [*Transcript of the Taupo seminar*]. Wellington: New Zealand Forest Service.
18. Morton, J., Ogden, J., & Hughes, T. (1984). To save a forest: Whirinaki. Auckland: Bateman.
19. New Zealand Forest Service (1952). Working plan for Pureora working circle. Auckland: New Zealand Forest Service.
20. New Zealand Forest Service (1972). Metric conversion tables for forestry. Wellington: New Zealand Forest Service.
21. New Zealand Forest Service (1977). Management policy for New Zealand's indigenous state forests. Wellington: New Zealand Forest Service.
22. New Zealand Forest Service (1983). Sawmilling in the King Country: An historical review. Wellington: New Zealand Forest Service.
23. O'Donnell, C. F. J., & Dilks, P. (1987). Preliminary modelling of impacts of logging on forest birds in South Westland. Department of Conservation, Science and Research Internal Report, 1, 1–30.
24. Poole, A. L. (1967). The future forest resource of New Zealand. Rotorua: Dominion Sawmillers' Federation.
25. Roche, M. (1990). History of Forestry. Wellington: New Zealand Forestry Corporation Ltd
26. Royal Commission on Forestry (1913). Report of the Royal Commission. Appendices to the Journals of the House of Representatives, 1913 C12.
27. Silver, B. (1994). The best of Bogor. Wellington: Silverculture Press.
28. Smale, M. C., Beveridge, A. E., & Herbert, J. W. (1998). Selection silviculture trials in North Island native forests: Impacts on the residual forest and their implications for sustainable forest management. New Zealand Journal of Forestry, 43, 19–30.
29. Smale, M. C., & Beveridge, A. E. (2007). Long term ecological impacts of selective harvesting on a New Zealand conifer-hardwood. New Zealand Journal of Forestry Science, 37, 3–22.
30. Smale, M. C., Fitzgerald, N. B., Arnold, G. C., & Bartlam, S. (2008). Basal area recovery of podocarp-tawa forest at Pureora 47 years after selective harvest. Lincoln: Landcare Research Contract Report LC0708/169.
31. Somerville, J. C. (1965). The timber industry of the King Country. MSc thesis, University of Canterbury.
32. Waitangi Tribunal (1993). The Pouakani report. Wellington: Waitangi Tribunal.
33. Wilkes, O. (2005). Transcripts of oral history recordings. Hamilton: Department of Conservation.
34. Wilkes, O., Gaukrodger, J., & Ritchie, N. (2011). Chronology of Pureora. Hamilton: Department of Conservation.
35. Young, D. (2004). Our Islands, our selves. Dunedin: University of Otago Press.

The Pureora Project 1940–78

7

D.J. Gaukrodger, N.A. Ritchie and C.M. King

Abstract

In the late 1930s, NZFS began to consider long-term management plans for the remaining native forests. The result as applied to the west Taupo forests, the Pureora Working Circle, was delayed by war but eventually published in 1952. It was designed to control the allocation of native timber blocks to sawmill companies, and invest in a staged transition to plantation forests of exotic species. This chapter describes the plan, the building and operation of the sawmills, and the allocation of cutting rights under contract to NZFS. It describes how the logging operations were conducted in the bush, from cruising (surveying and estimating the volume of standing timber), through felling the trees, hauling the logs out of the bush with bulldozers and logging arches, and loading the trucks for transport to the mills, to how the mills cut the timber. It summarises the histories, the sawing machinery and the annual production of the local mills up to 1978.

Keywords

Logging of New Zealand native forests • Logging contracts • Pureora Working Circle • Logging operations • Cruising timber sale areas • Bulldozers • Logging arches • Sawmill technology • Saw doctors • Sawmill closures • Barryville • Post-war histories of King Country sawmills

Origin of the Pureora Project

Until the 1930s, extraction of timber from indigenous forests had proceeded on Maori and private lands as well as Crown blocks [16]. The general attitude of the time was "cut out and get out", so planning was minimal. There was fierce competition between privately-owned timber companies for cutting rights in profitable and accessible forest blocks. As the easiest blocks were exhausted and the bush line moved further

D.J. Gaukrodger
3a Te Angi Street, Turua, RD4, Thames 3574, New Zealand
e-mail: beejaygauk@xtra.co.nz

N.A. Ritchie
Department of Conservation, Te Papa Atawhai, Hamilton Office, Private Bag 3072, Hamilton 3240, New Zealand
e-mail: nritchie@doc.govt.nz

C.M. King (✉)
School of Science, University of Waikato, Private Bag 3105, Hamilton 3240, New Zealand
e-mail: cmking@waikato.ac.nz

C.M. King et al. (eds.), *The Drama of Conservation*,
DOI 10.1007/978-3-319-18410-4_7

into steeper country, rising costs favoured the larger companies at the expense of the small.

Sawmills and the local communities that depended on them were short-lived (Chap. 5; Box 7.2) because, as the history of free enterprise amply demonstrates (never more clearly than in and since the 20th century), curbing powerful commercial interests for the sake of the long-term common good requires a political decision by an even more powerful authority. In the case of the sawmillers working the central North Island forests, that was supplied in 1938, when the Director of Forests A.D. McGavock proposed a 5 year plan that would control the allocation of cutting rights through official channels, and provide a more certain future for bushmen and sawmillers.

McGavock anticipated that the State Forest Service would become the "dominant forest owner" as privately owned forests became depleted, and that the future role of Forest Service operations would be to provide economic production units and to act as a testing ground for new machinery and techniques.

McGavock's plan was designed to ensure a more carefully rationed utilisation of the remaining native forests. It implemented Ellis' vision of a deliberate strategy to compensate for the inevitable exhaustion of the supplies of native timber over 40 years by establishing exotic plantations well in advance, that would mature in time to replace them. The plan specified the establishment of state-owned forest industries, including sawmills, and the upgrading of housing conditions to provide a better quality of life for forest workers.

In 1939, the new Director of Forests Alex Entrican continued to focus policy on the sustainable management of the forestry industry in the central North Island, Westland and Southland. Ever since Ellis, "a sustainable forest industry" had meant, in NZFS terminology, a progressive shift in timber supply towards exotic production forests (Chap. 6).

Entrican envisaged a combination of State-owned forests perpetually managed by selection logging, plus innovative, progressive timber processing, and attractive, good quality housing and villages for forest and sawmill employees. A significantly improved comprehensive system of assessment of native timber available for sale was developed and instituted, which, yielding more realistic values, helped to persuade sawmillers to improve their techniques and efficiency.

Pureora State Forest 96

East of the Maori-owned or private blocks which were already being milled (Chap. 5), there were huge areas of forest available for exploitation. A preliminary 2 % reconnaissance of this land had been made in 1929 by State Forest Ranger Whitehorn, who recorded vast stands of millable species comprising 43 % matai, 36 % rimu, 14 % totara, and 6 % kahikatea (Whitehorn, unpubl.). This area was designated as Pureora State Forest (SF 96) in 1935, with the intention that the creation of this new production forest would help supply the accelerating demand for timber and rural employment.

Post Splitting

After Pureora was gazetted as a State Forest, one of the first actions of the foresters was to survey the very extensive damage done sometime between 1895 and 1915, when a severe windstorm had swept the area of the future SF 96, uprooting thousands of trees. So the first revenue derived by the Crown from Pureora was obtained from post splitting operations, which provided much-needed employment during the Depression years. The working conditions were hard (Figs. 7.1 and 7.2).

More than 300,000 first class totara posts and many thousands of strainers were cut from the wind-thrown totara trees. The Lands Department held the splitting rights, and utilised all the posts and strainers on their land development schemes.

Fig. 7.1 A small tractor used by Ned Barrett to extract fallen timber used for post-splitting, and abandoned in the bush when it broke down in the 1940s. DOC now protects it as a historic artefact. *Crown Copyright, Department of Conservation Te Papa Atawhai. Photographer unknown*

Splitting was done on contract, and during the peak splitting period in 1929, 28 men were engaged. Splitters were paid 30 shillings (NZ $1.50) per hundred for first class 'broad-axed' 5 × 4 inch posts. By 1945 the price rose to £24 (NZ$48) per hundred, because of the diminishing supplies of totara suitable for splitting, and an increasing demand for posts together with a shortage of labour during the war years of 1939–45.

At first, bullocks were used to haul split posts to the fringe of the forest. From there they were carted by lorry over a roughly-formed road for 2 miles to E&B's Mangapehi tramline (Fig. 7.3) for transport to Mangapehi station on the North Island Main Trunk Railway (NIMTR).

In 1942 the Lands Department allowed the early splitting permits to lapse. In the same year a permit was issued by NZFS to J.B. Halcombe of Mangapehi, who employed Ned Barrett on a sub-contract. Ned used a Caterpillar two-ton tractor to pull split totara posts and battens from the bush, but the permit lapsed in 1945 and no further splitting permits were issued.

Later, Ned became the first local person employed by the NZFS at Pureora. When his tractor broke down it was left where it stood for the bush to reclaim (Fig. 7.1). The tractor, and a hollow totara stump converted to provide accommodation for a couple of splitters (Fig. 7.2), are both among the historical attractions accessible from the Timber Trail (Chap. 14).

The Pureora Working Circle

A survey of SF 96 made by the Te Kuiti office of the State Forest Service in 1940 [2: 126] mapped a total area of 14,644 acres (5925 ha) in eight blocks (Fig. 7.3). The map shows one small area of 297 acres (120 ha) in Block 2 (2392 acres) where the foresters conducted a 100 % appraisal, and reckoned from it that Block 2 alone carried 9.75 million board feet, or 23,007 m^3 (1 million board feet—2360 m^3, see Box 6.1) of timber. But the war interrupted the development of plans to manage this valuable resource, and diverted men and material to more urgent matters until the early 1950s.

The Pureora Working Circle plan of 1952 was developed by NZFS to guide timber operations in the forest blocks surrounding Pureora village and the four mills depending on them (Fig. 7.4).

Fig. 7.2 Two post splitters (Mac and Ken Phillips) working in the bush during the war years converted a hollow tree stump into a temporary hut (see Walker and Cooke, 2003: 134). *Crown Copyright, Department of Conservation Te Papa Atawhai. Photographer: John Mason*

The Working Circle comprised the whole of Pureora State Forest 96 and part of Pouakani State Forest 93, and the plan ran from April 1952 to March 1957. It covered 21,327 acres, which was estimated to carry 56,560,000 $feet^3$ (1,601,600 m^3) of merchantable timber. That was enough to offer permanent employment to 40 men.

The purpose of the Working Circle Plan was to "regulate the yield in such a manner as to maintain the forest on a sustained yield basis. It directs the method of felling and disposal of the cut, and prescribes the establishment of a new exotic growing stock and the protection and maintenance of the forest" [11].

The extraction of native timber rose rapidly after 1945, greatly assisted by new technology. Powerful bulldozers and power saws, heavy road trucks better able to carry logs from steep country than steam lokeys on tramlines (Chap. 5) and improvements in sawmill technology (Box 7.1) raised productivity to a peak in the early-mid 1950s. Around this time, nearly half the total output of sawn timber from the North Island came from the King Country [13]. E&B alone controlled five sawmills in the Pureora area by the end of 1945 (numbers 23, 45, 54, 55, and 209 in Box 7.2), which between them cut 9 million board feet a year (21,237 m^3/year) [2: 126]. In 1947 the life expectancies of these five mills were estimated to be only 4, 8, 13, 13, and 20 years respectively [2: 131]. It is hardly surprising that production fell steadily from the 1950s onwards, and the closing down of uneconomic mills handling only native timber (Box 7.2) foreshadowed a very different future for the industry—especially for companies like E&B that were slow to appreciate the need to embrace the future potential of pine plantations.

Logging and Sawmilling at Pureora

The major differences between the logging at Pureora and the earlier operations to the south and west, centred on Ongarue, Waimiha and Mangapehi (Chap. 5), were that by the time systematic logging started at Pureora in 1945, the allocation of logging contracts was managed by NZFS rather than by negotiations with Maori or private landowners, and mechanised transport (logging trucks) and log-handling machinery (diesel-powered bulldozers) were available. This new technology almost completely replaced the steam winches which had been used to haul the logs to the tramway skids, and the steam locomotives and rail tractors which previously had been the main means of hauling the logs to the mills (Figs. 5.10 and 5.11). Chainsaws had also begun to supersede man-powered cross-cut saws, although the chainsaw men were still called cross-cutters.

During 1943 the Forest Service announced plans to log Pureora SF 96 using its own staff, to

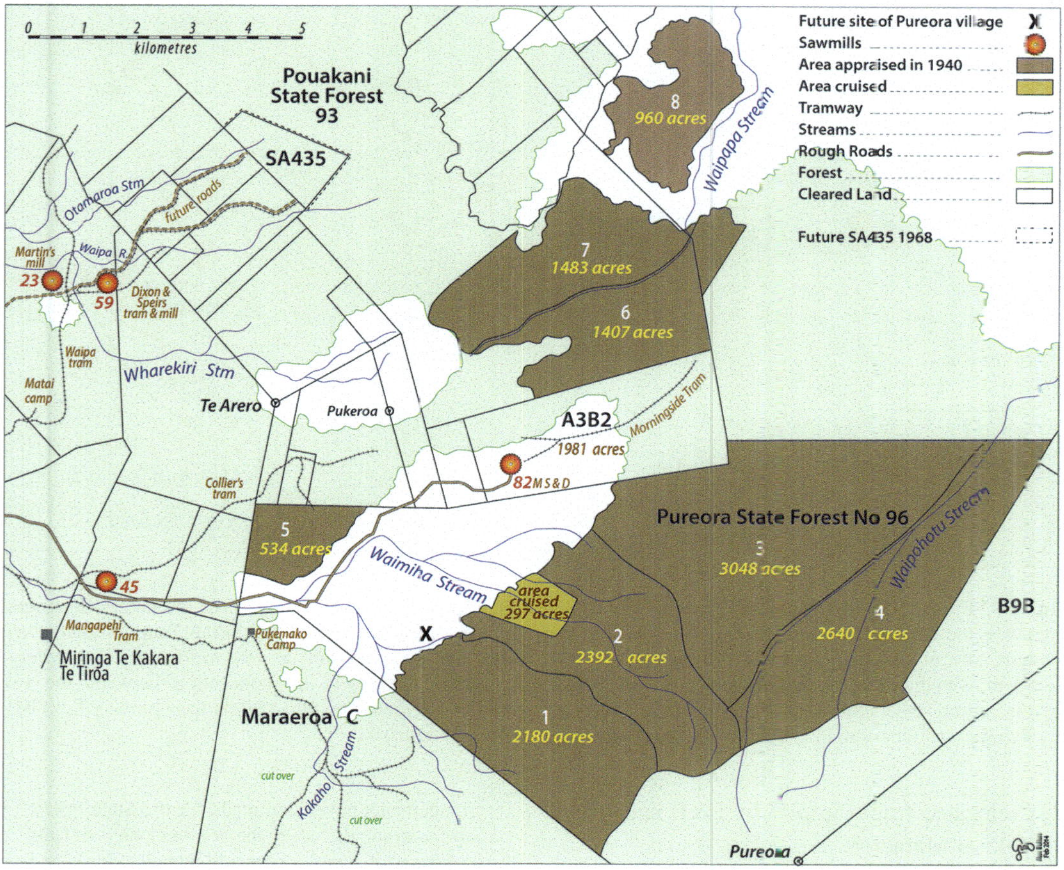

Fig. 7.3 The area surveyed by SFS in 1940 in preparation for the Pureora Project, scheduled to start in 1945. Four working sawmills are shown, with their tramways. The Mangapehi tramway crossing the Rangitoto-Tuhua block was extended into Maraeroa C, where Ellis & Burnand built their Maraeroa mill in 1945. There was no link eastwards over the ranges to Lake Taupo; the road from Te Kuiti reached only Benneydale, and thereafter was a rough track to Marton Sash & Door's mill no. 82. At top left is shown the pre-war Dixon & Speirs tramway, which was replaced by forest roads in the 1950s. These forest roads are significant to the Pureora story because they gave access to the future milling block SA435, allowing NZFS to grant Ellis & Burnand a 15-year logging contract there, 1958–1983. "Area cruised" shows the location of the 100 % survey made by SFS in 1940 (for a definition of "cruising", see text). X marks the approximate position of the future Pureora village. *Redrawn by Max Oulton from Anderson (2008: 56, 127 and 145)*

sell saw logs and peelers on the open market to private mills, and to construct a permanent logging village at Pureora. The following year the decision was made to proceed with the establishment of a logging scheme at SF 96. The largest logging areas close to the village site but not under NZFS control were the Maori-owned block of Maraeroa C and the A3B2 block, where logging had already started (Chap. 5).

By mid 1945, there were 48 mills working in the King Country, including many new ones being established on both sides of the NIMTR as the bush edge retreated. Two of the existing mills were near the future Pureora village site, both working on private or Maori land. One had been built by Marton Sash and Door Co. at the infant Barryville site in 1941 (Fig. 5.15), and the other

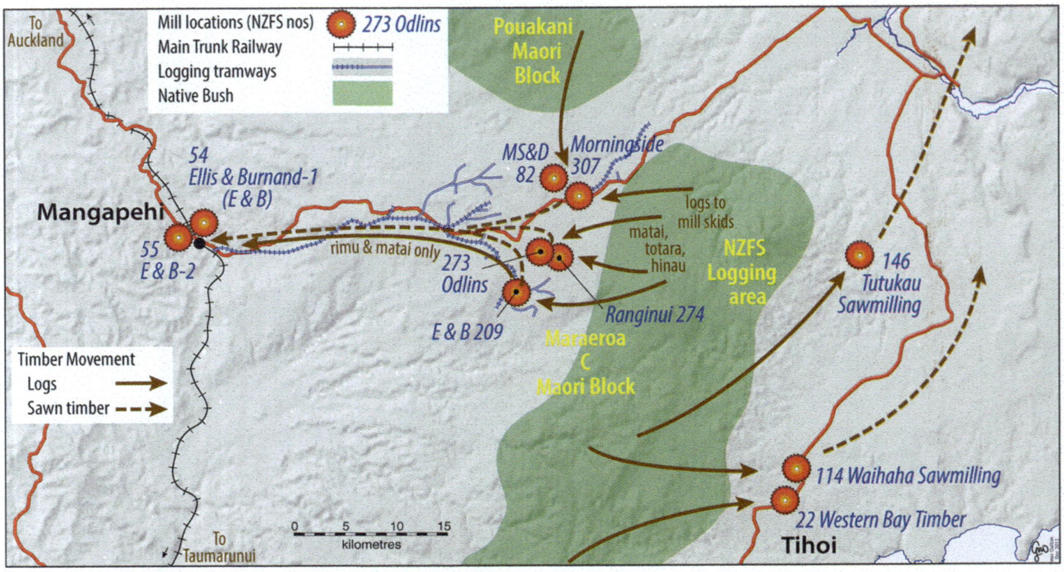

Fig. 7.4 From 1952 the Pureora Working Circle Plan organised the collection of logs from bush blocks on the western side of the Hauhungaroas for four of the five timber mills in the Pureora area (except MS&D 82, whose private cutting rights expired in 1955) and the delivery of sawn timber by tram or road to the Main Trunk Railway at Mangapehi. On the eastern side of the ranges, logs were sent to independent sawmills in the Taupo western bays area, and the sawn timber went to Putaruru and Rotorua by road. There was no connection between the two systems. *Redrawn by Max Oulton from Somerville (1965: Figs. 15 + 16)*

was relocated from Piropiro by E&B and rebuilt in 1945 at Maraeroa.

By 1946, Pureora village was taking shape, and three additional timber mills were being planned in the immediate area, as described in a contemporary announcement [4]:

> Greatly increased production in the timber industry in the vicinity of Mangapehi should result from the early establishment of three new sawmills. The Morningside Timber Co. plans to establish a new mill at the Marton Sash and Door's settlement, 18 miles east of Mangapehi, and the company has surveyors working there now. Negotiations are in progress to secure a piece of State Forest land for a mill site, and work will be commenced shortly on the erection of mill buildings and about 20 houses to accommodate the workers.
>
> The Marton Sash and Door Co. has made housing sites available and the Morningside company has acquired sufficient bush to provide for an estimated 30 years of cutting. When the project gets under way, about 40 men will be employed.
>
> A few miles away the Hutt Valley Timber Co. [=Ranginui Timber Co.] proposes to establish a new mill alongside the State forest, and arrangements have been made for the State Forest Service to supply logs to the new company. At least six houses of superior type will be built shortly and about 16 men will be employed in the initial stages.
>
> Close to the Hutt Valley mill site the C. and A. Odlin Timber Co. proposes to establish a further mill to handle State forest timber, and a big building programme is also planned by this company. About 20 men will be employed here.
>
> The three new mills are being established near to the new state forest village of Pureora, near Mangapehi, where 10 new houses of the State house type are almost completed, Extensive roading has been carried out in the area with modern machinery and telephone facilities are nearing completion.

The two new mills built at Pureora by C. & A. Odlins and Ranginui Timber Co. were completed in 1947, both situated about 1.6 km west of the present DOC field base at Pureora. Both mills were supplied solely by NZFS from State Forests.

The third new mill was built in 1949 at Barryville (Fig. 7.5) by the Morningside Timber Co. (acquired by Carter's in 1922), next door to the

Fig. 7.5 Carter's (Morningside) timber mill at Barryville, first established in 1949 and rebuilt after a fire in 1965. This is the mill that was forced to close down in 1978. *Crown Copyright, Department of Conservation Te Papa Atawhai (May 1977). Photographer: John Mason*

existing Marton Sash and Door's mill. In 1946 Carter's had negotiated directly with the private owners of A3B2 (Chap. 5) to gain the cutting rights for timber from this block for 20 years (1949–69) [17: 326]. The two mills operated side by side until MSD's mill closed in 1955. The Carter's mill would also have closed in April 1969 when it had exhausted its contracted supplies of logs from A3B2, but it had recently been rebuilt by the workers themselves after a disastrous fire in 1965. To maximize the benefit to the workers from their investment in their own jobs, arrangements were made to supply the mill with logs from Pureora SF 96 [5].

The three new plus two existing mills were located close to the bush edge (4–7 miles), which minimized the cost of carting logs to the mill. All five mills (four after 1955) carted their sawn timber another 15 miles by tram or road to the railway at Mangapehi (Fig. 7.4).

Until the mills in the village were ready, the first harvest from Pureora SF was processed by E&B's Maraeroa mill (Fig. 5.13). In April 1945, a turn-off was constructed at Delaney's Corner (Fig. 14.3), from the road which led from SF 30 to the Maraeroa mill, by dozer driver Bruce Archer. It is named after Bert Delaney, who was General Manager of E&B from 1942. He was one of few company Directors who wanted to prove the future potential of radiata pines. He planted a small demonstration plot of pines at the corner of the road leading to the Maraeroa camp, fenced off from marauding horses, with a sign proclaiming "Delaney's Corner" [2: 155–156]. They grew well, proving that pine plantations could be successful in that area. After the native forest was cleared from Maraeroa C, the owners replanted the whole block in pines (Fig. 5.14).

Bruce Archer also built a road to the proposed Pureora village site, and opened the way to the first skid built in June. On 5 July 1945, the first log was driven out of SF 96 to the Maraeroa mill by Sonny Hughes [22]. It was a totara log containing 157 feet^3 (4.4 m^3) of timber. By the end of 1945, logs were being produced and carted much faster than they could be processed at the Maraeroa mill.

Adverse climate conditions and insufficient skilled workmen, accommodation and equipment delayed the early stages of the project. Heavy machinery was limited to a GMC 6 × 4 truck, an old D8 dozer, a Ford 20 cwt truck, a Fruehauf trailer and a diesel-powered compressor for quarry work. But by 1947–49 more equipment had arrived, and the project was well on the way to being a profitable scheme, since supplying the national demand for building timber was a sellers' market.

The working plans were completed and signed for Waihaha and Pureora in 1952 [11], based on estimates that a maximum allowable cut of 1.25 million board feet/year (2950 m^3)(up to 3 million bd ft, 7079 m^3, in Waihaha only), would give the village and the sawmills a life of 50 years. That required some restraint in output,

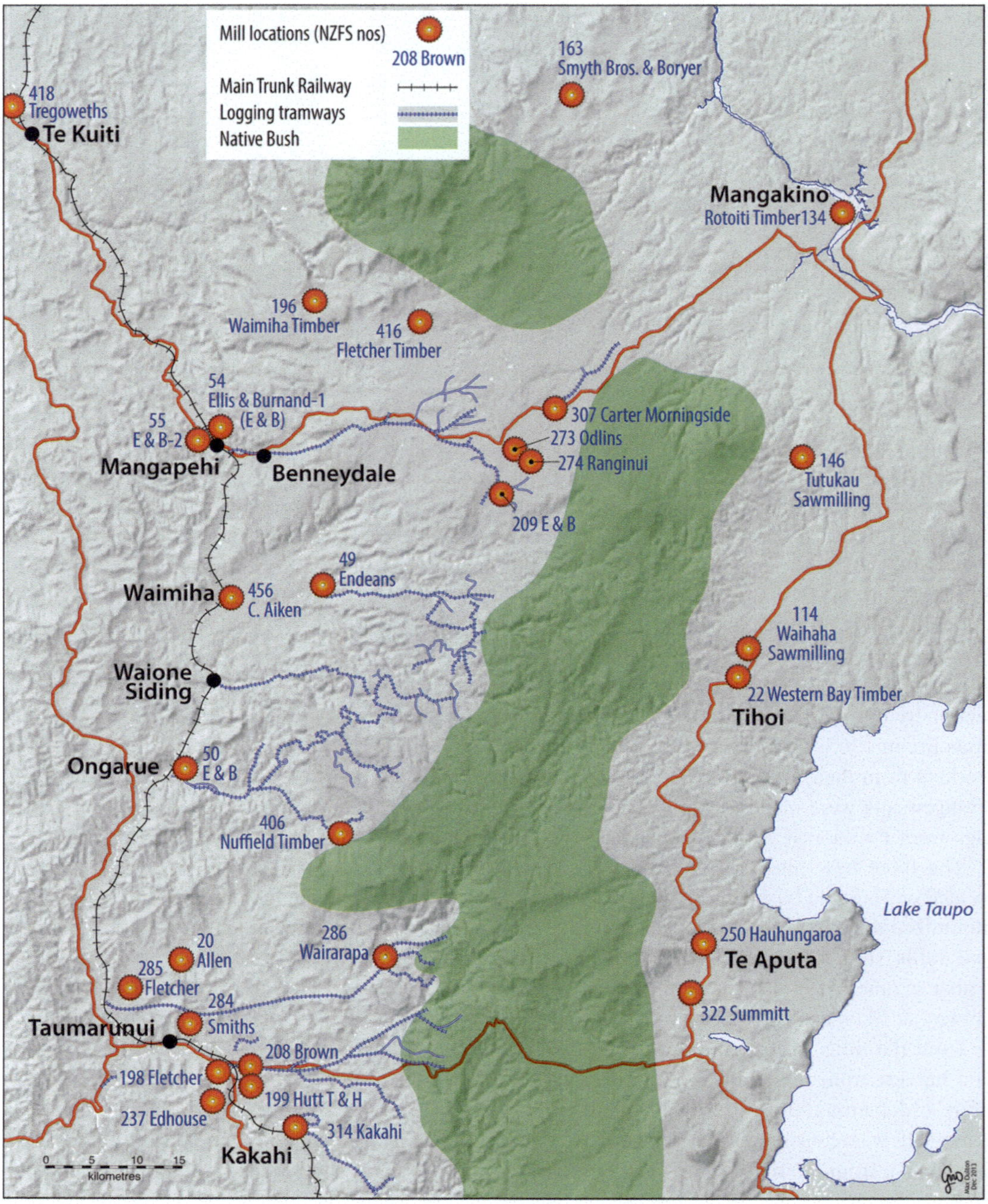

Fig. 7.6 Distribution of sawmills cutting timber from the forests of the Hauhungaroa and Rangitoto Ranges in 1964, when NZFS supplied logs to 32 of the 44 timber mills still operating in the King Country. Sawmill numbers are as allocated in order of the date of application by the State/NZ Forest Service register. Also shown are some of the routes of the old timber tramways as mapped by Yonge 1993, most abandoned by the 1950s. Distribution of forest diagrammatic only. *Redrawn by Max Oulton from Somerville (1965: Fig. 24 and Table 9)*

achieved in July 1952 when the Government removed the subsidy on overtime, paid to the timber industry since the last war.

By 1964, there were still 44 mills in the King Country, about the same number as in 1945, but only 13 of them were the same ones [18: Table 9]. Thirty-two of these mills drew their logs from the Hauhungaroa Ranges, of which 29 are shown on Fig. 7.6. Most of them used traditional machinery. For a summary of sawmill terms, see Box 7.1, and for the histories of local mills significant to the Pureora story, Box 7.2.

Box 7.1 Sawmill technology

The terminology used to describe sawmilling equipment and the processes of handling the logs varied but usually included the following.

Debarking: Heavy bark can foul the saws and transfer systems, and increase wear on the saw teeth, so mills handling relatively uniform exotic logs remove the bark before sawing, using a rotating ring of blunt knives or teeth. This is not possible with large and irregular native logs.

Headrig, head saw or **breakdown saw**: a powerful saw (or a pair of saws) that makes the first cut into a log on the breakdown bench. Older mills used thicker, wasteful twin circular saws, later replaced by thinner, more efficient vertical frame or bandsaws.

Kerf: the width of the saw cut, determined by the divergence between the teeth that protrude left-and-right from the blade. The wood occupying the path of the kerf is instantly converted to saw dust, hence the wider the kerf, the greater the proportion of a log wasted.

Pacific bench: a carriage for moving a log to the next process, mounted on small steel wheels and operated by a system of wire ropes and clutches.

Baulk: trimmed, unfinished log ready for further processing.

Breast bench: supports a single circular saw that cuts off one plank at a time from a baulk of sawn timber.

Slabs: the outside edges (facings) of the tree trunk, complete with bark, once used for firewood but now more often chipped for export.

Resaw: a secondary saw that produces smaller, finer cuts known as flitches. Gang saws can have two or more saws mounted in a frame, cutting in parallel to produce several flitches or planks at a time.

Docking saw: used to trim the flitch to length, remove defects and square the ends.

Grading: the process of sorting the timber into different quality classes to determine what the end-use will be. Grading maximises the value of the timber by allocating each class to its expected market.

Power: steam engines fired by timber slabs ran virtually all mills until the late 1940s, but were a serious fire risk. Newer mills were built to run on electric or diesel power as soon as these became available.

The Pureora mills were important players in the national figures for timber output at the time. According to the NZFS Sawmill Register for the year March 1965, the Maraeroa, Odlins and Ranginui mills all produced 1.5–2 million board-feet per year (3540–4720 m^3/y) each, and Carter's mill at Barryville over 3 million board feet (7080 m^3) per year. Even the smaller mills listed in Box 7.2 averaged at least 1 million board feet (2360 m^3) a year each.

In 1964 about 900 men were employed in the sawmilling industry throughout the King Country, producing rough sawn timber worth about £2.3 million, and the owners of the timber industry depending on the forests of Hauhungaroa Ranges reckoned the mills to have a further life of 15–20 years [18]. But they were wrong.

Box 7.2 Histories of some timber mills featuring in the Pureora story

Each mill was identified by a site-specific registration number allocated to it on the date of application by the Forest Service [18]. For locations, see Figs. 7.3 and 7.6. Data mainly from Somerville [18] and Anderson [2].

Pureora Sawmills Ltd at Barryville ran the largest mill, producing 43 m^3 of sawn timber a day. This was the third, newest and only electric mill on the site (of the two others, Marton Sash and Door's closed in 1955, and the first, steam-driven Carter's Morningside mill burned down in 1965), and it employed or contracted a total of 55 men (35 in the mill and yard). Logging and carting to the mill were managed by NZFS. The principal breakdown equipment was a flat top carriage with twin circular saws, and a twin circular head rig saw with a Pacific carriage. Further along was a 1.4 m band resaw and two breast benches. This mill was unusual for its time in having a small band resaw, but it still used inefficient circular saws as well, which meant that the average grade recovery was 48 %, and the mill recorded a high proportion of waste.

Owner	NZFS reg. no.	Location	Opened	Closed
Western Bay Timber Co.	22	Tihoi	1939	1976
Martin's	23	Horokino	1935	1952
Collier's /E&B since 1943	45	Tiroa	1925	1957
Endean's	49	Waimiha	1928	1996
Ellis & Burnand Co.	50	Ongarue	1913	1966
Ellis & Burnand Co.	54	Mangapehi No. 1	1904	1967
Ellis & Burnand Co.	55	Mangapehi No 2	1908	1968
Dixon & Speirs Ltd.	59	Horokino	1941	1959
Marton Sash and Door Timber Co.	82	Barryville	1941	1955
Waihaha Sawmilling Co.	114	Tihoi	1947	1979
Tutukau Sawmilling Co Ltd.	146	Arataki	1950	1979
Smyth Bros & Boryer	163	Ngaroma	1938	1975
Waimiha Timber	196	Kopaki	?	1957
Ellis & Burnand Co.	209	Maraeroa	1945	1967
C. & A. Odlin Timber Co.	273	Pureora	1947	1971
Ranginui Timber Co.	274	Pureora	1947	1976
Carter Merchants/Morningside	307	Barryville	1949	1978
R. H. Tregoweth Ltd.	418	Te Kuiti	1955	Still open

In 1946, there were 14 mills operating in the Pureora area as surveyed by Vaney & Gibson [19], of which only six survived by 1978, four of them with fixed logging contracts. These four mills (Fig. 10.1) had a combined theoretical capacity for sawn timber production of 170 m^3/day, in practice varying from 32,966 to 36,662 m^3 per year from 1972/73 to 1976/77, i.e., 19–20 % of all sawn timber production for the North Island. Their combined production comprised 73 % rimu/miro, 21 % matai, 3 % kahikatea, 1 % totara, 1 % tawa, 1 % other species.

R. H. Tregoweth Ltd's contract with E&B required them to log, cart and mill the timber from E&B's contract area (Fig. 7.3), which accounted for 90 % of their production, averaging 35 m^3 a day, until 1983. The company employed 25 men at the mill, and seven in the bush. The mill was paid by the grade of the sawn timber produced, giving it an incentive to achieve a high sawn timber grade output. It was equipped with a vertical breakdown saw for splitting large logs, located in a separate building close by. Inside the mill, the gear included a 1.4 m Fuji band headrig saw and Edwards carriage, two breast benches (one automatic and one manual pin) and a Fuji band resaw, plus two twin docking saws. This more efficient equipment enabled them to record an average grade recovery of 62–70 %. The documents recording E&B's decision to subcontract Tregoweth's mill to handle logs under their 1970 NZFS contract are held in the Fletcher Trust Archives, file FA0055/1/2.

Neither of these two mills was cutting as much native timber as their contracts allowed. Barryville was permitted to take 35,543 m^3 a year until April 1985, and Te Kuiti, 20,736 m^3/year (total 56,279 m^3/year), but the total cut was only 46,000 m^3/year, according to NZFS figures.

On the eastern flank of the Hauhungaroas the two small mills had contracts expiring in 1979 and no further private resources. Both operated old technology (twin circular break-down saws and traditional breast bench and docking saws), and were supplied by bush operations (logging, loading) run by NZFS; both were expected to close down when their contracts ended. Tutukau Sawmilling Co Ltd at Arataki was established in 1950 and employed 10 men to produce 13 m^3 sawn timber a day; Waihaha Sawmilling Co Ltd at Tihoi, was established in 1947 and employed 9 men to produce 11 m^3 sawn timber a day [19].

In addition, there were two more small mills without contracts, employing 21 men at Te Kuiti (17 at Waiteti Sawmills Ltd, cutting on private land, and 4 men at Woodturners NZ Ltd), and four small contractors employing a total of 14 men on felling and transport infrastructure.

The official statistics show that the total output of native sawn timber by all timber companies in the NZFS Auckland Conservancy combined was steeply declining, from 35.8 million board feet (84,479 m^3) for the year ended March 1967, to 22.6 million board feet (53,330 m^3) by March 1972 [17]. These figures meant that the long-term prospects for companies dependent on native timber alone were not good. The senior executives of those companies must surely have read the figures, and come to the same conclusion.

The 1968 and 1970 15-Year Contracts

On the other hand, even after the most easily accessible timber resources were approaching exhaustion and the mills were already running down, there was still a lot of native timber for sale to companies that could reach it. NZFS was required by the Government of the time to place the native sawmilling industry on a more permanent basis. So in 1968 and 1970, NZFS called for applications for long-term cutting rights, which led to two separate 15-year contracts to supply a total of 56,279 m^3 of timber a year. It was of course standard practice at the time, but the existence of these two contracts proved very significant to later negotiations.

When the Forest Service took control of all timber sales, it introduced a system under which sales were contracted for "appraised quantities of timber standing on defined and demarcated blocks" known as sawmill areas (SA) [13: 11–12]. SAs were pre-planned according to timber volume, lay of the land, access and other logistical considerations. Where access roads cut through blocks of standing timber (Fig. 7.7), wind damage often followed.

Fig. 7.7 Perham Avenue, a picture-perfect tourist road lined with tall standing podocarps near Pureora Forest Park HQ. See map and associated NFS survey lines in Fig. 13.6. Roads cut through standing forest to access timber cutting areas commonly created wind-tunnel effects, especially if, as here, aligned to the prevailing wind, so increasing the risk of damage to otherwise untouched forest. *SCION 8042505 (1979). Photographer: J. Johns*

On 1 December 1968 a contract was granted to Ellis & Burnand [2: 145], allowing them to cut at the rate of up to 700,000 feet^3 (20,736 m^3) a year until 1983 [10]. Timber from the block on offer, SA435 (Fig. 7.3), was accessible by existing forest roads off SH30 previously serving Dixon & Speirs' mill (closed in 1959).

The 1968 contract was supposed to help defer the closure of E&B's No.2 mill at Mangapehi on the NIMTR (their larger No. 1 mill had already closed in 1967). But NZFS had specified that the logs had to be cut in one of the new and more efficient bandsaw mills (Box 7.1). E&B's No. 2 mill was still using the old circular saws, and could not be modernised. E&B had considered plans to build a new bandsaw mill in Benneydale, but (greatly to the disappointment of local men needing work) eventually decided there was no need to invest in a new mill when R. H. Tregoweth's mill in Te Kuiti had the equipment required to handle the logs to NZFS's satisfaction [14]. So E&B sub-contracted Tregoweth's to mill the logs from SA435 [2: 149–150].

Part of the contract was to clear-fell SA435 ready for replanting in exotics. E&B engaged the Valley Logging Co. to produce the logs, and Tregoweth's carted them to the mill on their own trucks [2: 145]. The timber resource was expected to be exhausted by December 1979, well before the contract expiry date of December 1983.

The 1970 contract was originally granted by NZFS to Odlin's mill in Pureora village. The papers held in the Fletcher Archives (Chap. 10) include the agreement in 1971 that established a new joint venture, Pureora Sawmill Ltd (PSL), and the legal transfer to it of the Odlin's contract. The newly established company then took control of Carter-Holt's Morningside mill at Barryville [12], complete with access to an annual commitment of 35,543 m^3 of logs a year until

1985. The cutting area was not specified in the contract, but could have been in any of the blocks of SF 93 or 96 shown on Fig. 7.3. The contract was signed off for 15 years even though the New Zealand Wildlife Service was already doing surveys showing that these forests were of high wildlife value (Chap. 9), and the timber supply was expected to be exhausted by 1981 [10].

Both contracts were still current in 1978, but neither was completed. The 1973/74 annual report of Pureora State Forest had already seen lots of trouble ahead.

> Because of all the reserves set aside [under recent changes in NZFS policy] there will be a shortfall in contract supplies, with 11 years of the 15 year agreement still to run. There have evidently been discussions between Auckland and Rotorua Conservancies about the possibilities of drawing timber from the Waihaha block and transporting it via a proposed Tihoi—Pureora road. There is little doubt that the 15 year agreement was made only a matter of 2 or 3 years too early—i.e., before emphasis on conservation arose [22].

Trouble did indeed arrive, with consequences taken up in greater detail in Chaps. 9–11.

Logging Operations in the Bush

The timber industry included two quite separate operations, which were usually done by different crews with different specialist skills and under different employment conditions. Felling trees in the bush and delivering the logs to a sawmill was the job of the logging crews. Cutting the whole log into planks ready for sale to timber merchants was the job of the sawmillers. Men doing both jobs lived in Pureora Forest village, though not under the same conditions (Chap. 8); most of the residents of Barryville were sawmillers.

Logging crews and their controlling officers were usually trained and employed by NZFS, so were public servants subject to the same regulations as other public servants. Sawmills were usually private enterprises that employed mill workers under their own, usually minimal, regulations.

A typical early logging crew of 12, as listed in the station diary for 9 February 1950, could produce 3000 "cubes" (cubic feet: see Box 6.1) of logs a day with one D8 bulldozer and driver, a power saw and three men, one scarfer, two crosscutters, one breaker-out, three on the skid, and one foreman. By the late 1960s, better mechanisation allowed OC Bill Drower to report that a typical crew comprised two crosscutters, one breaker-out, one tractor driver, one truck driver plus a staff trainee and a logging officer.

Logging crews were on the job by 7.30 am, and worked until 4.30 pm. They got two smoko breaks in the day, one at around 10 am and the other in the afternoon. They boiled the billy both times, and sat in a shelter, often just a tent fly—"nothing sophisticated about Pureora!" commented Jack Fyffe, logging officer during the 1950s [21].

The local mills being supplied from Pureora forests were all commercial companies in competition with each other, so there was an advantage to whichever company could persuade the crews to deliver the best logs. But when asked if the mills could specify what kind of timber they wanted, Jack Fyffe replied that they just got what was delivered to them.

When Roy Winwood (manager at Odlin's Pureora mill) complained that he was getting all the hollow-butt matai, Jack pointed across the road to the Ranginui yard where a skid was full of hollow matai—"everyone gets a fair share of them", said Jack. The best and highest value logs were the "peelers", which had to be smooth and round and suitable for sending to a specialist veneer-cutting machine.

Cruising

Cruising is the term used to describe the standard method of surveying a standing forest in order to calculate how much timber it could be expected to yield. This inventory enabled the forest owner to set a realistic price for the sale (the stumpage rate), and the logging contractors and mill owners to estimate their potential costs and profits.

A cruising team comprised a cruising officer and two men who would go out with a Senior Ranger and sometimes a sawmill manager to identify the block to be surveyed, and consider the access and haulage routes. Then the crew would set up a camp or a hut as a base for the job. Cruising teams had to be prepared to live rough under canvas in the bush for the duration of their assignment, usually 10 days on, 4 days off. They had poor living conditions, and absolutely no official provision for what we would now call Health and Safety precautions. They risked hair-raising escapes from dangerous jigger accidents and charging wild bulls—but they also enjoyed the beauty of bush and wildlife, off-duty hunting, practical jokes, camaraderie, and lifelong friendships. The general opinion was that it was a great job for a fit young man who enjoyed the outdoor life [6: 158, 169].

Ken Seymour was a young NZFS ranger when he was sent off cruising for Ellis & Burnand in the Rangitoto Ranges in 1954. He commented that few employers could entice young people into such a lifestyle now, but back then it was all part of the job.

After the SA boundary had been cut and surveyed, "control lines" were marked through the block with a compass and surveyor's chain, the standard unit of land measurement at that time (1 chain = 66 feet, or 20.1 m: see Box 6.1). Then, representative samples were taken through the block at set intervals. Parallel control lines were marked at 10-chain intervals (every 201 m). To get a 10 % measure of the volume, a 1-chain strip up either side of each line was fully surveyed, and all trees were measured and counted by species, diameter, and merchantable height.

After all the lines were run, the result was a complete map of the block, including the expected access. Back at the office, the field team would use log-volume tables to work out total timber volumes by species, the cost of production and the predicted realisation price.

Cruising rangers had very good bush skills, and were able to describe all aspects of an area. They were adept at surveying land areas and drafting their own SA maps—including the challenge of locating old land survey pegs established more than 50 years previously. Many of them rose to become District Rangers or to more senior positions in NZFS. Dennis Harris's collection of stories about life of former NZFS staff in the bush in the 1950s [6] relates the early cruising adventures of several people who later appeared in the Pureora story, some as OCs—including Darbie Perston, Ivan Frost, John Gaukrodger, Bill Drower, Kitch Pedder, Jack Walker, Ken Seymour, Buster Seager, and Gavin Molloy.

The NZFS cruising manual stipulated that a 10-chain (201 m) strip had to be left along highways and rivers, which added up to a lot of forest saved from logging by NZFS policy throughout its history. NZFS never got any credit from environmentalists for that, commented former OC Ivan Frost, or for the early reservation of a belt of lowland rimu along the slopes of Pureora Mountain—and that "makes my blood pressure go up", he added [21].

After the logging of that block was completed, a Forestry officer would do a check appraisal in the field to ensure that the real harvest more or less matched the prediction, and that the logging company was satisfied. If there had been an under-run he would go through the block and find out if the stumps had been cut too high or if any good trees had been missed; over-runs were usually simply recorded and used to compensate for the next under-run. Ivan Frost did this job for some time, and found that mostly the estimates worked out pretty well. The same applied to checks made at the mill, required to confirm the royalty payable by the mill to the owners of the forest that had granted the mill's cutting rights, calculated from a residual formula (i.e., less costs).

Timber cruising in the future PFP began in Taringamotu SF 121 in 1925, and the first cruising team entered Pureora Forest in 1930, although Pureora SF was not gazetted as a permanent State Forest until 1935. As the Pureora project was being planned in the early 1940s, cruising data from a sample block of 297 acres was used to calculate the yield of the future Pureora Working Circle (Fig. 7.3). Kitch Pedder was one of the first cruising officers to be based at Pureora in 1947.

Asked if he still worked in the rain, he replied: "Oh, yes! Gosh, yes! Field books were of ordinary paper, keeping them dry was the biggest curse. You used a rain coat and crouched over the book —it still got wet, but you'd have a good stock of blotting paper, until it was too wet to write, then it was time to go home" [21]. The last commercial timber cruise at Pureora was done in 1981 at Waihora by John Gaukrodger and Ron Cumming.

Scarfing and Felling

The process of felling the tree began with scarfing, done by a specialist axeman who cut a triangular wedge out of the trunk on the side it was expected to fall. Then the traditional two-man crosscut saw (the standard technology up to the late 1940s) was used to cut into the trunk from the other side (Fig. 7.8), until the weight of the head (the canopy and upper branches) pulled it down. Sawyers had to be skilled in judging the right moment and direction to run as soon as the trunk began to move, or run the risk of being killed as it fell.

The old two-man cross-cut saw was a powerful tool in the hands of a skilled crew, but the arrival of power-driven saws stepped up production per man to unprecedented levels. A Danarm chainsaw, a massively heavy tool by today's standards, was used at Pureora for the first time in November 1947, and it was still there in December 1950.

Early estimates (1948, apparently by Geoff Hammond) were that power saws could increase output per man/day by at least 1400 board feet, up to 6400 board feet (3.3–15.1 m^3) per day (Fig. 7.9). They could replace two men out of every nine, and save £47 in wages per fortnight. Although there was considerable resistance to the introduction of chainsaws by the 'old time' cross-cutters working for E&B on the Tuhua block, Ivan Frost reckoned that there was no resistance to the introduction of power saws from the bush crews at Pureora. "Well, I guess if you'd been pulling on a cross-cut all day…." [21].

But OC Eric Johnstone was of the opinion that the cross-cutters could keep up production, and he would prefer not to use chainsaws. He was not alone in his reservations. Buster Seager, who worked with logging crews in the 1950s, remembers that the Danarm worked only in the vertical position, so stalled if turned to cut horizontally; the Disston could be turned but had a

Fig. 7.8 Felling a tree with a cross-cut saw and driving wedges. *Hamilton City Library, image 1468*

Fig. 7.9 Using an early two-man Mercury chainsaw to fell a rimu. Safety helmets were being introduced, but were unpopular with the bushmen. *Photographer unknown, copyright assumed SCION image 8043889*

big ugly motor and motorbike handles, a monstrous thing. Lugging such a heavy machine through thick bush to the working area was a challenge in itself. (Trying to lift the specimens of both that can be seen in timber museums, one can see his point.) Buster reckoned that the early power saws were not used much for felling, because the much lighter old "peg-and-drag" saws were better for that.

Arthur Grimshaw was the full-time saw doctor at Pureora during the 1950s (Fig. 7.10). It would take him half a day to sharpen one saw by hand. It would then last two days or less, depending on what the crews were cutting.

Under the new OC Kitch Pedder, the use of chain saws increased. The 1953/54 Annual Report mentions that the two Mercury 11 hp, 6 foot Disston chainsaws used at Pureora allowed the felling gang to be almost halved. The two-man Disston could be used single-handed by a very strong guy, but normally a helper's handle was clipped on at the far end. Chain oiling and tightening was done at the motor end. It was a great loss when in 1955 a tree with a badly decayed butt fell on one of the Disstons, wrecking it.

Disston chainsaws went out of production in the early 1960s, by which time the Pureora gangs were changing to McCulloch or Homelite gear driven saws, and then to Pioneer 620s. But the old two-man Disstons were still running, and bushmen familiar with them swore that it would be a long time before a better saw is made.

Logging Arches

The logs had to be hauled to a skid site, where they could be loaded on to a truck that would take them to the mill. First, the end of the log was sniped (rounded), a D-shaped groove was cut to take the cable (Fig. 7.11), and then the log was hauled along the ground to a collecting area. This was inefficient, because steam haulers needed up to a kilometre of heavy wire cable, the log generated a lot of friction with the ground, and it often caught on obstacles such as rocks and stumps.

The introduction of logging arches, heavy steel frames towed by a tractor, was a great improvement. They were designed to lift one end of a log clear off the ground. Dragging the log through the bush on the other end reduced friction and saved time, and the logs arrived at the sawmill much cleaner and less damaged. Logging arches worked best where the ground was sufficiently hard and level to drag logs slung on chains to the skid. They also saved a lot of labour, since a steam hauler needed a crew of five men whereas a tractor hauler needed only two.

Le Tourneau in Australia was making a rubber-tyred logging arch in 1945. The Director

Fig. 7.10 Arthur Grimshaw, saw doctor at Pureora SF 96 in the 1950s, with a cross-cut saw. *Alexander Turnbull Library, Wellington, New Zealand. F164545-1/2, 28 October 1948. Photographer unknown*

of NZFS Alex Entrican ordered that one be imported and sent straight to Pureora for trials. By 1949 the OC's comments make clear that the rubber tyres were fine in dry weather, but often got bogged in the deep mud produced by Pureora's wet climate and soft pumice soils.

In 1951 a 25-ton logging arch was assembled at Pureora, but it worked well only on flat to easy going terrain. The towing tractor needed plenty of room to back the arch or manoeuvre it around obstacles, so it was restricted in what it could do or where it could go towing an arch behind it. The more manageable Hyster arches with caterpillar tracks (Fig. 7.12) arrived at least by 1956.

Logs were hauled to a collection area, where a skid or platform was built from large logs notched together, to create a stable base for loading logs on to bogies or log trucks. The skid logs would be laid at the right height and angle level with the truck. The truck access would often turn into a sea of mud, and the skids wasted a lot of good timber, but in the hands of experienced bushmen in the 1950s, even the largest logs could be loaded by a system called parbuckling (Fig. 5.10), as described by Buster Seager [21].

A rope would be wound around a log three times, the hook banged in, and the log rolled gently from the skid onto the truck. Some would

Fig. 7.11 A log being prepared for hauling. The leading end has been sniped (rounded), and a "D" groove cut to secure the steel cable with which the hauler will pull the log from stump to skid. *Crown Copyright, Department of Conservation Te Papa Atawhai. Photographer unknown*

not roll, so had to be slid, with the rope attached to each end in turn. If necessary the position of the log on the truck had to be adjusted with timber jacks, which could lever heavy weights an inch at a time, much as a modern bottle jack can lift a huge truck off the ground. It was hard and risky work, especially if the log rolled unexpectedly, or when the men themselves got bogged in the mud and had to be pulled out with a winch.

In a typical day at the skids, the large logs would be dragged out of the bush one at a time, and then winched on to the skids and loaded on to Mack trucks, 5–6 loads a day. There were always at least two men working the skids, one on the winch and one on the hook, and usually a ranger or bush boss as well. A scaler worked full time at the bush skid or at the sawmill skid, measuring the logs, writing a number on their ends and filling in a docket for each load.

Bulldozers and Tractors

At least by 1941 it was known that hauling logs with American-made crawler tractors (imported by NZ Caterpillar agents Gough, Gough and Hamer) was better than using wire cable haulers, but they were expensive, so not always available. Much depended on the skill of the driver. The best operators, such as Scotty Pihama and Tai Hona, were widely acknowledged as the most expert drivers in the business.

The value of a heavy-duty tracked bulldozer was not really appreciated until Tracey Gough demonstrated what it could do at Mangapehi. E&B engaged a contractor to use a Caterpillar Diesel 75 tractor to supply eleven million board-feet of timber over an eight month period. The machine started the job in January 1935, and finished it within the allocated time, in fact it pulled so much timber that the mill ran out of log storage space [2: 97–98].

After this conclusive demonstration, the use of bulldozers became more accepted, and they soon became the main method of hauling logs from the New Zealand bush [1]. Caterpillar D8s became indispensable for efficient handling of large logs in inaccessible country, although at first they were available only second-hand. Two of these are described in the station diary for 1949 as "in poor condition after service in the Pacific".

Great must have been the rejoicing, therefore, when two new D8s arrived in 1955. They served about ten years before they were replaced by D7Es, smaller but with power shift transmissions and hydraulic blade and tilt controls, far superior to the cable-controlled blades of the D8s. The various models of Caterpillar bulldozers became

Fig. 7.12 A D8 tractor tows a logging arch, a heavy steel lifting arm on its own caterpillar tracks, at Pureora. The arch lifts the leading end of the log off the ground, reducing drag and damage to the log on the way to the skid. *Crown Copyright, Department of Conservation Te Papa Atawhai (1950). Photographer unknown*

the standard equipment for logging operations and road works throughout the King Country. The mills and logging gangs that used them, the hard lives of both men and machines, and the frequent need for maintenance, are well described by Anderson [1].

The minimum efficient fleet was three, two for logging and one for tracking and roading, and allowance was needed for machines to be serviced. Maybe that is why the last, old D8 remained on the books at Pureora until the 1973/74 year. By the time they got rid of it, it had had 72 % down time. The last dozer left in 1986 was a D7G.

NZFS bought a Caterpillar D7 for Pureora in October 1963 [1: 108], but the logging crews needed more. Logging officer Dave Yanko tells a story about how the Pureora staff managed to get them, even though they were by far the most expensive option [21]. Head Office purchasing officers wanted to get Komatsus, because these worked well in exotic forest at Kaingaroa and they could get 2½ Komatsus for one Caterpillar. Yanko and his team, knowing the conditions they had to work in were much wetter and muddier than at Kaingaroa, objected. So it was agreed to run a comparative trial.

In May 1972 three large machines (a Komatsu, a Caterpillar D7, and a TD20 International based at Pureora and normally used for roading) assembled for the trial. All three were in the

Fig. 7.13 Ace tractor driver Tai Hona with his machine. *Crown Copyright, Department of Conservation Te Papa Atawhai. Photographer unknown*

150–180 hp range. Head Office specified that a certain tree should be handled in a certain way, using a Hyster logging arch, by all three tractors in turn.

The D7 just walked away with it; the Komatsu got stuck just dragging the arch without the log on it, and then dug into the ground and couldn't get out. It looked as if Yanko would win his point, until the Head Office guys said, OK, let's try the International. It was driven by Tai Hona, a real expert (Fig. 7.13), who could make any machine perform very well, but the Pureora team did not want any more Internationals, so Dave stopped Tai and persuaded him to drive it like a learner. They got their D7s.

A famous official mishap, remembered in great detail by Bill Drower, happened in 1968 when the subframes and all the running gear (tracks etc) from one of the first D7s were railed from Mangapehi for overhauling by Stevensons, an engineering company in Auckland. The company were sent the train, wagon and consignment numbers, but next day they reported that the parts had not arrived.

Mangapehi confirmed that the wagon had left, and Auckland that it had arrived and had been shunted, but it was not in the siding. The NZFS Conservancy Mechanical Officer investigated, and found that the wagon had been forwarded to the steel mill along with four others containing scrap steel. He made a quick trip to the mill, but was too late. A few components were saved, but most of the rest had already been cut up ready for melting down. Many weeks later, new parts arrived from overseas, and NZ Rail received a claim for more than NZ$30,000 [21].

Bulldozers have now been in use for nearly eighty years in the New Zealand bush, and although they are not nearly as popular as they used to be, it looks likely they will pass a century of great service in 2037. Indeed the only tools which have served New Zealand's bushmen for longer are the humble axe and the felling wedge.

Trucks

Pureora was established relatively late in the history of the New Zealand timber industry, so never had to rely on bullocks or steam engines, the traditional forms of log transport used in the early days of logging operations (Chap. 5). The Pureora mills used road transport from the start (Fig. 7.14), starting with army surplus trucks and tractors. They were much less of a fire hazard than were the old steam engines spitting sparks, but not always good bargains.

In 1948 ten Mack trucks arrived in NZ from the UK, and Pureora put in for their share. They were ex US Army, very heavy and immensely powerful, fully capable of coping with the conditions, but they had petrol engines with about 16 different adjustments to the carburettor, and were not cheap to run. One averaged 4 miles to the gallon on a

Fig. 7.14 A Leyland Hippo, one of several makes of heavy trucks used to carry logs from skid to mill. *Crown Copyright, Department of Conservation Te Papa Atawhai. Photographer unknown*

return trip to Auckland, and this during a period of post-war petrol rationing! A picture taken of one of these Macks involved in an accident in March 1950 (Fig. 7.15) gives an idea of their size. Otherwise, serious accidents involving work vehicles were few and seldom caused injury—not even in the collision in 1968/69 between a Leyland Hippo and a Bedford roading truck, which caused the Bedford to be written off.

By 1953 there were two Macks with Hayes trailers giving excellent service, until they were retired in 1956 because of chassis cracks. They were replaced with Leyland Hippos, and later with a Foden (which itself had a sad mechanical history). Bill Drower describes how a Foden engineer visiting New Zealand was taken to the bush to watch the loading procedure, followed by unloading at the mill, and was horrified at the extent to which the truck keeled over at the skid as the log landed on it. After a long discussion the company agreed to upgrade the truck with heavier-duty parts, and after that it performed well [21].

Trucks became indispensable for getting logs to the mills over the rough bush roads, but their high maintenance requirements were hard to meet because the workshop was always short of skilled mechanics. The truck maintenance problem was solved in 1977 when NZFS decided it would no longer cart logs from the skid itself, but require Pureora Sawmills Ltd to employ a contractor from Benneydale. Ten years later the NZFS roading unit (George Paul, foreman, with Jim Heta, Tom Waka, Dave Ihaia and Kira Hughes Snr) was itself disbanded, and from 1987 all roading was done by contractors.

Highlight of the 1972/73 year was the arrival of a Hough 90 mobile loader, which did away with both the dangerous loading methods of the past, and the need to waste timber constructing skids (Fig. 7.16). At first the bushmen were concerned it might not be able to handle very large logs. By the time it had been in use over a year, it was difficult to imagine how the logging crews once did without it.

The long-term effect of mechanization was, of course, to reduce jobs for bushmen. By 1983 the Pureora logging crew was down to two: Tony Hartley the machine operator, and Joe Ngatai the feller, crosscutter and breaker-out.

Safety Gear

Safety helmets came in for bush operations during the mid 1950s. They were very unpopular with the bushmen, who made every possible excuse not to wear them. Dave Yanko never forgot the drama of his very first tree on his first day with a logging

Fig. 7.15 When a Mack truck drove off the pumice road, the log it was carrying shot forward and pinned the driver, Bob Abraham, in his cab, breaking his leg. It took a Hyster logging arch and D8 tractor two hours to rescue him and his mate. *Crown Copyright, Department of Conservation Te Papa Atawhai (March 1950). Photographer Keith Webster*

crew in 1956 [21]. Dave watched in horror as a branch of a rimu broke off as it fell, and hit Tom Ruki on the head. Tom was not wearing his helmet, and was knocked out cold. But he woke up, and they sent him home for a week.

Another accident mentioned in the 1962/63 annual report showed the value of helmets even more dramatically. A senior trainee, D.W. (Des) Bergman, had just scarfed a tree, and turned his back on it as it fell. It snapped a small tree as it went down, which caught him on the back of his head. The crew found him lying face down in a pool of blood, and feared the worst.

The logging officer, Athol Ferguson, called OC Ivan Frost, and they got him into the ambulance and raced to Te Kuiti. Bergman was unconscious for two or three weeks, but regained almost full use of his faculties except for the loss of feeling on one side of his face. His aluminum helmet was bent at right angles, but it undoubtedly saved his life. (It probably helped, added Frost, that he was a very fit young man and a good rugby player).

At the Mill

Early sawmill technology was wasteful, deafening and very dangerous. Advances in the design of the saws and log handling equipment have achieved great improvements in safety and minimisation of waste, but they inevitably disadvantaged the older mills. Both static and working models of many of the machines described here may be seen at the specialist timber museums at Putaruru and Matakohe.

Logs arriving at the mill were rolled or hauled from the trucks onto the mill skids in the yard, cleaned of mud and stones that could damage the saws, and then transferred on rollers or bogeys to a carriage taking them to the first stage of processing, the breakdown bench (Box 7.1). The carriage supported both the logs and the jacks or dogs that held them in a fixed orientation, and moved them through the headrig (the breakdown saw) at a speed adjusted by the rate of cutting. Wedges driven into

Fig. 7.16 By 1977, logging operations at Pureora were fully mechanised. A D7 tractor has brought rimu logs to a collection area, where a loader stands ready to lift the logs directly on to a waiting truck. The wasteful and dangerous skid loading operations of the past were no longer needed. *Crown Copyright, Department of Conservation Te Papa Atawhai (1979). Photographer D.J. Gaukrodger*

the cut prevented it from closing and jamming the saw blades, and water cooled the saw blades.

For the first cut through large-diameter native logs, the headrig needed to have saws capable of making very deep cuts. All three mills at Pureora and Barryville used twin circular saws (Fig. 7.17) which cut from above and below simultaneously. E&B's Ongarue mill used a vertical breakdown saw (Fig. 7.18), fixed at both ends and mounted in a frame which oscillated up and down, driven by the main engine via cranks and belts under the floor [9: 307]. Vertical breakdown saws were slower but could handle larger logs. In either case, the two halves of the log then moved on to a second breakdown bench equipped with a pair of twin circular blades.

The men had no protection against the ear-splitting noise inside the mill—the saws, the steam engine, the belts and pulleys, the transfer chains—and they could not talk, so communication had to be by a series of hand signals [3: 102].

The sharpening and maintenance of the saw blades was an essential and highly skilled job. Circular saws were made as a flat metal plate which had to be thick enough to support its own weight and resist bending and tooth damage on contact with hard timber, so the cut made (the kerf) was relatively thick and produced a lot of waste as saw dust.

A large spinning saw becomes a flywheel, generating centrifugal force that causes the outer edge to stretch more than the inner. This distortion and heat expansion could buckle the blade and make it wobble during a cut, widening the kerf and demanding more power to keep running [9: 320].

Fig. 7.17 A breakdown bench with a headrig equipped with twin circular saws, used by Odlin's mill at Pureora (built in 1947) and also by many other older mills. *Crown Copyright, Department of Conservation Te Papa Atawhai. Photographer unknown*

Fig. 7.18 A breakdown bench with a headrig using vertical saws fixed in an oscillating frame, at Ellis & Burnand's Ongarue mill in 1966. *Hamilton City Library, image 69*

For circular saws to cut true, a region of the blade had to be tensioned. The saw doctor's job included hammering the blade in a zone about half to two-thirds of the way from the centre until the metal is slightly thinner and pre-strained, and checking the result (Fig. 7.19). When the saw was running, the pre-strained forces counteracted the distortive forces, maintaining the shape of the blade and the efficiency and accuracy of the cut.

The idea of a band saw, a continuous loop of flexible steel saw blade running around a top and bottom wheel, has been attractive since it was first patented in 1809. The lower drive wheel pulls the bandsaw blade down through the log as

Fig. 7.19 Saw doctors regularly maintained both circular and vertical saws. The task of sharpening a circular saw included tensioning the blade. *Hamilton City Library, image 1576 (1929)*

it is fed into the saw. A band saw blade can be thin and can cut logs of any width with a narrow kerf, which drastically reduces wastage to sawdust (Box 7.1). Band saws were installed in three New Zealand timber mills by 1905, but at first they were not a success [8].

The problem was that the structural variation inherent in most native timbers caused the flexible band saw blades to wander from a straight line. A crooked or inaccurate first cut affected all subsequent cuts from that log, which ruined its sales value in a market that demanded straight-cut timber. Hence the traditional vertical and circular saws remained unchallenged in New Zealand until nearly the end of native timber milling.

Since the late 1960s, improvements in bandsaw technology have corrected the earlier problem of inaccurate cutting, and mills handling only native logs have closed down or converted to exotic timber. Circular and vertical saws have been phased out from the headrigs of all mills in favour of bandsaws. The thinner band saw blades cut logs more efficiently and faster than could circular saws, and much faster than vertical saws.

Like circular saws, bandsaws also have to be tensioned and sharpened regularly. Special machines were developed for automated sharpening of a bandsaw, which hold the blade on a frame, teeth up, and advance it by one tooth at a time past an automatic grinding stone. The stone drops down to sharpen each tooth in turn at the correct angle, then lifts up out of the way with each forward step. (Working models of these and many other timber handling machines can be seen at the Matakohe Museum).

When the outside sapwood and bark has been cut off, the heart wood is split into squared baulks. These are then moved on to smaller re-saws that cut the baulks into narrower flitches (unfinished planks) and edges (irregular cuts to be rejected). Sawdust and shavings were dragged out of the way along wooden channels by an endless chain fitted with bars every few feet, to be dumped outside or burned in a conical incinerator.

The final product was a neat stack of rough sawn timber, graded by quality and species, ready to go to market. All further processing was done by the parent companies or customers of the mills.

Mill Closures Before 1978

Within the three decades after 1945, about a third (18,000 ha) of the western flank of the Hauhungaroa Range was intensively logged. About a third (6010 ha) of what is now the North Block of PFP was logged for podocarps, mainly rimu, between 1976 and 1978, leaving a logging-induced tawa-dominated canopy [7]. In total, some 83 % of the huge west Taupo forests were logged between 1945 and 1977 [24].

One by one, the older mills around Pureora village closed down or were taken over as the bush blocks supplying them were exhausted (Box 7.2). First to go in 1955 was Marton Sash & Door's mill at Barryville [23]; then Collier's mill (owned by E&B) at Tiroa in 1957; then E&B's mill at Maraeroa in 1967. Next was Odlin's at Pureora, after C. & A. Odlin bought out the Ranginui mill next door and closed their own in 1971. They kept the old Ranginui mill going for a few years, modified to cut slab into firewood. Eventually the Ranginui mill became dilapidated, its heap of sawdust looked unsightly, and its mill houses at the western edge of Pureora village were worn-out. It closed in 1976, but survived as an empty shell until the early 1980s. The old sawdust heap still lies there, part overgrown but still huge (Fig. 10.2).

By 1978, only four mills were drawing timber from State Forests in the Pureora district (Fig. 10.1). Two were on the eastern flank of the Hauhungaroa Ranges, run by Tutukau Sawmilling Timber Co. at Arataki and Waihaha Sawmilling at Tihoi. On the opposite side of the ranges, in far-away Te Kuiti, was the relatively new mill of R. H. Tregoweth. The only survivor anywhere near Pureora village was Carters' Morningside mill at Barryville, itself the second of Carter's mills on that site (Fig. 7.5).

The prospects of getting any further independent cutting rights had for a long time been poor, so in August 1971, Carter's at Barryville and Odlin's at Pureora had agreed to work together. They signed long-term contracts enabling an amalgamation to form Pureora Sawmills Ltd, managed by Magnus Russell (resident in Pureora village since 1948). Because this joint venture run by the Carter-Holt and Odlin's companies at Barryville was the last timber operation offering employment to local residents in 1978, it was the one that later became the focus of the famous conservationists versus loggers dispute (Chaps. 9 and 10). Logging in the Pikiariki area of SF 96 (South Block) continued to supply it—at least for the moment.

The most efficient mills were phasing out heavy circular saws (Fig. 7.17), like the ones used by Odlin's at Pureora, for thinner and more efficient band saws like those at Tregoweth's in Te Kuiti, which could get more boards from every log in less time, halve the waste to sawdust, and adapt to milling exotic timber in future. Mergers and further closures were a good way of channelling resources into keeping at least the better mills going. This logic became a significant part of the decision-making process in 1978 (Chap. 10).

The profitability of the logging business as a whole depended in part on the value put on the logs, and hence the price paid by the sawmills to the forest owners. Before the formation of the State Forest Service (SFS) in 1919, timber sales had been based on output from the forest, regardless of how much timber was wasted at the mill. SFS changed the system to charging for volume input, whereby every log that went into the mill was measured at the bush or mill skids and charged for individually. Thereafter it was in the mill owners' interests to process logs as efficiently as possible, with minimal wastage [6: 174]. But the prices set by the forest owners, at first by the Maori landowners and later by Forestry officers, were often so low that experienced logging officers like Jack Fyffe reckoned "they were giving the stuff away" [21]. Jack was not the only one to notice, but unlike Jack, others had power to change things.

The underpricing of such an important and essentially irreplaceable national resource shocked Geoffrey Palmer, Attorney General in the Fourth Labour Government that came to power in 1984. He commented in his memoirs that " In 1986, NZFS was selling logs on the East Coast [at prices] that barely covered the cost of cutting and transport….[this] cannot continue"

[15: 506]. It was one of many factors that led Palmer to help draft the State Owned Enterprises Act 1986, with drastic consequences for Pureora.

The following year, responsibility for non-commercial State Forests was passed from NZFS to the new Department of Conservation, and commercial plantations were handed over to the New Zealand Forestry Corporation. The Government suppressed the strong objections of senior NZFS staff, who had invested decades into building the exotic forests into a very large national asset, but could not prevent them expressing their outrage to professional conferences overseas [20]. The rest, as the saying goes, is history.

References

1. Anderson, K. (2000). Servicing caterpillar. Taumarunui: CAS Publications.
2. Anderson, K. (2008). Maoriland sawmillers. Ellis & Burnand Ltd, Sawmillers and timber merchants of Mangapehi. Manurewa: Ken Anderson family.
3. Anderson, K., & Cooke, R. (2014). "Tringamoot": The story of Taringamutu Totara Sawmills Ltd in the Taringamotu Valley. Taumarunui: Taumarunui and Districts Historical Society Inc.
4. Auckland Weekly News, 16 October 1946.
5. Groome, J. G. (1978). The role of Pureora sawmill in the King Country economy. Taupo: J. G. Groome and Associates.
6. Harris, D. (Ed.) (2009). Timber cruising and other forestry stories. Wellington: FirstEditionPublishers.com.
7. Leigh, A. H., & Clegg, S. E. (1989). Maniapoto District wild animal management plan, 1989 to 1999. Hamilton: Department of Conservation.
8. Mahoney, P. (2011). Vertical saws in sawmills: A heritage comparative analysis in global and New Zealand context. Wellington: Department of Conservation, unpublished report.
9. McMillan, L., & Walker, A. (2005). Ongarue: A place of the heart. Ongarue: Ongarue School Centenary Committee.
10. Molloy, G. J., Rockell, J. D., & Ure, J. (1973). Introduction to the state forests of the Rangitoto and Hauhungaroa ranges. Wellington: New Zealand Forest Service.
11. New Zealand Forest Service (1952). Working plan for Pureora working circle. Auckland: New Zealand Forest Service.
12. New Zealand Forest Service (1982). Report on Pureora forest. Auckland: New Zealand Forest Service, Auckland Conservancy.
13. New Zealand Forest Service (1983). Sawmilling in the King Country: An historical review. Wellington: New Zealand Forest Service.
14. New Zealand Herald, 29 November 1968.
15. Palmer, G. (2013). Reform: A memoir. Wellington: Victoria University Press.
16. Roche, M. (1990). History of forestry. Wellington: New Zealand Forestry Corporation Ltd
17. Simpson, T. E. (1973). Kauri to radiata. Auckland: Hodder & Stoughton.
18. Somerville, J. C. (1965). The timber industry of the King Country. MSc thesis, University of Canterbury.
19. Vaney, J. C., & Gibson, L. J. (1978). Current and potential use of timber from the state forests of the Hauhungaroa and Rangitoto ranges. Wellington: New Zealand Forest Service.
20. Whyte, A. G. D., & Poole, A. L. (2001). New Zealand forestry politically compromised. Fremantle, Australia: Commonwealth Forestry Association.
21. Wilkes, O. (2005). Transcripts of oral history recordings. Hamilton: Department of Conservation.
22. Wilkes, O., Gaukrodger, J., & Ritchie, N. (2011). Chronology of Pureora. Hamilton: Department of Conservation.
23. Wilson, R. A. (1957). The first fifty years of the Marton Sash, Door and Timber Company Ltd. Palmerston North: Marton Sash, Door and Timber Co. Ltd.
24. Wright, L. W. (1980). Decision making and the logging industry: An example from New Zealand. Biological Conservation, 18, 101–115.

Pureora Forest Village and Its Community, 1945–87

8

D.J. Gaukrodger, N.A. Ritchie and C.M. King

Abstract

Pureora Forest village was designed by the New Zealand Forest Service (NZFS) as a planned logging settlement, to provide a stable long-term future and good housing conditions for the workforce. This chapter describes the establishment of the village from 1945; the people and the social organisations they created; and the officers of the milling companies and of NZFS who were in charge. The stories of life in the village during its prime, and of how the community developed in isolation, of the children, the schools, the Football Club (both the players and the bar), the volunteer fire brigade, the sly-groggers, the ambulance, are all previously unpublished, genuine solid-gold social history, contributed by people who lived there and who recorded their first-hand memories on tape.

Keywords

Planned logging settlements · Rural community life · Barryville · Pureora Forest village · Social history of the timber industry · Prohibition in the King Country · Pureora Volunteer Fire Brigade · Sawmill fires · Wood chopping competitions · Inia Te Wiata · New Zealand House pouihi · Logging accidents

The most vivid first-hand descriptions of a place can come only from the people who actually lived there. Fortunately for anyone interested in the history of Pureora, the archive at the Department of Conservation (DOC)'s Hamilton office contains boxes of tapes and transcriptions of interviews, organised and conducted in the late 1990s by former OC John Gaukrodger and researcher Owen Wilkes with many of the people who lived at Pureora from its earliest days—including four of the longest-serving OCs [18].

In addition, NZFS required each OC to keep a detailed Station Log and to make Annual Reports

D.J. Gaukrodger
3a Te Angi Street, Turua, RD4, Thames 3574, New Zealand
e-mail: beejaygauk@xtra.co.nz

N.A. Ritchie
Department of Conservation, Te Papa Atawhai, Hamilton Office, Private Bag 3072, Hamilton 3240, New Zealand
e-mail: nritchie@doc.govt.nz

C.M. King (✉)
School of Science, University of Waikato, Private Bag 3105, Hamilton 3240, New Zealand
e-mail: cmking@waikato.ac.nz

C.M. King et al. (eds.), *The Drama of Conservation*,
DOI 10.1007/978-3-319-18410-4_8

on their operations. Owen and John summarised relevant details from these and other sources (e.g., file notes and official correspondence) into a single document. All this, plus additional information from DOC archaeologist Neville Ritchie and John Gaukrodger, amounts to a detailed and authentic chronology of the birth, life and decline of the Pureora Forest village [19]. The descriptions in this chapter are based on extracts from this wonderfully rich and revealing material, most of it never before published.

Construction of the Village

In 1940, the Te Kuiti Forestry office began to plan the establishment of the village at Pureora designed to house timber workers and forestry officers, plus roads and all the other necessary facilities. The site chosen for the village was inspected in November 1944 by the Housing Department. The following year Forest Service architect Lew Hahn arrived in Te Kuiti to manage the project [2: 132]. Approval was given by Head Office to locate seven 8′ × 10′ army huts at the E&B Pukemako camp (Chap. 5; Fig. 7.3) as temporary single mens' quarters to accommodate builders working on house construction at Pureora.

A roading survey was completed by Geoff Collett, NZFS District Ranger at Te Kuiti, and by March 1945, 18 miles of line had been cut, and the main extraction routes for working the first 2500 acres of SF96 were defined. Crawler tractors were unloaded, and in May the rough road to the village site and No. 1 skid were completed. Logging started almost immediately, to help alleviate a critical post-war shortage of building timber.

From the outset there were labour shortages. Expert dozer drivers and qualified tradesmen, especially mechanics and builders, did not want to work in the backblocks when easier work was readily available in the cities and towns. Labour for general forestry work in the seedling nursery and tree crop teams was also hard to get.

By the end of October 1945, the village area had been levelled and roads formed by dozer driver Bruce Archer; paddocks had been tilled and grassed; trees planted for ornamental and shelter purposes, fences erected, a telephone link to Benneydale established, permanent single men's quarters and a cookhouse for the building contractors had been constructed, and a water supply for the village was in operation. Contracts were let for the construction of 12 houses for staff in December 1945. Construction began in early 1946 and was completed over the next two years. Water and drainage reticulation were completed in 1949.

Fig. 8.1 Pureora Village five years after establishment, showing the early single mens' huts, the new State houses for NZFS staff, and (in the distance) the mill workers' houses at the west end of the village nearest to the Odlin's and Ranginui mills. *Crown Copyright, Department of Conservation Te Papa Atawhai (1950). Photographer: Buster Seager*

Fig. 8.2 Pureora Village on a crisp winter day in about 1978. The flat-topped house on the right was typical of those occupied by mill workers' families. It was then occupied by the NZFS/DOC goat hunting team from 1978 to 1989. A similar house (the Blue House, across the road) was converted into a field base and accommodation for visiting researchers. *Crown Copyright, Department of Conservation Te Papa Atawhai. Photographer: John Mason*

For the next 40 years the village settled into the landscape (Figs. 8.1, 8.2 and 8.5).

Location and Isolation

Rusty Russell's father Magnus was one of the carpenters sent to finish building the Ranginui mill at Pureora in the winter of 1947. He recruited Rusty to help, picking him and Alec Reid up from Rotorua and driving to Pureora via Te Kuiti—"a three-hour trip if you didn't get stuck".

They arrived in the middle of the night, and fell into the only accommodation available, a group of single mens' huts with two beds squeezed into each of them. There was no hot water so "First job in the morning was to get the axe and break your way through the ice on the drum outside. So you can imagine there wasn't much washing done".

Even when the village was well established, there was no electric power supply from outside until 1956, but a diesel-driven generator started operating at 6 am and ran until 10 pm. If the generator cut out, say during a dance at the village hall, the chap on duty would have to run down to the shed and start it again. If it cut out again, it was usually because one or more of the houses was overloading. That meant isolating part of the village until the electrician could come, and it was usually not difficult to guess which part to cut out. Next day the electrician usually found that someone had been putting a heavy gauge wire into their main fuse.

The water supply came from an unlogged area behind the village. It was untreated, so the Health Department used to come regularly to check it. Sediment was minimized by keeping the logging well clear of the stream, but "the bug count", as OC Ivan Frost called it, "used to depend on how far up the creek the last dead deer was".

Village life was especially difficult for women unused to country life Helen Russell remembered arriving in Pureora in 1953, a city girl from Rotorua who had never seen a mill or lived in a

Fig. 8.3 A typical NZFS logging crew. *Left* to *right* Sonny Anderson, John Evans, Digger Tane, Joe Ngatai, Ben Kane, Laurie Flay, Bill Reti. *Crown Copyright, Department of Conservation Te Papa Atawhai. Date and photographer unknown*

logging village before. There was no store, no power at night, and a bus only three times a week, which arrived in Te Kuiti at 11 am and left at 2 pm. So you had to get all the shopping plus a haircut, or a tooth filled or a doctor's appointment within those times. The gravel roads were unkind to her high-heeled shoes, and she was homesick at first. Her children began to arrive after 1954, and all of them in due course attended the Pureora School. Most of the men had grown up close to the bush, and relished the camaraderie of the tight-knit working teams (Fig. 8.3).

Telephones

The lack of telephones seriously accentuated the isolation. The 14 mile phone line from Benneydale to Pureora, an old earth circuit, was finished in 1947, and a tiny post and telegraph office opened on 15 December 1947. It was run by Ivy Grimshaw, daughter of the resident saw-doctor, Arthur Grimshaw (Fig. 7.10).

By May 1950 there were still only two manual phones available in the community at night for emergencies. Even in 1956 there were only three in offices—at the Forestry HQ, the Ranginui mill (manager Magnus Russell), and the Odlins mill (manager Roy Winwood). The only home that had a phone was that of the NZFS OC, and the office phone was switched over to it at night and at weekends. Ivan Frost remembered that he and Pauline would light their fire as late as possible on a Sunday morning, because as soon as the smoke appeared from the chimney, someone would be over to use the phone.

The summit of Pureora is the highest point north of the central volcanoes, and offered a tempting possibility for TV companies extending their coverage in the early 1960s. For a while there was a risk that they might build a huge mast there, and a road to construct and service it. Fortunately for the landscape values of the future PFP, they chose Te Aroha instead, but the outside world crept in eventually: in 1972 a microwave mast was built on Rangitoto, and a repeater in 1983. NZFS ensured that the large cleared construction area was restored by 1985.

Roads and Buses

The central location and rugged surrounding landscape (Chap. 1) explain why Pureora Forest

Fig. 8.4 Barryville mill and village in the late 1970s. The circular structure is a grade sorting table, and the conical wastewood burner stands next to the mill at *top left*. The single street of mill worker's houses was completely demolished after the mill closed in December 1978. SH 30 crosses the picture at the top (right to Benneydale, left to Mangakino). *Crown Copyright, Department of Conservation Te Papa Atawhai. Photographer unknown*

Village was isolated for so long. When the two mills at Pureora were built in 1947, the only way in or out was the road from Te Kuiti, which virtually stopped at Benneydale. From there to Pureora village there was just a rough mud track. For the first few years, the 53 km to Te Kuiti felt like 100 km of dodging potholes among slushy pumice and gravel. The road still included a section of unsealed gravel in 1979, when Dawson visited Pureora in the course of his research [5].

The road between the village at Pureora and the nearest neighbouring village of Barryville (Fig. 8.4) was at first only a walking track of 4 km, until after 1956 when the Public Works Department used road metal (gravel) from the Pureora Quarry to form a vehicle track. The rock used tended to form sharp edges when crushed, which shredded tyres. It had to be covered with pumice, making it prone to potholes.

There were no links eastward to Lake Taupo across the ranges (Fig. 7.4). In March 1947, it took Lew Hahn three days to get from Pureora to Tihoi on foot, but he concluded that it would be possible to push a connecting road across the saddle between Pureora and Titiraupenga one day. The present Link Road was built in 1979.

The road to Mangakino was finally finished in 1955, again using crushed metal taken from the Pureora Quarry. Before the Whakamaru dam was built, the only way to cross the Waikato River was a Bailey bridge across a narrows where Maori used to jump across.

The NZR Road Services bus to Te Kuiti started running in 1950, carrying the older children to secondary school and the housewives to do their shopping. The bus driver occupied a house in the village, which enabled him to be based at Pureora rather than at Benneydale. By 1971 the improvements in the roads and the increasing number of private cars meant that the service was running at a loss, and it stopped on 6 July 1979 [19].

Housing

The existence of new houses and the beginnings of a community were quite an attraction for potential workers: one of the earliest residents, Kitch Pedder, came to Pureora in 1947 only because he was offered a share of a state house.

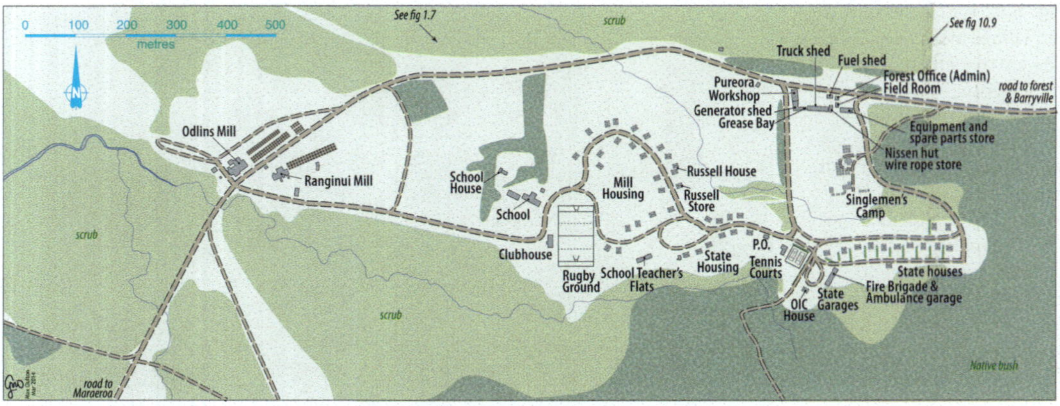

Fig. 8.5 Map of Pureora village in 1953, during the heyday of the timber mills. *Drawn by Max Oulton from aerial photographs*

It had no open fire and the winters were cold, so they used to undo the stove front and pile the wood in, making the kitchen black with smoke.

The "flat-top" houses for the mill workers were built of prefabricated materials at the E&B factory in Hamilton, all to the same flat-roofed design regardless of destination, then trucked onto the site and bolted together. There were hundreds of them all over the country, all with malthoid roofs and painted in pastel colours. The 1947/48 Annual Report from the Te Kuiti office of NZFS recorded 22 sawmill scheme houses: Ranginui 5, Odlins 6, and Morningside 11.

By 1952 there were over 30 homes at Pureora, half established under the Timber Workers Housing Scheme (Fig. 8.5). All the houses had wood ranges with wetbacks, and radios. There were no fridges, since there was no power at night, no insulation despite the cool climate (Box 1.2), and only seven houses had vacuum cleaners [13]. But social life in the community was growing, and active sports clubs were developing.

The single mens' quarters comprised a row of individual huts (Fig. 8.1), plus an ablutions block. and a cookhouse run by Gertrude Grimshaw. In 1947 this redoubtable lady charged the bushmen £2 a week for meals. Years later, Peter Archer (Gertie's grandson-in-law) once met an old bushman who was there at the time, who still raved about Gertie's cooking. Regrettably, the consolations of the cookhouse did not prevent the single men from becoming notorious for damaging their huts and causing a nuisance to the adjoining married quarters.

The difference in size and quality between the gabled state houses allocated to forestry workers by NZFS and the flat-tops provided for timber workers by the mills was always a potential friction point that had to be carefully managed by the OC. Critical comments about this and related social issues [13] were the main reason why, when Kitch Pedder first arrived as OC in late 1953, he was specifically instructed by the DG of Forests to deal with it. All the same, most reports showed good relationships, allowing for a bit of over-indulgence by the single men from the mills on paydays and at weekends.

Inevitably, rough use and cold wet weather began to tell on the houses, particularly the flat-tops occupied by the timber workers. By 1962/63, several were in a poor state of repair because of leaking roofs. By the 1968/69 year only three old-style malthoid roofs remained in the village (one of them appears at the far right of Fig. 8.2).

The 1972/73 annual report stated that 14 timber workers' houses and 32 huts would soon have to be replaced unless hundreds of dollars was spent on each for weather proofing and replacing worn out plumbing, fittings and joinery units. By then only 8 of the original 18 sawmill houses were occupied, and one of these was by the local shopkeeper. The population was 55 men, 37 women, 99 children, total 191, and the school roll was 102.

The next year, the sole carpenter's time was "being dissipated on maintaining old drains, septic tanks, and replacing decayed material on some houses", because accommodation for both married and single men had continued to deteriorate. The following year, 1975, NZFS policy changed to allow workers to live in nearby urban areas, and travel from there to their jobs in the village.

Collecting the Pay

Staff were paid every second Thursday, in cash. In Kitch Pedder's time the OC and pay clerk had to drive to the bank in Te Kuiti (53 km), which meant half a day lost for two men. "We kept a revolver in the little suitcase we carried the pay in", Pedder remembered. This was standard NZFS practice in the 1940s and 50s [6: 145], but not necessarily much of a comfort. "We often wondered what we would do if we were stopped—would we be able to get it out in time? Would we have actually used it?" [18].

Bill Drower tells a story about a real panic over the worker's pay which arose when an NZFS staffer was dispatched to Te Kuiti to collect stores, post mail and collect the pay bag from the bank. He did not realise until standing at the bank counter that he had accidentally posted the envelope containing the cheque for the bank along with the other mail. It took several frantic calls to the Postmaster, the Forest office, the bank and the Chief Postmaster in Hamilton before the mail box could be opened, the vital envelope retrieved and the business at the bank concluded. That staffer found a thousand reasons not to be sent on the same errand again.

When SH 30 eastwards was opened, the pay account was transferred to a branch of the BNZ bank in Mangakino (33 km). BNZ would send a secure truck to deliver cash to the bank every other week, and it would be open 10 am to 2 pm.

Dave Yanko (Fig. 8.7) enjoyed doing this job, because it gave him and the clerk the chance to drop into the pub for a few beers after picking up the suitcase. They took the suitcase full of money in with them, and tucked it down behind the foot rest in front of the bar. "One day we must have been there a couple of hours, we took off, and remembered halfway home that we didn't have the money. We ducked back to the pub and the suitcase was still there, and the relief was …well, we sobered up that quick" [18]

When the precious suitcase with its enormous quantity of cash arrived back at Pureora, the pay clerk and the driver would lock themselves in an office for a couple of hours and divide it all between the employees' pay envelopes. Then everyone would file into the office on Thursdays about 5 o'clock and pick up their pay.

Direct credit was practically unheard-of, and transferring cash into the post office savings bank was too much trouble, so all the local NZFS staff went out with their two weeks' wages in cash, and a lot of them would go straight to the nearest bar. They'd have some really good sessions on Friday and Saturday, and then the money would start to run out during the week. The weekends between paydays were never as good as the weekends after a payday.

The worst problems with management of money and time came up over Christmas, as described in detail by John Mason. From Christmas to New Year, NZFS (like most Government Departments) shut down. Then, everyone had to take their two weeks' annual leave before the mills started up again in mid January. That neatly solved any problems over leave for the rest of the year—there wasn't any. But in December NZFS paid two pay days at once, so people went away for their holidays with a month's pay in their pockets, and two weeks later they were broke even though there was still two weeks to go before the next pay day.

That created a lot of social problems, until OC John Gaukrodger brought in an unofficial system by which people were given only three weeks' pay before they went off for their four weeks off, and then when they got back he gave them the fourth week's pay in arrears so that they had something to live on before the next regular pay. The same problems plagued the mill workers, but the mill managers never changed their system.

Workshop

The need for constant maintenance of ageing machinery meant that the workshop was an absolutely central part of the NZFS project, and retaining good staff was a perpetual headache for OCs. The isolation of the village meant that it was always difficult to recruit and keep enough skilled mechanics to maintain the fleet. A-grade mechanics were worth their weight in gold, but living at a remote place like Pureora required a special type of family, especially when the children got to high school age. Those who stayed did so mainly out of family or tribal loyalty.

In 1953 there were two 8-year old Caterpillar bulldozers and two Mack trucks on the books, plus the gang buses and the OC's car. Within two years, chassis cracks made one Mack unroadworthy and the other marginal. Jack Fyffe, logging officer of the time, was always bitter about the way the bush crews were starved of equipment. He was supposed to have access to four tractors, but "you were lucky if you ever had two going at once".

Jack was especially annoyed when orders for two new tractors were approved, and they were given to another forestry operation. By the time Ivan Frost arrived in 1957, there were supposedly two trucks and three dozers available, but only one truck and no dozers were operational, and staff morale was low.

Equipment breakdowns often figured large in the annual reports compiled by successive OCs, but no action was taken even in 1960/61 when worn-out equipment caused a loss in production. Ivan Frost's dismal conclusion was that this was "A black year for breakdowns, it seems we will have to grind to a halt before anything will be done. We have had no new equipment since 1956". The next year, a new logging truck arrived.

The workshop was not a comfortable place to work, especially in winter. That all changed in 1968/69 when, to general astonishment, underfloor electric heating was installed in the workshop. It meant that a mechanic would no longer have to leave his work to thaw out his hands around the stove, his tools would not be icy to the touch, and even his overalls would be warm to put on if left on the floor overnight. Bill Drower reported that time-and-motion studies of several of the longer repair jobs recorded a 20 % improvement in output—sufficient to justify the installation of underfloor heating.

One of the biggest problems was getting spare parts. All major stores and parts had to come by rail from Auckland, but were often put on the train before the advice note had arrived, so the Pureora staff had no idea when to expect the goods. Neither did NZR staff consider it part of their job to tell customers that their stuff had arrived, so it might lie at the Mangapehi rail yard for days before anyone went to look for it. Over the years, commented Ivan Frost through gritted teeth, these high costs and delays in obtaining parts must have added thousands of pounds on to the costs of their operations, or would mean having to assemble a new machine out of bits of broken ones.

In 1956 the workshop store was run by Buster Seager. The main items he dealt with were tractor parts and track gear, because the pumice roads were very hard on tractors.

Relationships between the mills were usually friendly, and included various forms of mutual help, but soured quickly if damage was not immediately made good. The Pureora station diary for September 1951 records that "Ranginui broke one of our 6′6″ fishtail saws. [NZFS] refuses to loan another".

NZFS Officers in Charge

Over the 40 years (1947–87) that Pureora village was run by NZFS as a planned forestry operation, it had eleven permanent or acting OCs, as follows: Lew Hahn 1948; Eric Johnstone 1948–53; Keith (Kitch) Pedder 1953–57; Ivan Frost 1957–62; W. H. Robinson/Athol Ferguson (Acting) 1962–64; Bill Drower 1964–69; Darbie Perston 1969–1974; Jack Walker 1974–80; Dave Yanko (Acting) 1980–81; John Gaukrodger 1981–87.

The life of the OC in a small and isolated forestry community required the expertise, not only of a qualified forester, but also the wisdom and keen perception of a sociologist. When Ivan Frost arrived as OC he already knew what he was coming to, as he had first worked there as a young Leading Hand on timber cruising in 1947. Nevertheless, he recognised that he'd have been a better OC if he'd had some sort of social training.

Technical problems, he said, were simple, because you could either solve them or you couldn't, but social problems were never ending. The people he had to deal with were not only employees, but also the guys he drank with. He was gregarious and had to socialise, and was on the committees of all six local clubs, but although the community included some people he would like to invite to his house, there were others he wouldn't, and he had to be the same to everyone. Outsiders don't realise that managing such issues required skills you were never taught, he said, so you just took it out on your wife when you got home.

For example, one episode reported in May 1951 arose when the OC Eric Johnstone refused to provide a mill worker with a Forest Service vehicle, since he could have used one belonging to the Odlins mill where he worked. The Timber Workers Union became involved in the subsequent stoush, and Eric was exonerated. The station diary dryly comments that the episode illustrated the general attitude of the community, who tended to see Forest Service equipment as public property. Two years later, another comment noted that "Community relations difficult but with slight improvement. Mill companies not helping".

An even more illuminating example of the troubles of an OC was revealed when Margaret Smith, a student from Wellington, undertook a social study of life in Pureora village [13]. Her analysis was based on interviews with 35 families, mainly the wives. She observed a worrying distinction between the state and mill employees, which arose in part because NZFS looked after their own people better than the mill owners looked after theirs. Most of the state houses at one end of the village were in good repair, but virtually all the mill houses at the other end needed maintenance and painting. The topsoil had been replaced around the state houses, but not round the mill houses, so gardening was mainly confined to state house families.

Some families in Pureora village kept poultry, until wild NZ falcons discovered in the 1950s that domesticated birds were easy game [3]. Falcons were of course also protected, but in pioneering days, pragmatic considerations tended to dominate. The falcons were largely exterminated by village residents, although they have since made a remarkable recovery.

The mill workers drew £20–35 (NZ$40–70) per fortnight, before deductions for rent of 15 shillings to 17/6 per week (NZ$1.50–1.75), and 3 shillings for the single mens' huts. The men also often asked for advances in wages, which drastically limited the contents of the next pay packet. Odlin's average total wage cheque was £380 (NZ $760), of which only £193 (NZ$386) might be issued as take-home pay, after much of it was retained for rent, tax and wages paid in advance, mainly to finance the mill workers' heavy drinking. A married bushman entitled to £25/6/6 per fortnight lost deductions of £10/4/–; a single man earning £22/19/7 paid deductions of £6/17/10.

The mill managers justified the situation by saying that unless wage advances were allowed, the employees would not remain. But the system meant that four out of five mill workers were living hand to mouth, including those with several children. State employees were better off because no such system applied to them.

In December 1953 the new OC Kitch Pedder was called to a special meeting in Wellington with the DG of Forests Alex Entrican and other senior staff including W.J. Kinlock, stimulated by the implications of Smith's thesis. The meeting was to discuss the future of Pureora village, and how Forestry authorities could meet the challenges identified by Smith's work. Whilst recognizing that Smith's was a special study for a special purpose, which did not include forest management, nevertheless NZFS accepted that they had a special responsibility to develop a good community spirit at Pureora.

Pedder was told that the OC's job did not start and end with supervising log deliveries according to schedule, but it involved, among other things, ensuring local leadership and inspiration from Forestry staff, breaking down the class distinction between the forestry and mill workers, looking after the mill houses to exactly the same level as other buildings, ensuring tidy lawns and frontages in the village, encouraging churches and other outside interests, and helping to establish a local committee capable of showing interest and ability to manage the proposed new community hall. "Forestry authorities are liberal with providing recreation room and camp amenities", said Kinlock, "and intend to remain so, but evidence of some appreciation from villagers was required".

Pedder was instructed to make early contact with the mill managers, get them to clean up the disgraceful appearance of the mills and their surroundings, and to cease giving sawmillers advances on their wages, which was spent only on beer. NZFS instructions (preserved in a file note in DOC archives) end by pointing out that "how Pedder was to carry out his onerous tasks in the village was up to him, but he could be assured of every help and understanding from HO. Signed W.J. Kinlock 9/12/53". Most of these problems were already known to successive OCs, but they had been identified in startling detail by Smith's analysis of village life.

Far from resenting such a burden, Pedder agreed that providing for the social needs of the village was a very important part of the job, and attributed the idea of the canteen business and the Football Club bar to that meeting.

It was not easy, and sometimes the social distinction between the OC and the rest had consequences for his family. Pedder's wife Nellie taught at the Pureora School, and one day their small son Ron came home covered in bruises. Mr. Patrick the headmaster asked Nellie if she knew whether Kitch had recently had to discipline anyone? The answer was that yes, he had recently had to sack a man, and it was that man's son who had knocked Ron about.

Pedder was no doubt pretty annoyed, but he had to try to remember that sawmillers were always under tremendous pressure to produce the required quantities of timber, and the mill managers were not always willing to make allowances for isolated working conditions.

One of Ivan Frost's proudest moments came in 1962, after he had left Pureora. He was asked to return, in company with logging officer Athol Ferguson, to choose and help fell a perfect totara tree for a very special purpose. It had to be straight and long enough to provide a 52-foot log to be carved by Inia Te Wiata into a pouihi ("totem pole") for New Zealand House, then being built for the new NZ High Commission in the Haymarket, London.

They found three possible trees by lunchtime and organized a logging crew to fell the best one, but to put it down on flat ground without breaking it meant cutting against the direction the tree was leaning, and all the scarfing and wedges they had available would not move it. By 5 pm they were sitting down considering their next move when, very slowly, the tree began to fall. It went down within inches of the prepared site and landed without the slightest crack (totara trees often shattered when they hit the ground).

Inia spent seven years carving it, in a workshop in the basement carpark (Fig. 8.6), and it was finally erected in 1972, a year after he died

Fig. 8.6 Inia te Wiata carving the pouihi for NZ House, London, from a totara felled in Pureora Forest. *Alexander Turnbull Library, Wellington, New Zealand. ATL 1/2-190156 F*

[15]. Years later (in 1989), Ivan visited London and proudly photographed "our" tree and the spectacular carving work on it.

The heyday of native timber milling passed with the 1950s, and after that the progressive closing of exhausted mills (Chap. 7) inevitably led to a gradual reduction in the number of mill workers living at Pureora, and an associated decline in the school roll. The OC in the 1970s, Jack Walker, saw his job in the village as being a combination of mayor and father confessor, a hard ask during the height of the anti-logging controversy that took them all by surprise in 1977–78. He occasionally also had to act as tour guide for important overseas visitors. The station diary for May 3 1974 notes that "A Mr. Suharto from Indonesia came to see Pureora logging".

Fig. 8.7 Dave Yanko, active Pureora resident, long-serving logging officer and acting OC in 1980–81. *Crown Copyright, Department of Conservation Te Papa Atawhai. Photographer unknown*

Staff

In 1957, the Forestry staff numbered 12 handling logs in the bush, 12 growing exotic seedlings in the nursery and planting them out, two in the office, three mechanics if available, one or two on roading, one in the rope store splicing strops and sharpening saws, and a storeman. Plus a camp cook and his off-sider, and a camp sergeant to keep the place tidy.

The OC had to coordinate all these different tasks and specialities, helped by senior staff including a logging officer, a planting officer, and various technical advisers. It wasn't always easy, yet Ivan Frost remembered Pureora as a good community with few hassles or frictions among men who worked, drank and played sport together regardless of rank. He reckoned that the occasional disciplinary action required of the OC usually caused more trouble among the wives than among their men.

Typical of the long-serving Forestry families of the village were Dave Yanko (Fig. 8.7) and his wife Mauven. They lived at Pureora for 30 years, until he was transferred to Tairua Forest, Whangamata, in November 1984. Dave had been in charge of all bush operations until the moratorium, and then he took responsibility for roading. Both were fully involved with the community; all but the eldest of their four children were born there, and all attended the school; Mauven was on the committees of the CWI and the school for years, and later worked in the school office; Dave was a stalwart of the Bush United rugby club both as player and secretary; and both were involved in Christmas parties, fire brigade competitions, rugby socials, the lot.

The place and its back-of beyond reputation were well known to the senior staff, and they understood the way that isolation tended to accentuate the problems of recruiting and retaining skilled labour. The official employee statistics for the 1952/53 year quoted by Smith [13] showed a 70 % turnover in 12 months, with a net loss of 16 working men.

Staff made great efforts to improve social life in the village, but almost a decade after Smith's survey the 1960/61 annual report was still complaining about the shortage of skilled labour, by then due to a different problem: a commercial company, NZ Forest Products, was offering better wages that were enticing men away. Fortunately, someone higher up took action. The next year, a new bonus scheme improved production, and an increase in mechanics' wages meant that the workshop could, for a time, hold a full crew.

Some of the labour shortages of the postwar years were met by British assisted immigrants. They found 1950s New Zealand in general, and Pureora in particular, very different from anything they had previously experienced. When Kitch Pedder got a call from Wellington, saying that a group was coming up on the overnight train, arriving at Mangapehi in the small hours of the morning, he got huts ready, laid down mattresses, and drove down to meet them. By daylight, one had disappeared already; by contrast, one (Gordon Gillespie), who couldn't get over the OC himself going to pick up his new men, stayed for years.

Probably around half the village community were Maori, and the stalwarts among them were invaluable people to have in the village. Some families, like the Pihamas, could stabilize the whole group. Frost could sometimes help combat the lure of better money elsewhere by some under-the-table deals.

Such payments were illegal of course, but necessary to keep key staff like ace dozer driver Scotty Pihama. Scotty could drive one of those big bulldozers "like a sewing machine", as Ivan Frost put it. The downtime of his machine was half that of all others, because he knew it so well and was quick to spot anything wrong.

Like many men working around noisy machines without any ear protection, as was usual then, Scotty became hard of hearing, but would never admit it, so he had to second-guess what the OC was saying. He was still a key member of any logging crew, and if he could not solve a mechanical problem, nobody could.

In general, the Maori of those days led very hard lives, observed OC Ivan Frost; they were tough as old boots, and got into some horrendous fights, but they were gentlemen. They never used weapons, and always turned up for work on Monday regardless, yet once they got to the end of their working life they were likely to lose their homes, and never had any money to buy one of their own. Life expectancy was shorter for Maori than for Pakeha, and there were a lot of tangi (funerals), all of which were attended by the OC.

Mauven Yanko had a story about one of her neighbours, Bev Rudsit, who had a real struggle living at Pureora—with five kids and a wood stove, cutting her own slab firewood to fit the firebox to keep the hot water and the food coming—she said they were the happiest years of her life, because she was surrounded by good neighbours, almost all Maori.

Bill Drower, OC in the mid to late 1960s, was in charge of village residents working in a variety of jobs (including 12 forestry staff of all grades plus mill workers employed by two different companies plus teachers, store keepers, and a bus driver). Drower saw the result as providing sufficient diversity among the locals for energetic, resourceful and helpful community, and the few exceptions as making no difference.

At the other extreme of social judgement, Jack Fyffe, in charge of logging crews, and Jack Walker, a later OC, attributed part of the reason for the eventual demise of NZFS to the policy of the Department of Social Welfare to bring unemployed people to work in the forest, whether they were capable of doing the job or not.

Fyffe (born 1913, and a life-long bushman right down to his boots) was equally scathing about "all those BSc jokers…they were a law unto themselves" whose crackpot ideas wasted money that should have been shown as a profit. It says a lot about the Pureora community that, even with so many different private opinions, the village was usually described as a friendly place.

All the small communities round about had a hall, and on Saturday night there was always a do somewhere. Rusty Russell would go round the village saying "I'm leaving for…." And if you wanted to go you just hopped on his truck and away he'd go. At 2 or 3 or 4 in the morning you'd all hop on the back of the truck and come home. They were a lot of fun, those days remembered Dave Yanko.

By the mid 1970s, the village housing and amenities were showing their age, and the depletion of the forests was causing anxiety about job security. Firm decisions about the future of the village were obviously needed, but the top brass at NZFS Head Office were caught up in national-scale political arguments about West Coast logging proposals, which had begun to affect decisions on the central North Island forests too (Chap. 6). The 1973–74 Annual Report commented that the last few years had been a very unsettled period for Pureora, and morale amongst the residents had been declining steadily.

In response, NZFS began positive moves towards re-establishing Pureora village on a permanent basis, by letting a contract for three new houses and a new recreation room, and starting a programme of re-roofing some houses. In the short term, these, and the recent improvements to the new school and teachers' flats, were "…. helping to allay flagging morale amongst residents", said the OC. In the longer term, these improvements only made the residents more determined to resist the final blow that was to change all their lives only four years hence.

Some of the longest-serving key staff who contributed their memories to the oral history programme were able to supply a succinct summary of life in the village, mainly positive and fascinatingly different from those of the OCs. In Dave Yanko's words, "She was an excellent life, let's do it all again".

Pureora Forest School

The Education Board were planning to build a school in 1945 at E&B's mill at Maraeroa, but Lew Hahn suggested Pureora would be a more central site to place a school to serve the number of children expected to be living in the area within a few years. Pureora residents missed out on a school at the time, but in 1947 they tried again.

The school at Maraeroa, by now rather ramshackle, was to be enlarged, perhaps because the school committee was made up of mostly Maraeroa residents. Pureora residents opposed the plan, and this time they won. During the Easter break in 1948 the school was moved to Pureora, by a team including Geoff Hammond, Roy Callaghan, Henry Simpson and Bruce Archer (son-in-law of Arthur Grimshaw), for a contract fee of £250 (NZ$500) [2: 175]. Two D8s and other Forest Service equipment were used to widen the road in places, and some damage was done. The station diary comments that "The exercise was officially investigated", and one of the guys involved lost a salary grade because of it.

The village community continued to grow as more houses were completed, and by 1951 the Education Department was putting up extra rooms as the post-war baby boom and the height of local logging operations helped to boost the school roll to 155 pupils in 1952/53 [19]. The children of all the local sawmill communities (Pureora, Barryville and Maraeroa) went to the Pureora Forest School (Fig. 8.3). A bus carried the 45 primary school children and 9 pre-schoolers from Barryville to Pureora (4 km) and back every day.

One of the first questions asked by prospective employees in the mid 1960s, reported OC Bill Drower, was "What is the schooling like?" The answer was that the teaching was excellent. Helen Russell remembered a long line of dedicated staff—Jack Green, Alan Ingram, Rod Nielsen, Bill Murray, Vic O'Rourke, George Keown—who all did not only a great job with the kids but also gave a lot to the village social life.

Bill Murray was a Maori teacher who understood his own people and knew how to work with them, remembered Mauven Yanko, then a teacher's aide. If Maori kids tried to skip school he would ask the others where they were, and if he was told "Oh, they have to babysit 'cos their Mum's sick", he would not settle for that. He'd stalk up the road, break up the card game and tell them, "Here, you, look after your kids, and you, back to school". Only he could have done that. At one stage, Bill had 46 kids in his classroom.

Unfortunately, the school buildings did not match the teaching. A new member of the School Committee, Lew Read, teamed up with Bill, and with the support of the rest of Committee prepared a comprehensive proposal to the Hamilton

Fig. 8.8 Pureora Village School, across the road from the football field. *Crown Copyright, Department of Conservation Te Papa Atawhai (1986 or 1989). Photographer: John Mason*

Education Board requesting an upgrade in building. Plans to build a new, relocatable 4-room school began in 1968, including insulation and heating better than the norm for a rural school, and it was ready by February 1970.

In 1972, the school roll was 102, and for the next eight years the school was graded for four staff, with rolls of 104 in 1976, 107 in 1977, 114 in 1978 and 115 in 1979 [9]. It was still very difficult to find staff, especially as two houses were no longer available for teachers.

The Post Office and General Store

The first post and telegraph office opened on 15 December 1947, but for many years there was no other shop of any kind. Smith's survey of the inadequate facilities in the village up to 1953 specifically recommended the establishment of a general store, to give the women living in different areas of the village a common meeting ground.

Another of Smith's recommendations was that NZFS should establish a canteen. The first meeting of various staff interested in helping to make this happen was called by Kitch Pedder in 1954, fresh from his meeting in Wellington. It elected a Committee, including Colin Sutherland, Harry Bunn, Kitch Pedder, W.C. Johnstone (the NZFS clerk), and Ken Seymour to organize it.

The idea was to encourage a social hub for the village, run by volunteers after hours, and its stated intention was to generate proceeds to be spent on assisting the organisation of the community. Within a year it was in operation, in a little shed at the back of one of the houses. Stock was bought wholesale from Bond and Bond, who allowed a month to sell it. Harry Bunn described the canteen in great detail.

> We each put in a fiver as a capital base. We bought high turnover, high profit lines like chocolate and cigarettes, and we sold those in a week, spent it again, then spent it again, before we had to pay the bill. I was treasurer and I remember being quite startled that the first bill was £300 or so but I was able to pay it! We set up in a spare garage and put in shelves. It became the focal point of the village. We decided to have ice cream for the kids, and we bought a 15 gallon fridge. …we could sell 15 gallons in 4 days. Eventually we bought another 15 gallon unit. The suppliers were astounded at how much ice cream we sold…and supplied us with a 50 gallon unit. They showed us how to roll a hollow ice cream to double the profit. We found there was an enormous profit on it anyway so we piled it on and made enormous ice creams – we became famous for the size of our ice creams and

people came from all over the place! And we had soft drinks of course, and they delivered them and picked up empties …the kids would bring in half a dozen bottles and immediately spend the deposit on other high-profit items. It flourished, so much so that people said, hey there is enough [custom] to run a shop. …We made so much money …we completely refurbished the hall, £400 worth, we supported the fire brigade, all the sports clubs, they all got handouts. The profit that first year was way above what any of us were earning, even Kitch. It ended when we all got transfers [18].

Within a couple of years NZFS sold the canteen at cost to Mrs. Ruby Alexander, wife of a fireman at Odlin's mill, and she ran it as a shop with basic supplies from 1956. Helen Russell used to help her in the shop, and stood in for her when she wanted to have a day in her beautiful garden. In 1964, Ruby's husband died suddenly, and a year later she decided to move on.

Helen and Rusty Russell took it over in 1965, moved it to a new building and developed it into the Pureora General Store under the IGA franchise (Fig. 8.9)—just another responsibility to add to Rusty's cartage business and their new baby. They catered for everything, from ice cream and cigarettes to working boots, swannies (bush shirts) and ploughs. Watties and Lever Bros would rail orders of ten cartons or more of tinned food for free to Mangapehi, and Rusty would pick them up at the station.

Helen got to know all the local children sent down to the shop by their parents with a fistful of money to pay off their bills, and heard a lot of stories from them. A potbelly stove in the store made it a warm social centre in winter—anyone who was there, eating a pie or warming their gumboots, would stoke the fire or go get wood for it, and Helen could even get the kids to nip outside and bring in her washing for her while she was busy doing orders.

Most clients were honest, as indeed one needs to be in a small community where everyone knows everyone else's business. Once a couple of brazen lads pinched some swannies, and wore their new gear to work next morning, "still with the creases in" said Helen, still surprised years later. Jack Walker sorted them out, she added. He made them pay for them or bring them back. It is not clear whether her conclusion that there were "No hard feelings" refers to the lads or to herself, but Jack's attitude seemed to ensure it wouldn't happen again.

Burglaries were rare, and usually due to outsiders; Helen mentioned two. Once, "they just backed their tow ball in through the window; [next morning] the glass had gone out and so had a lot of other things".

A much more serious burglary happened when Rusty was away. Helen went down one morning with the cash box under her arm and found the front door open, the padlock on the

Fig. 8.9 Pureora General Store, and the Russells' house. *Crown Copyright, Department of Conservation Te Papa Atawhai. Photographer: John Mason*

ground, the freezers open and the shop a total shambles. The burglars had taken all the school t-shirts, soft drinks and sweets, torch batteries and three axe handles ("those would have been the first things they'd have hit you with", said the police when they came to investigate). The same night, NZFS lost all their chain saws. Eventually the police tracked down a suspect to Te Awamutu, and in Court a local woman recognized him as someone who had been in the village that day. The Russells ran the revamped shop until GST came in in 1987.

The long awaited fully fledged Pureora Forest Post Office opened in what had formerly been the visiting doctor's room, by the parking bay in the middle of the village, in July 1972. The Post Office and the Store were vital links to the outside world for the villagers. Between them they offered not only a mail service but also a bank and a communal meeting place, an asset to the community with a local significance out of all proportion to its size. They provided credit from payday to payday, and the only petrol pump for 13 miles (22 km).

The Volunteer Fire Brigade

Hot machinery combined with flammable fuel and accidental sparks (not to mention lightning strikes) always makes fires a perpetual hazard in forestry and timber operations, and a big fire could cause massive economic damage. The Forest Journal kept by forestry staff at Te Kuiti also commented on two other hazards common in the early days: many fires were started by sparks from log-hauling steam engines, or by pig hunters [12].

Between December 1945 and March 1946, a period of unprecedented fire hazard, 62 fires burnt over 16,330 acres of State Forest land, including 390 acres (157.6 ha) at Pureora. The scorched timber sustained damage amounting to 25,000 bd ft/acre (146 m^3 per ha).

In response, a Fire Emergency Army radio truck was stationed at Pureora, a fire hazard monitoring station established and, because there was then no telephone, a radio-telephone was installed to link Pureora with the Te Kuiti office.

The Pureora Fire Brigade could be called either an industrial unit under the Forest Service, or a volunteer unit under the Fire Service. Which label was used depended on who was asking, since at that time only the Fire Service had a country-wide liquor licence. The Pureora Brigade was a Volunteer unit.

The Pureora Volunteer Fire Brigade was formed in 1951, financed by grants from NZFS, but it was also needed at the mills. Timber mills burned down with depressing frequency. It was the 1938 fire that destroyed the Mapara Timber Co.'s mill at Poro-o-tarao, and the 1939 fire at the Marton Sash and Door (MS&D)'s mill at Waione, that enabled the MS&D company to take over Mapara's cutting rights and expand their operation to Barryville (Chap. 5). MS&D also lost two other mills within a few months in 1940 (National Park and Erua) [20].

Mills near the Main Trunk Railway were especially vulnerable to fires started by sparks from steam engines. The Waione and National Park fires both began soon after an express train went by in the small hours of the morning, reports Wilson [20]. The National Park mill was burnt down twice more, and E&B lost and rebuilt their No. 1 mill at Mangapehi in 1945 and the smaller No. 2 mill in 1947 [2: 107–110].

At Pureora the Odlin's mill was burnt down in the middle of the night in 1957. Rusty Russell was the first to see the flames leaping up from it —he had had a bout of hepatitis so was not doing much drinking in the evenings. There was no siren, so he drove round the village with his hand on the horn to wake up the volunteer fire brigade. After the mill was rebuilt, Rusty kept his job until 1966, when he finished working for Odlin's and sold his logging truck.

On 15 October 1968 the Morningside mill at Barryville burned down, and was rebuilt by Christmas. Another fire which started on 10 February 1970 in the Ranginui timber stacks, and crossed the road to damage the Odlins sawdust and slab piles, was attributed by Forestry staff to the accumulated result of years of bad housekeeping at the Ranginui mill. Minor chimney and kitchen fires in the village were common.

The Pureora Volunteer Fire Brigade members worked hard at their practice drills, and entered into the competitions between local brigades organized by the national Fire Service. They always did well, and won trophies several times. The Volunteer Brigade club house had a bar which was opened on Wednesday nights after practice, plus Thursdays and Fridays, supposedly open to members only. The regular members included many of the mechanics from the workshop, in part because the Chief Fire Officer was the mechanical overseer, Bruce Tricklebank.

At first the volunteer brigade was, of course, staffed entirely by men, but after the NZ Fire Service changed its rules in 1979 to permit women to train as fire-fighters, three women from Pureora joined up at once. Who else, they asked, could tackle fires in the village when the men were all out in the forest? The longest-serving members were Mrs. Belle Packer, Mrs. Eileen Ihaia, and Mrs. Sharon Tricklebank, whose husbands were also in the brigade. They were joined a couple of years later by Mrs. Bella Heta and Mrs. Patricia Hall. All five received long-service awards in 1982.

The women trained every Wednesday night, along with 12 male colleagues. Practical fire-fighting exercises were available on old forestry buildings, although normally their duties were limited to smaller fires in hedges and chimneys. The Chief Fire Officer was glad to know that if some of his men were too deep in the forest to get back in time, their wives would always be at the ready.

The Volunteer Fire Brigade worked only in the villages and at the mills. Fire control in the forest was the responsibility of the NZFS silviculture and production forestry teams, and was often very difficult. Early firefighters in remote areas had to carry water in 4-gallon back-packs, remembered Ivan Frost. The portable Paramount pump was a big advance—it could be set up at a convenient stream, and would pump water through a one-inch pipe for at least a mile. The only problems were that, at 85 pounds, it was heavy to carry into the bush [6: 85], and the Canadian centrifugal motor often refused to start, which must have been desperately frustrating for the firefighters.

By the 1970s fire-fighting equipment was more sophisticated and also more necessary, since burning was becoming the standard method of clearing logged areas at Pureora, but was still a hazardous operation. The accumulation of nearly 25 years of cutover slash that had been root-raked and planted provided plenty of fuel right next to highly flammable standing non-merchantable trees. The 1973/74 Annual Report from the Pureora OC describes an occasion when a helicopter was called into douse a fire, but was late arriving. The pilot had seen a fire, found a pond, and deposited a load on the blaze, but got rude gestures from the men on the ground. He eventually realised he was putting out the wrong fire [19].

The Volunteer Fire Brigade fulfilled a social need as well as providing a local fire fighting force, so the arguments for keeping it were more than merely economic, the OC told a NZFS review commissioned in 1974/75. It survived, but not for long. After 1981 the village began to empty, and some of the most dilapidated vacant houses were burned down for fire brigade practice.

Prohibition

When Ngati Maniapoto first agreed to allow the opening of Te Rohe Potae for railway surveys in 1882 (Chap. 5), it was on condition that the Government agreed to declare the King Country "dry". As historian Michael King put it: "If the presence of the Pakeha was allowed, then there had to be some protection from his vices" [7: 253]. Perhaps this explains why Maori offences related to alcohol in the King Country at that time (1935–44) were so low (0.745 per 1000 people, compared with 1.345 per 1000 in the Waikato, and 3.743 in Gisborne).

Ngati Maniapoto were in a strong position to negotiate, since they knew how much the Government wanted to push through the Main Trunk Line. They had largely distanced themselves from the Waikato tribes who, still smarting from post-war land confiscations, were less inclined to be co-operative. However, an official review of the "solemn pact" made between the Government and the chiefs of the King Country came to the conclusion that there was no bargain agreed for a proclamation of prohibition in return for concessions on the railway.

The main driver of the proclamation was a petition signed by 1400 Maori from Te Rohe Potae district, proving that prohibition had the almost unanimous assent of the local people. Moreover, at the time the petition was circulating (September 1884), the decision to use the King Country route for the railway, made on 24 October 1884, was still another month away [10].

Temperance organisations fought hard to bring about and maintain the prohibition order, as might have been expected. The Gospel Temperance Mission strongly influenced the Ngati Maniapoto chief Wahanui, who travelled to Wellington in November 1884 to plead the case. The proclamation was duly made on 3 December, under section 25 of the Licensing Act 1881. The Government agreed with Wahanui on humanitarian grounds (the desire to save the people from a proven evil), so he won his point with no reference to the railway, but perhaps neither party fully realised the potential of railway communications for undermining the decision in future years.

> There can be little doubt that the Maoris intended to keep intoxicating liquor entirely out of their territory. Unfortunately, it was permitted to follow up the workers constructing the railway, and when the railways began to run, liquor was transported upon them, and Europeans have been permitted to import intoxicating liquor ostensibly for their own consumption. .... but ...despite illegal and discreditable trading by Europeans, the Maoris have been very considerably protected by the prohibition of the legalized sale of intoxicating liquor in the King Country [11: 200].

Many ingenious ways were found to get round the alcohol ban. Goods sheds at stations within the dry area were said to handle more consignments of sly grog than any other rail stations between Auckland and Palmerston North; one horse collar made frequent trips from Taumarunui to Auckland for repairs, until police found inside it a copper inner tube, which came back full of whisky [4: 118].

Another story concerns a magistrate who often travelled to Te Kuiti on the train, arriving as usual the night before he was to hold court. On one occasion he came to hear one of the more important of the ever-increasing number of local sly grog cases. It was the custom for the arrival of the magistrate to be celebrated with a little private but convivial party with his fellow solicitors, although this one was rather better supplied than usual. Next morning he arrived at court with a somewhat sore head, and the accused sly-grogger duly stood before him. The case had to be dropped after a solicitor informed His Honour that, "Sorry, sir, that was Exhibit A that we drank last night" [4: 116]. The small country courthouse in which this memorable event was recorded still stands behind its kauri front pillars in Queen St, Te Kuiti (Fig. 8.10), and is still regularly used for its designated purpose.

Europeans going to live in the King Country knew before they went there that it was a dry area.

> It is therefore to their dishonour that, after having settled there, a section of Europeans has consistently endeavoured to secure the repudiation of the terms of the Covenant with the Maoris....In 1926, an exceptionally vigorous effort was made to secure a poll in the King Country on the question of whether license was to be admitted. A petition from Europeans in this sense was signed by some 5000 people. A petition signed by some Maoris was also got up, asking for a referendum of all the Maori people as to whether license should be granted or not. Sworn declarations were made by a number saying that they had refused to sign, but their names were nevertheless found on the petition. On the other side, thirty-five leading chiefs in the King Country issued an Ohaki, or solemn testamentary declaration, reminding the people of the original Covenant, warning them of the evil that strong drink does, and urging them to remain in the path marked out by their forefathers [11: 200–202].

Country clubs operating on the locker system (locker owners were supposed to declare that they were white, over 21 and held liquor locked up and solely for their own use) made sociable drinking legal long before the 1946 Royal Commission recommended the lifting of the ban on alcohol sales if 60 % of the population voted for it.

Fig. 8.10 Te Kuiti courthouse, scene of many sly-grogging cases during the era of prohibition in the King Country, and still a functioning Magistrate's Court today. *C.M. King (2013)*

At a referendum in March 1949 a strong Maori "No" vote kept the result below the 60 % threshold. But later in 1949 legislation was passed to allow licences to charter clubs, and after 1954, to pubs. By August 1954 the Football Club in Pureora village was operating a licence under restricted hours.

The first pub at Benneydale opened in December 1955. Rusty Russell was determined to be first in when the doors opened, so he and his mate (Boy Nathan) unloaded 4000 feet of timber in half an hour and got to the pub with two minutes to spare.

They found Paddy Hallen, a big coalminer from Benneydale, stretched across the door with his arms folded. Rusty accepted he could not be first in, but he was determined to be last out. There were speeches and celebrations all day, including at least one declaring that the Maori people would fight on for ever against liquor in the King Country (according to the *Otorohanga Times*, reporting the opening). Others were apparently willing to accept the new regime, since the paper also reported that 392 gallons of beer were consumed that day. Finally Rusty and Boy, no doubt among others, were thrown out by the local constable.

The Football Club

When Pureora village was first established in 1945–46, flat areas were cleared not only for houses, a school, and roads, but also, right from the start, for a football field (Figs. 8.5 and 8.11). The field was bulldozed and graded in their own time by machinery operators Bruce Archer, Don Howe, Sonny Hughes and Hoppy Benbrook. It had a natural embankment, but was rather narrow outside the sidelines. The first local team started playing in 1948, and a year later they took a seven-a-side team to a competition in Benneydale and were the runners-up. Bill Watene was one of the more successful coaches during the 1980s.

A junior rugby team was entered in the Maniapoto competition in 1955–56 under coach Jack Fyffe. Other early coaches were Mike Mason in 1959, and Allan Ingram, headmaster of the school and a former Hawkes Bay All-Blacks triallist, in 1960.

The significance of rugby to Pureora social life was illustrated by a frank comment by former OC Bill Drower, that the two questions always asked of prospective employees were, do you

Fig. 8.11 The Bush United Football Club in action at Benneydale. The clubhouse was built at Pureora in 1960, and was moved to Benneydale after the Pureora club closed in 1987. *Crown Copyright, Department of Conservation Te Papa Atawhai. Photographer unknown*

play rugby? And if so, in what position? Drower admitted that the answers sometimes just about determined whether a man got the position or not. The Ranginui mill manager Magnus Russell would employ only young men who played rugby [17].

Many of the mill workers employed at the mills in the surrounding area (Box 7.2; Fig. 7.6) were good rugby players, but they worked on Saturdays so they played on Sundays in Mangapehi or Benneydale. The Benneydale senior team included a lot of strong coal miners, until the mine closed (Chap. 5) and then they didn't have enough Benneydale men for a full team. Those that were left would not join another club so long as it had a name too closely linked with Pureora or Forestry, said Rusty Russell. So the name was changed to Bush United, and everyone joined.

The Bush United rugby team got only one or two training sessions a week, but was always up there with the leaders in local competitions, in part because the players were all so hard fit from the heavy manual work they did during the week. Unfortunately for them, that was the period when Colin "Pinetree" Meads, later a famous All Black, was playing for Waitete, the Te Kuiti club. Henry Flavell was picked for the Maori All Blacks shortly before his sudden death in a logging accident in February 1967.

But in 1973/74 they had an extremely successful season, winning the Maniapoto-Otorohanga Senior B competition with 16 wins in 16 games. In 1976 the club went on a trip to Australia, and four members won selection to represent the King Country (Sonny Anderson, Pae Wynard, Gordon Hill and Brian Cressy). In the 1977 season they won 17 of their 20 matches [17].

The history of the Bush United Rugby Football Club was summarized at a function celebrating 25 years of competition at Pureora, as reported by the local newspaper [8]. Club loyalty was impressive: six of the seven life members elected over the years returned for the club's silver jubilee: Graeme (Rusty) Russell (president for 18 of the 25 years), Alec Watts, Dick Porini, Dave Yanko, Bill Reti and Sonny Anderson.

There had always been a close relationship with Waitete Football Club off the field, at the same time as intense rivalry on it. So Waitete accepted the invitation to play the jubilee match, which Waitete won 13-6. The day was finished off with a Jubilee Ball at the clubrooms, catered to a very high standard for 220 ex-members and supporters by the Pureora women.

The Football Club Bar

The first club house was built by volunteer labour ("we never had any trouble organizing working bees", said Rusty), using logs given by NZFS and cut by the Ranginui mill. Organisations like NZFS, the Ranginui Timber Mill managed by Magnus Russell, the Odlins Timber Mill managed by Jim Russell, and E&B Maraeroa managed by Bill Morton, all helped to keep things going.

Rusty Russell remembered the first club building as "like one of those old wild west hotels, guys flying out the windows…we [were all] young guys, we needed rugby, we needed sport, we needed our beer". Unfortunately for the men, Kitch Pedder, OC at the time, was legally obliged to maintain the ban on alcohol in the King Country. After one particularly wild party that ended with the theft of all supplies and all funds, the first clubhouse was forced to close down [13].

Within a few years, the Football Club had built a new, marginally legal bar in an area that was otherwise still dry. The bar was opened in 1954 and operated on restricted hours, with the knowledge of the Te Kuiti Police (better than sly grogging and house parties, or men driving home drunk from Benneydale, they said). It opened on Tuesday and Thursday after practices.

Kitch Pedder sometimes encountered consternation from the wives about their men returning home late from work. Sawmill managers were not helpful in backing him up by ensuring the bar closed on time, so that was yet another job for the OC. Ladies were not admitted to the club, which was in some ways a pity [13].

At first the accounts were not very well kept (because, said Jack Fyffe, "the guys running it were taking out more than was going in") and the Club ran up a £500 debt to Waikato Wines and Spirits. Then Rusty and Bluey McLennan took over the Committee, notified WW&S that the club was under new management, and undertook to pay off the debt if they would resume deliveries.

For nearly 20 years after that the club bar operated at a profit averaging about 30 % a year, which was ploughed back into football trips, the school, or any other community activity that needed support. The only problem was that the liquor store was vulnerable to break-ins, until it was rebuilt in concrete blocks with stronger locks.

In 1960 a new rugby clubhouse was built for the renamed Bush United FC, which cemented its role as the focal gathering place for the village community, both mill workers and forestry workers. Rusty remembered that the guys from different workplaces tended to sit in their own corners, but rivalries between Ranginui and Odlins people tended to settle down once they broached the bar. Anyone who still wanted to do battle would go outside; no-one would interfere, and the next minute they'd be back inside drinking again, each with a black eye.

Bruce Tricklebank, stalwart of the Volunteer Fire Brigade and its own bar, also played for Bush United so often drank at the Football Club bar after practices, and at weekends after games. There was no official connection between the two bars, and few people were actually members of both clubs, they just went to whichever one was open longest.

There were some people who simply didn't drink, but most who had been resident for a while ended up drinking more than they had before—including women—because there was not a lot else to do. Getting drunk was socially acceptable, except when it ended up causing neglected children and domestic assaults. John Mason remembered a regular stream of battered women turning up for comfort on the OC's doorstep, and some youngsters started drinking early because they saw their parents doing so much of it.

Nevertheless, the men were very discriminating in their drinking habits, and would not drink beers they disliked, such as the new brand "Lucky" beer, even when it was offered free in large quantities during a promotion campaign by NZ Breweries. The Barryville staff, regarded in Pureora as "rogues and vagabonds", had their own separate club, but would come over to Pureora at times.

According to Jack Walker, there must have been a few of those in Pureora too, judging by the number of disappearing NZ$20 bills [IOUs] that were put down to "the rats". But NZFS staff began to take a closer interest in the Club's accounts, and eventually they instituted a system of bush justice to restrain "the rats". It was so effective that, if Rusty Russell had come back from Mangapehi with a truck load of beer (several hundred crates with 2 dozen bottles a crate) too late to unload it, he'd leave it parked in the driveway of his house, and nothing would be pinched while it was under his care.

The same did not apply to accidents supplying what the guys thought of as fair game. Rusty tells a story about a rail wagon that got derailed during shunting, and was stuck in a siding with broken crates of beer visible at both ends. There were "6–8 dry truck drivers down there a coupla times a day with a few cut fingers, [so] there weren't many full bottles left by the time the insurance assessors arrived. Jack Kenney had beer hidden in a culvert, he dived in and picked up a few bottles on every trip back".

Especially after the logging industry passed its 1950s–60s peak and the future of the village began to cloud over, successive OCs recognized the importance of encouraging the football club at Pureora, because it was the only sports and social organization available to the community. They supported it for years, but in the end, the mill closures and the departure of many NZFS staff in Pureora inevitably had a drastic effect on the club, and it could not field a team for the 1987 season. But it survived, because the club, complete with its clubhouse, moved to Benneydale and reformed with players from the new freezing works. The club celebrated its 50th Jubilee in 2005 [17].

Other Sports

Women were ardent supporters of the efforts of their men on the rugby field, but they also wanted access to a sport they could play, and which did not involve travelling to facilities elsewhere. Tennis was widely popular with women and children as well as men, so in November 1950 the village tennis courts were officially opened (Fig. 8.5).

By 1951 the villagers were organizing successful sports days, with wood chopping (Figs. 8.12 and 8.13) and athletic events plus a dance in the evening, and hosting tennis tournament teams against Mangapehi at Pureora. Five years later there was a profitable working men's club and a billiard table at the fire station.

In the mid 1950s social life within the village was thriving. The Committee running the Football Club bar kept to a strict schedule for closing

Fig. 8.12 New Zealand has a long history of forestry, which makes it a leading country in the international sport of wood chopping, and it has produced two world champions. Here a standing block competition is in progress at a Pureora sports day. *Crown Copyright, Department of Conservation Te Papa Atawhai. Photographer unknown*

Fig. 8.13 This portrait of Sonny Bolstad, a champion axeman from a Rotorua club competing against Pureora men at an event in Benneydale, illustrates the strength and concentration required for competitive wood chopping. *Graeme Reinhardt*

hours (not always maintained in later years); complaints arising from Smith's survey about the state of the road frontages of the sawmill houses had been acted upon; and a canteen was being organized. Sandy Sunnex ran a cricket club and helped to build the first club rooms.

Helen Russell and Mauven Yanko were keen members of the Country Women's Institute, which at first meant getting someone to drive them over to Maraeroa. In time they established a branch in Pureora, where they met first at the Football Club and later in the school staff room. They hosted other CWI groups and visited others, held competitions in cooking and floral arrangements, and organized wonderful Christmas parties. Every woman in the village was invited to go, and it was a very friendly group. Dorothy Seager would have gone more often if she had not been pregnant most of the time, she said.

NZFS provided a Quonset hut (a lightweight prefabricated building of corrugated galvanized steel, semicircular in cross section) to be used as a village hall, where films were shown regularly every Friday night by an independent projectionist. He paid 30 shillings a week for the hall and in turn charged 2/6 a ticket. The hall was invariably full, so this was apparently a profitable venture, and the entertainment was not limited to watching the film. Smith's sociological eye appreciated watching the audience more [13], especially the cat calls, the ribald comments on the charms of the heroine, and the audience's responses when, after every few hundred feet of film, the projector would break down. The hall was also used for dances and various other functions.

Medical Care and the Pureora Ambulance

Logging villages need rapid access to medical facilities more than do most rural settlements. So Lew Hahn, the first OC of the Pureora project, started asking for an ambulance in 1947. An established service, provided by an old ex-army Dodge, sprung like a bullock dray, appeared in the records only in about 1954. It was staffed by

Forestry men on a roster, but part-funded by Carters and Odlins, since it served their employees too. It remained in service for 15 years, when it was replaced by an International ambulance. The old Dodge was transferred to other duties, and lasted until 1987.

The Labour Department employed a St John ambulance officer who went round all the logging camps training staff who wanted to obtain their First Aid certificates. All the Pureora staff had to know something about first aid, and all the ambulance crew had their certificates, plus occasional lectures from the visiting doctor on how to handle emergencies. Several Pureora women with nursing skills, including Mauven Yanko and Helen Russell, looked after minor problems in the village and acted as "patch-up person" after accidents in the bush.

The most frequent commissions required of the ambulance service were to take expectant women to Te Kuiti (53 km northwest) or Mangakino (33 km northeast) to have their babies. Ivan Frost's first outing as ambulance driver for this purpose ended after 33 km with a puncture at Kopaki. Ivan struggled to change the wheel, but the nuts would not come off. After a superhuman effort he had got only two of the five off, complete with studs, when suddenly out of the dark came a voice saying "Hey, I think you turn them the other way, eh". No-one had told Ivan that Dodges had left hand threads on their wheel nuts.

Sometimes the ambulance and its passenger got as far as Benneydale (22 km) before events overtook them, so to speak. At least one Pureora baby was born on the floor of the doctor's surgery there (after which the doctor ordered a brandy for the driver, Kitch Pedder, saying he needed it more than the mother did).

Another woman reached Mangapehi Post office (28 km), where the postmaster and his wife delivered the baby. The driver, Graham Bell, a single man, swore he would never again have anything to do with pregnant women. Fortunately, some women operated a mutual-help "you go with me this time and I'll go with you next time" system so they could often leave the guys (other than the driver) out of it.

Other mothers did not even get that far. Ivan Frost tells a story about a woman from Barryville whose husband could not leave their other children, so he had to drive down the road like a mad thing with her alone in the back. As they ran down the last hill to Benneydale she called out "I think it's coming!" Ivan debated whether to stop and deliver or keep on driving, but recalled the doctor saying that if it happens on the way it's usually straightforward, so he kept on driving.

In another couple of minutes he pulled up outside the doctor's house in a flurry of stones, and there was the baby already out and crying. "We cleaned the baby up, then back to the ambulance ….she's sitting there smoking a cigarette with the baby in her arms, and we're off to Te Kuiti….ah, one way and another it was rich living!" he chuckled [18].

Considering how inherently dangerous logging was, and that medical care was confined to weekly visits by a doctor and district nurse [16], Pureora had a commendably short casualty list for forty years' work (Box 8.1).

Box 8.1. Safety record

Only three deaths of NZFS personnel were recorded in the official annual reports.

15 September 1950: a bushman was killed by a log rolling on him.

16 January 1958: bush boss Kaipara Harris was killed when a matai he had just felled hit a dead tree which didn't break. The matai bounced back and caught Kaipara full in the chest. When Jack Fyffe got there he was lying beside the log with his Disston chain saw still running on the other side of it. Any such accident is dreadfully shocking for anyone, especially one that involves a very experienced bush man, but this one happened in front of Kaipara's teenage son Tom, who had just left school and was helping the logging gang.

23 February 1967: dozer driver Henry Flavell was killed instantly when a large tree, whose roots had rotted away, fell

across the cab of his D7 crawler tractor and crushed it. Immediate investigation showed no evidence of negligence by Flavell or any fault in NZFS procedures. Vibration from those heavy tractors on the fragile pumice soil was not a manageable problem, and no cab could have been made strong enough to resist such a blow.

Nevertheless, on 16 March Mrs. Flavell's lawyer came to inspect the site, and he initiated a claim for compensation. The case later got to the Supreme Court in Hamilton, and lasted more than three days before a jury. Mrs. Flavell got compensation for the loss of her husband and his income, but NZFS was exonerated. The Caterpillar company and their agents were concerned enough to ensure that D7 cabs were strengthened from then on [1: 174].

Non-fatal accidents were also relatively few, but included one man hit on head with loading hook, another with a broken leg, and a third (Des Bergman) who was hit by a falling tree, and would have been killed if he had not been wearing a helmet.

The most dramatic accident happened on 18 March 1950, when a heavily loaded Mack truck carrying a huge log from Pureora to the Maraeroa mill went off the pumice road into the soft verge. The log shot forward, demolishing the cab, jamming Bob Abraham, the driver, in his seat and breaking his leg. His mate was also hurt. It took a D8 dozer with a Hyster logging arch two hours to pull the log off and free them (Fig. 7.15). Both ended up in hospital.

Forestry is still a dangerous occupation, even now. No fewer than ten forestry workers throughout New Zealand were killed in 2013 [14].

References

1. Anderson, K. (2000). Servicing caterpillar. Taumarunui: CAS Publications.
2. Anderson, K. (2008). Maoriland sawmillers. Ellis & Burnard Ltd, Sawmillers and timber merchants of Mangapehi. Manurewa: Ken Anderson family.
3. Beveridge, A. E. (2013). Personal communication.
4. Craig, D. (1962). South of the Aukati Line: A history of the King Country. Te Kuiti: King Country Chronicle Ltd.
5. Dawson, J. A. (1979). Pureora Forest Park: Costs and benefits. MA thesis, Victoria University of Wellington.
6. Harris, D. (Ed.) (2009). Timber cruising and other forestry stories. Wellington: FirstEditionPublishers.com.
7. King, M. (1977). Te Puea. Auckland: Hodder & Stoughton.
8. King Country Chronicle, 28 March 1980
9. King Country Land Use Advisory Committee (1977). King Country land use study. Te Kuiti: King Country Land Use Advisory Committee.
10. McLintock, A. H. (1953). The King Country and No-Licence—An examination of the facts of the pact made between the Government and Maori chiefs. Wellington: New Zealand Alliance: Papers relating to licensing in the King Country, H-25, in Alexander Turnbull Library, Wellington.
11. Murray, J. M., & Cocker, J. (1930). Temperance and prohibition in New Zealand. London: Epworth Press.
12. New Zealand Forest Service (1944). Forest Journal entry for 21 July 1944. Te Kuiti: New Zealand Forest Service.
13. Smith, M. W. M. (1953). Some aspects of family and community life in a New Zealand saw milling village. Diploma of Social Science, Victoria University College.
14. Sunday Star-Times, 15 January 2014.
15. Te Wiata, B. (1976). Inia te Wiata, most happy Fella: A biography. Auckland: Hutchinson Group (NZ) Ltd.
16. Waikato Times, 28 November 1987.
17. Waitomo News, 11 August 2005.
18. Wilkes, O. (2005). Transcripts of oral history recordings. Hamilton: Department of Conservation.
19. Wilkes, O., Gaukrodger, J., & Ritchie, N. (2011). Chronology of Pureora. Hamilton: Department of Conservation.
20. Wilson, R. A. (1957). The first fifty years of the Marton Sash, Door and Timber Company Ltd. Palmerston North: Marton Sash, Door and Timber Co. Ltd.

Conflict: Protest Words to Action in the Forest 1970–78

9

C.M. King, D.J. Gaukrodger and J.R. Hay

Abstract

This chapter summarises the background to the refusal of NZ Forest Service to accept the NZ Wildlife Service's recommendations to halt logging in the parts of Pureora State Forest occupied by the kokako, a threatened endemic bird, and the public protests and tree-sitting action that followed. A high-profile seminar in Taupo in March 1978 debated the issue at length, and stimulated 1735 public submissions.

Keywords

NZ Wildlife Service surveys · North Island kokako · Habitat valuation · Yellow-crowned parakeet · Kakariki · Native Forests Action Council (NFAC) · Anti-logging actions · Stephen King · Tree-sitting protests · NZFS seminar in Taupo March 1978

The Royal Forest and Bird Protection Society (RFBPS) and its predecessor, the New Zealand Native Bird Protection Society, and the Waipoua Forest campaign led by Professor W.R. McGregor in 1948 [22], had been drawing official and public attention to the continued decline of native forest fauna since 1923, but with limited or only local effect. Conservation did not become a public issue in New Zealand until remarkably recently. Suddenly, in the mid-late 1960s, the sleepers awoke.

> With an abundance of clean water, with clear and clean air on all but the worst days, and with mountains and bush-covered hills within easy access of almost all parts of the country, there were many who believed that New Zealand was indeed "God's Own" country [so, with no worries about nature] …..diversification of the economy, which was still almost entirely based on primary industries, became a national goal. It was in this climate that the decision was taken to raise the level of Lake Manapouri to provide electricity for an aluminium smelter…..the government….imagined that New Zealanders would accept the loss of the lake as the necessary price of progress….they were wrong [32: 7].

C.M. King (✉)
School of Science, University of Waikato, Private Bag 3105, Hamilton 3240, New Zealand
e-mail: cmking@waikato.ac.nz

D.J. Gaukrodger
3a Te Angi Street, Turua, RD4, Thames 3574, New Zealand
e-mail: beejaygauk@xtra.co.nz

J.R. Hay
Department of Conservation, Te Papa Atawhai, Private Bag 4715, Christchurch 8140, New Zealand
e-mail: rhay@doc.govt.nz

C.M. King et al. (eds.), *The Drama of Conservation*,
DOI 10.1007/978-3-319-18410-4_9

As historian Michael King put it,

> New Zealand's first national conservation campaign... heightened awareness of the need for conservation of natural resources in general and turned the country away from its pioneering phase of simply 'quarrying' those resources into extinction. ...The National [conservative] Government of the day - and the opposition, whose ranks still included Hugh Watt, who had been so proud of the [Manapouri] project, was taken wholly by surprise [10: 378].

The success of the Save Manapouri campaign of 1971 changed the outcome of the 1972 General Election. It made conservation groups realise for the first time that they could put a hot conservation issue centre stage in front of the voting public, and generate the sort of response, from thousands of previously disinterested people, that politicians could no longer ignore.

Hence, the Manapouri campaign continued to influence events for years after its own immediate battle was won. It was the first to clearly identify a wider, fundamental problem that had hitherto been ignored in the rush to develop the national economy. Michael King put his finger on it.

> Some State agencies were committed to developmental policies that took no heed of environmental considerations. This became the basis of another set of disputes that broke out in the early 1970s over the use of native forests. The protagonists were senior executives of the New Zealand Forest Service, who wanted to continue to log mature native trees for timber, and environmental groups such as Forest & Bird who argued that the natural and ecological values of such forests outweighed the commercial gains from harvesting - and in the process destroying – them [10: 381].

This new set of disputes concerned the west Taupo forests, which had covered some 100,000 ha of the central North Island since time immemorial. Up to about 1950, they had seemed inexhaustible. Then, in the early 1970s, RFBPS and many other conservation groups began to read reports from the New Zealand Wildlife Service (NZWS) that described the current progress of clearfelling, so rapid and drastic that by 1978 only about 17 % of this huge forest remained unlogged [32, 33].

Over the protests of NZWS, the New Zealand Forest Service (NZFS) was continuing to log native forest—with predictable consequences for the last few known populations of North Island kokako, an especially beautiful and severely endangered endemic forest bird. Although the kokako and most other native birds had been legally protected since 1953, it was not illegal to destroy the only habitats in which they could live —even though many of these species could live nowhere else in the world but in New Zealand. Local extermination of kokako in logging areas began to seem inevitable, to be followed, very probably, by total extinction (Chap. 13). Even Shakespeare could have predicted that risk, when he put into Shylock's mouth the words: "You take my life when you do take the means whereby I live".

How could this be happening in New Zealand, a country which is a recognised pioneer in conservation legislation? New Zealand had established one of the first National Parks in the world—Tongariro, donated to the nation by Te Heuheu Tukino Horonuku in 1887 and reserved by Act of Parliament in 1894 (Chap. 5). By the end of 1960 there were eight more national parks and more than 1000 gazetted scientific and scenic reserves; and NZFS had created 16 State Forest Parks by 1978 (Chap. 6).

The answer suggested by Wright [33] was that first, the protection offered to these reserved areas was, as Manapouri showed, not cast-iron. Second, many important habitats were not represented in the existing reserves system, including the dense podocarp forests where some threatened endemic species such as the kokako can reach high densities. Third, decisions were being made based on information that was being withheld from some of the interested parties. Finally, environmental groups were divided among themselves, dispersed around the country, generally subjective in their assessments, far removed from decision-making processes, and sometimes unaware of how to fight effectively for their convictions.

The Manapouri campaign went on for years, and gave the environmental groups valuable training for their next big challenge in the central North Island. On the other hand, unlike the Manapouri issue, the argument over logging the west Taupo forests was not simply about scenery and birds versus profits for big business; it also had important socio-economic dimensions for local sawmilling communities (Chap. 8).

Wildlife Service Surveys and Recommendations

During the 1960s, NZFS began to consider the ideas of multiple use and recreation in indigenous forests (Chap. 6), which eventually led to the development of the forest park concept [25: 272]. Independently, NZWS had begun to develop techniques for systematic bird surveys on the mainland in the late 1960s [8].

In 1970 they got together, and NZFS requested NZWS to survey forest areas scheduled for multiple use management, starting with a 40,000 ha block on the Mamaku Plateau. The surveys aimed to estimate the effects of timber production on native fauna and to provide data from which to determine the sizes and locations of reserves. An experienced NZWS officer, Ian Crook, commented with some astonishment that this was the first time that systematic studies of wildlife distribution had *preceded* planned forest operations [2].

The Mamaku experience helped establish the methods later used to survey over 80,000 ha around Pureora. NZWS concluded that a small, mobile unit of 3–6 teams of two people each could collect detailed information on the numbers and habitats of most native species in 40,000 ha of forest per month. But, Crook warned, handling these data exceeded the processing capacity of the small NZWS staff (they were then only a sub-section of the Department of Internal Affairs), so interpreting them would be as difficult as collecting them.

NZWS surveyed the west Taupo forests in the winters of 1970 and 1971, and again in 1974 and in 1976, including a total of 15,000 ha sampled at >600 bird-counting stations and 3000 substations of vegetation samples [8]. NZWS mapped their results to show the presence (or apparent absence) of a given species per 1000 yard square of the then (non-metric) National Grid. Teams followed transects by compass, 1000 yards apart, recording the results in 1000 yard blocks (Fig. 9.1).

Interpretation of positive records is often easy, but, as in most scientific endeavours and especially in bird surveys, absence of evidence is not evidence of absence. Negative records have to be corrected for differences between species in conspicuousness (colour, volume and pitch of song), time of day, weather, behaviour (migration, foraging height) and many other variations in the probability of detection.

Bird survey methods have been refined since the 1970s, but nevertheless the early data showed clearly that there were important differences in the distribution of some key species with habitat. In forest of Class L (see Box 2.1), dominated by podocarps 27–40 m tall at 75–125 trees per hectare, the average number of pairs per station in 1973 was 1.7 kokako, 0.7 yellow-crowned parakeets and 1.1 robins, whereas in forest Class M, where fewer tall podocarps (25/ha) emerged above a broadleaved lower story, these figures were 0.3, 0.3 and 0.6 [2].

Despite all their uncertainties, these data left no doubt about the obvious conclusion: some classes of forest support more threatened native species than others. NZWS concluded that the most valuable areas must be protected from further logging. They produced detailed reports clearly expressing professional concern about NZFS management policy for these forests.

The problem was that the concept of attaching value to natural resources was and is famously controversial, because it depends so much on the prior assumptions of the valuer. The need for semi-objective and partially numerical valuation systems was clear from the start.

That requires some means of comparing the costs and benefits of alternative strategies in the same terms. Christoph Imboden, then with NZWS, attempted to develop one for New Zealand [8]. He proposed a scale of wildlife value categories defined as in Box 9.1, and mapped them across the surviving forests of the Hauhungaroa and Rangitoto Ranges. By the time Imboden's valuation map was presented to the Taupo seminar of 1978 (Fig. 9.1), a large swathe

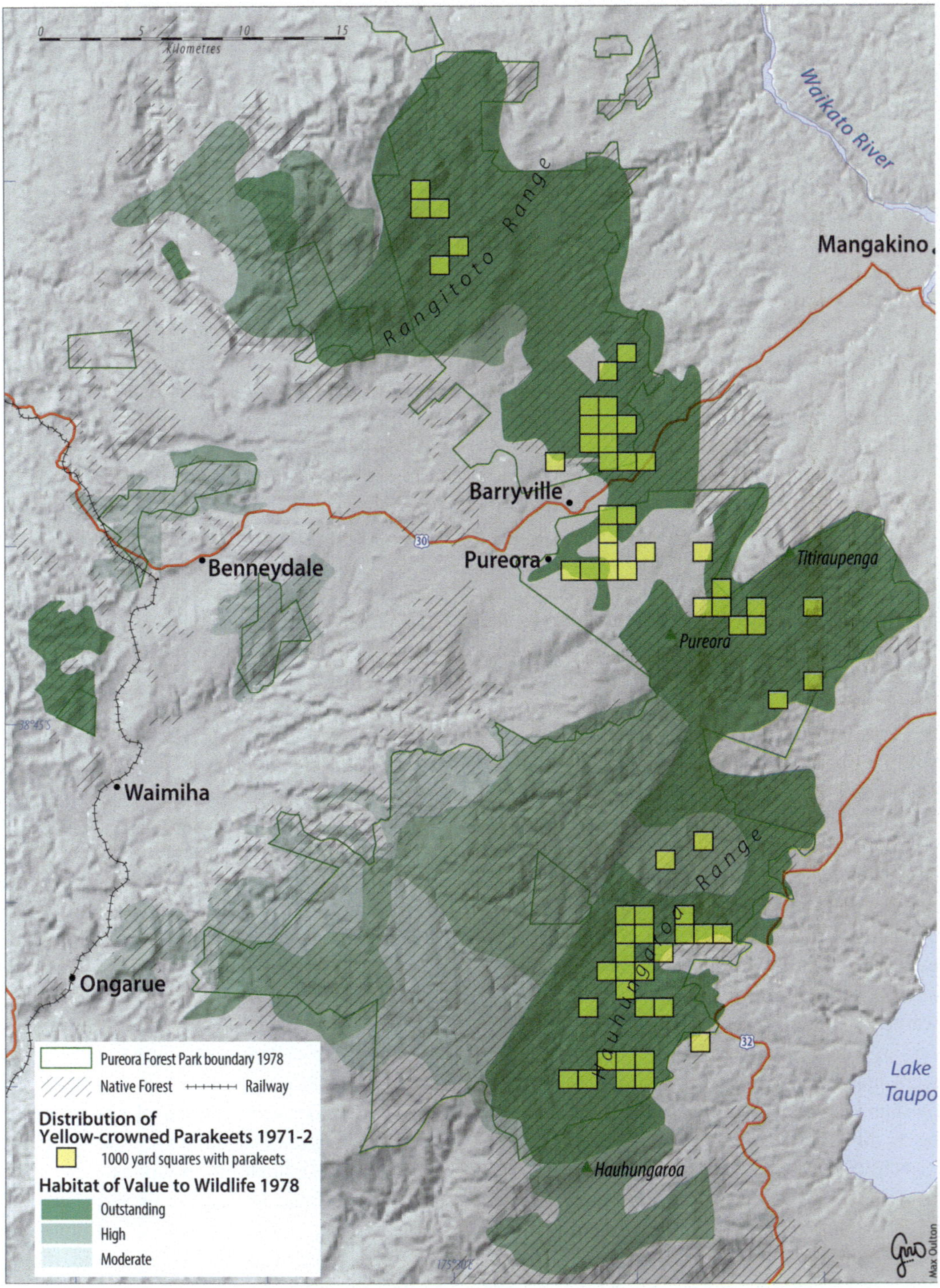

Fig. 9.1 Distribution of yellow-crowned parakeets in the west Taupo forests in 1000-yard squares surveyed by the NZ Wildlife Service in 1971–1972, superimposed on a map of Imboden's habitat valuation categories of 1978 as applied to the same areas and presented to the Taupo seminar. Note that in the years between the survey and the seminar, a large area of formerly continuous forest of outstanding value in the Pikiariki Road area east of Pureora, known to support parakeets (and kokako), was clearfelled and planted in pines. *Redrawn by Max Oulton from Crook (1973) and Imboden (1978: Fig. 5)*

of high-value forest that had been occupied by yellow-crowned parakeets in 1971/72 was already gone. How did that happen?

Box 9.1 Imboden's Valuation Scale for Wildlife Habitats [8]

Outstanding:

Presence of an endangered species listed in the Red Data Book of IUCN;

Presence of an isolated viable population of an endemic species with restricted distribution and limited abundance;

A largely unmodified habitat type not represented elsewhere to the same extent and large enough to support self-sustaining populations of all plant and animal species natural to this community.

High:

Presence of an uncommon, discontinuously distributed species not adequately and safely represented elsewhere in the region;

Presence of a species that has been significantly reduced in abundance and distribution elsewhere by human-induced habitat change;

A large example of a relatively unmodified habitat typical of the region and much reduced elsewhere.

Moderate:

Areas supporting good numbers of common wildlife species typical of the region;

All forest and wetland habitats not otherwise classified.

Imboden anticipated criticism of the broad definition of "moderate" value by pointing out that, because such huge areas of native forest and wetland have already been lost, *all* surviving remnants are valuable to some extent, so there is no need to add a further category for "low" value.

NZFS was sympathetic to the concept of "ecological areas" (EAs, proposed by John Nicholls of Forest Research Institute, FRI), and partially accepted the factually supported recommendations NZWS had made to NZFS in 1971. NZWS argued for the prohibition of logging in three areas of outstanding wildlife habitat (two in the Ranginui Road area (Fig. 9.2) and one on Pikiariki Road) that should be reserved as EAs.

Fig. 9.2 An example of forest of outstanding value, in an outlier in the Ranginui Road area. It had been scheduled for clearfelling, but was eventually reserved as part of the Waipapa Ecological Area. *Crown Copyright, Department of Conservation Te Papa Atawhai. Photographer unknown*

The problem for NZFS was that the three areas contained about 734,000 m^3 of timber out of the total of 1.07 million m^3 available in the forest, and ruling them out would leave a deficit of 136,000 m^3 [14] short of the total required to meet the two existing 1968 and 1970 logging contracts described in Chap. 7.

The sticking point concerned how such large areas could be protected in the Pureora forest in relation to national demands for native timber and to the current NZFS logging contracts. NZFS informed NZWS that all three could not be reserved without further information on the importance of each of the three areas to the survival of the kokako.

Late in 1972, NZWS supplied more information, suggesting that the Ranginui Road block was the most significant, so it was extended and set aside as the Waipapa EA. In the Pikiariki Rd area, logging had already started, and in order to meet commitments, NZFS required it to continue. An Ecological Area was eventually established in the Pikiariki area, but not until much later, and even then, it was much smaller than the ideal, and its irregular shape made it very vulnerable to wind damage [15].

Then as now, proposals in favour of "non-profitable" conservation land uses (for anything other than for protection of soil and water) could make progress only via hard bargaining and painful compromises. The NZFS officer then in charge at Pureora, Darbie Perston, commented in his 1972/73 report:

> For the first time, considerations affecting logging have been other than economic…Environmental protection, though irksome, will pay dividends and we must learn to modify our practices to conform to them [31].

NZWS welcomed the proposed Waipapa reserve, though not without pointing out how much more could be achieved if those reserves were made much larger. Imboden even proposed that the west Taupo forests as a whole be given IUCN Biosphere Reserve status, so providing New Zealand's contribution to a UNESCO programme creating representative conservation areas in all 193 of the earth's biogeographical provinces [8]. The idea was supported by the Auckland branch of the United Nations Association, but regrettably, it was too far ahead of its time, and now it is too late.

By contrast, a vocal non-Government conservation pressure group, the Native Forests Action Council (NFAC, formed in 1975) was adamant that the proposed Waipapa EA was an inadequate "token sample of the broad vegetation type represented by the once-extensive mixed podocarp rainforests…the stage has now been reached when all milling in virgin forest …should cease" [16]. NZFS could not agree at the time, although it was forced to change its policy later.

NFAC stimulated much public concern throughout the 1970s about conservation generally, protection of forests, and especially the survival of native birds in logged areas. The kokako aroused this concern far more than any other bird, for reasons clearly explained by Sir Charles Fleming, at the time of the launch of the NFAC campaign to save the remaining west Taupo forests [5].

> Next to the kiwi, the kokako is the most ancient and interesting bird on the mainland…Along with the tuatara, the native frog and the now-extinct moa, these are the last relics of the ancient time when New Zealand was part of the southern hemisphere continent, Gondwanaland…Most of the things we have thought of as our national culture, such as rugby football, have been imported from overseas. But our feeling for the bush-clad range behind the place where we were brought up is much more formative of us than the derived culture.

The Reasoned Case for Stopping the Logging

By 1974, the determined attempts by NZWS and others to halt logging in the giant totara stands in the Pikiariki area of Pureora forest had failed, but the general public was becoming more aware of these and the rest of New Zealand's dwindling native forest resources. Visits to Pureora by conservation groups and individuals, keen to see the situation for themselves, were becoming more frequent [31].

Scientists pointed out that the decline of kokako was clearly correlated with the reduction and fragmentation of forest area. The Pureora population was one of the largest remaining, yet NZFS agreed to protect only part of the Waipapa area, and refused to halt logging of other large areas of its core habitat. This combination of high iconic value, well-known and extreme but manageable threat, and great beauty was enough to trigger a powerful public reaction against what was perceived as heartless official intransigence.

Needless to say, the reality was far more complex than that. Some of the bitter complaints against NZFS management were quite wrong and unjustified, and many individual NZFS staff were the very opposite of heartless, but the public movement soon became unstoppable.

The 1978 Protests

Since reason had failed, more direct action was needed. It arrived in early 1977 in the form of Auckland-based conservation activist, Stephen King. King had been a passionate advocate of protection for native forests and Maori language since childhood, and for the next couple of years he was a well-known and controversial figure in the news media, instantly recognisable by his long hair and bare feet.

King was chair of the Auckland Branch of NFAC, which had been systematically identifying the best remaining North Island podocarp forests and visiting them. At Pureora he described the vista beyond the unlogged vegetation strip, left along the road edge to conceal the logging that was going on behind it (Fig. 9.3).

> As far as I could see was wasteland – smashed tree stumps, some 8-10 feet in diameter, broken branches, churned earth. Flocks of kaka were screeching in protest, and from the few remaining totara still standing in that raped landscape the song of the kokako poured out like a great lament mourning the fate of the huia, mourning the desecration of their ancient home [1: 46–47].

King did not simply stop at descriptive reporting, and he was not alone. He also got together a team of a dozen colleagues to plan what to do if reason could not prevail. NZFS used comparable tactics, based on the threat to the lives of the local community and on how they would mourn their lost jobs.

That was just the sort of controversy most loved by the news media. In response, on 17 May 1977 a camera crew and reporter from TV2 filmed a backgrounder on selection logging, plus a visit to Pikiariki and Tony Beveridge's forest plots. Such intense adversarial debates are not always conducted on reasonable grounds, then or now.

A high-level delegation from RFBPS followed on 12 August 1977, including Tony Ellis (President), Dave Collingwood (Conservation Officer), and representatives from the Waikato and Taumarunui branches [31].

Fig. 9.3 Inspecting the result of logging the forest of outstanding value in the Pikiariki Road area. The logging debris is drying out ready for burning. Left to right: FRI scientist John Herbert, NZFS District Forester Erle Robinson, FRI scientist Tony Beveridge. *Crown Copyright, Department of Conservation Te Papa Atawhai (February 1975). Photographer unknown*

Late in 1977 King and NFAC prepared a detailed, scholarly 100-page parliamentary submission and presented it to the Minister of Forests, Venn Young, seeking a stop to the logging and burning of the last area of unprotected forest at Pikiariki. They emphasised that this area supported a unique forest association including giant totara, which had previously been identified by NZWS as of especially high wildlife value (Fig. 9.1). They argued that Pureora offered the last outstanding opportunity for a mainland wildlife sanctuary including some of the most distinctive life forms of the pre-European New Zealand rainforest, and a good population of kokako in an unlogged habitat [16]. NZFS ignored the submission.

The Minister's refusal to stop the logging in response to this request was, to him and to the NZFS bosses and senior scientists advising him and to the government behind them all, simply reasonable: NZFS was bound by two rigid commercial contracts (Chap. 7), legally valid until 1983 and 1985. The NZFS position would be explained, he announced, and the various options could be discussed, at a seminar in Taupo in March 1978.

To conservation groups, his refusal was disappointing and unreasonable. Logging in the nominated areas around the giant totara would not be stopped before the proposed seminar. The giant totara themselves might not be cut down, but removing the trees around them, crushing their roots with heavy machinery, and burning the remains before re-planting in radiata pines (Fig. 9.4) would cause damage that would by then be irreparable [32].

After the urgent appeals put up by RFBPS and NFAC to the Minister of Forests in 1977, plus a final last-minute request in January 1978, were all rejected by the Minister, conservation groups felt they had sufficient reason and support to take drastic action. The ensuing famous and well publicised protest campaign generated a huge controversy, which made rational negotiation both more necessary and also more difficult.

Meanwhile back in the bush, the whole argument had come as a considerable surprise to men

Fig. 9.4 After clearfelling and burning, near Pikiariki (1977). Three small pine seedlings can be seen in the foreground, planted in the ashes. *Crown Copyright, Department of Conservation Te Papa Atawhai. Photographer unknown*

who worked every day in the forest and almost never saw a kokako, and to their bosses who saw their job as conserving the forest, i.e., managing it for *future use* [31]. The old hands like Buster Seager and Harry Bunn commented later that they never knew much about kokako when they worked in the bush, and could not identify one if they saw it by chance. They all knew and commented on the flocks of kaka and pigeons they often saw, and they knew the tui and lots of other birds, but the few who ever saw a kokako had trouble identifying it.

Tony Beveridge tells a story about Jack Fyffe who, in "one of his amiable moments in our shared field room, gave me an accurate description of an unusual bird seen near the logging gang on the slopes of Pureora Mountain; it could only have been a kokako. Jack was quite excited". Among very few written records that survive is a File Note written by Bill Drower in April 1968, describing a falcon attacking and killing a bird he did not recognise. Only much later, when the controversy ensured that pictures of kokako appeared in the newspapers, did he realise what he had seen.

Jack Walker, OC at the time, admitted that he could not understand what had triggered off the protests on his particular patch, considering there were other forests where the same kind of logging was going on and which had never attracted so much attention. Jack and many NZFS staff of his generation did not think much about rare birds, and he was not alerted by the increasing number of NFAC visitors because he had never worried too much about who was entering the forest. There was a permit system that was not rigidly enforced, in part because NZFS staff knew all the regular hunters and had many other problems to think about. So it is not surprising that they saw the whole controversy as being mainly about the threat to their livelihoods.

Crunchpoint: Sitting in the Tree Tops

Logging in the Pikiariki block was due to start again after the holiday break on Monday 16 January 1978. The issue featured on radio and TV on January 9, and attracted the attention of yet more high-profile visitors. On January 11 Ian Shearer, National MP for Hamilton East and his wife; Tony Ellis, President of RFBPS and his wife; Stewart Gray, member of Waikato NFAC and his wife; Ian Prior, president of ECO (an association of NZ environmental groups) and his wife; and NZFS staff Ossie Kirk (District Ranger at Te Kuiti), Erle Robinson (District Forester Te Kuiti) and Dave Yanko (2iC at Pureora), all came to inspect the ancient totara (reputedly 1000 years old, but probably more like 600–700 years) that were supposedly being felled. Shearer was himself a member of NFAC, and had a long history of criticising his own party's forestry policies [27] because, as he liked to say, he was an environmentalist before he became a politician.

NFAC leaders pointed out to the visitors that the logging area included the last significant area of 2 m plus diameter totara in public ownership and the largest surviving; some were trees equal in significance to the famous Tane Mahuta kauri of Waipoua Forest in Northland. NZFS were unmoved. The Assistant Conservator of Forests, Auckland, David Black, had met King on site in October 1977, but could not see what the fuss was about. These were just hollow old trees, he said—"a prevailing attitude among foresters who write off hollow trees as if they were useless", replied Stephen King. Black saw the clearfelling and conversion to exotics as an improvement of land use. "New Zealand is too small a country to have land lying idle" were his words. To which Stephen King replied that "by the same logic the Auckland Town Hall should be put to productive use and converted to a warehouse and the golf paddocks in Auckland should be ploughed up to grow spuds" [9].

Black was not alone in his opinion—it was the common attitude among foresters whose normal culture at the time, understandably, concentrated on producing timber rather than on the life of the whole forest community. Much later, Stephen King, speaking at a 1980s public meeting of 400 in Auckland at which NZFS managers were present, likened the timber community's disdain for hollow trees to writing off anyone with grey hair in a human community [9]. It was when trees reached the hollow stage that they entered middle age and began to contribute the most to their arboreal society, which we call an ecosystem. By contrast, in a forester's ideal world everyone with a sign of grey hair is removed. That's fine for timber cropping models, he said, but disastrous for a balanced ecosystem or a protected heritage forest.

Such different viewpoints (jobs versus birds, or human benefit versus natural processes, or more broadly, exploitation versus protection) are certainly not recent or confined to New Zealand —and nor have they disappeared since. Decades previously, a similar "yawning ideological chasm" separated the attitudes of pioneering Oxford ecologist Charles Elton from those of the foresters charged with managing Elton's famous study area, Wytham Woods. The conflicts of interests between foresters aiming to remove all unhealthy trees and ecologists interested in the rich wildlife of decaying timber were never far from the surface for at least 40 years of Wytham's history [26: 201]. A decade after Pureora, the conflict between logging and wildlife was fought out again in the old-growth forests of the Pacific northwestern states of the US. Doak [4] used population models to predict the effect of logging on the survival of the northern spotted owl, and concluded that the proposed US Forest Service plan would extinguish the owls.

Some of the Maori bushmen working in the forest were caught between these two quite different perspectives. In one area scheduled for clearfelling, Maori crosscutters assigned to fell about 50 huge ancient totara were anxious to leave them standing because they were rangatira (noble), and viewed in Maori tradition as tupuna (ancestors) (Chap. 3). The bushmen were not alone: they had the support of local Maori kuia (elder) Martha Hepi and the whanau (community) who were employed at the sawmill. But, as employees of the Forest Service, the bushmen were ordered to go back and cut the big trees down. Either they did it, or they lost their jobs and income for their families [9].

To be fair, the apparently intransigent attitude of NZFS senior managers was not because they did not realize that many old totara are hollow, but simply that the South Block forest in question was scheduled for clearfelling and conversion to exotics. Most men who had worked for decades in the forests developed a highly educated eye for a tree. A story is told about Francis Carter, co-founder of the Carter Holt Harvey empire,

> ...when experts were arguing over how many board feet of timber there might be in a giant totara that had been discovered in the Pouakani Block... 130 feet tall and 37 feet round at breast height. Francis quietly walked around it, looked up at the branches in its head and studied its base. "There's no timber in it", he pronounced. "It's hollow, full of kaekak [honeycomb timber]. Leave it to the pigeons and posterity" [23: 50].

Ironically, years later it was Carter's own company, then run by Francis' son Alwyn and co-owner of Pureora Sawmill Ltd, that was insisting on clearfelling the giant totara of Pureroa and the surrounding rimu, matai, miro and kahikatea. The argument by this stage was more about authority, exotic conversion and contracts than about the native forest. One un-named NZFS officer was reported to have said privately that "for the amount of timber involved in the southern part of the forest, the department [NZFS] could well have left the native trees alone" [12].

On Saturday 14 January 1978, a group of more than 100 NFAC members and supporters from Auckland, the Waikato and Bay of Plenty arrived in two busloads (Fig. 9.5), accompanied by fascinated reporters and TV crews from both news channels [33]. On the same day, the Director-General of Forests Malcolm Conway was reported as saying that any physical action by NFAC would be "ill-advised", adding that "they don't worry me, the logging would go ahead no matter what NFAC tried to do", because it was being done under legally binding contracts [27]. Protest action could affect neither NZFS's obligation to fulfil its contracts, nor save future forests, he said, since no similar contracts would ever be let because NZFS's former clearfelling policy had now been phased out. And, he probably added under his breath, NZFS senior managers don't like being told what to do by outsiders.

Instead of arguing, NFAC spent the weekend planting 100 native tree seedlings across the

Fig. 9.5 Native Forest Action Council (NFAC) protesters returning to their bus along Bismarck Road, January 1978. *Graeme Reinhardt*

access road, but two days later NZFS staff uplifted them and transplanted them out of the way.

Last-ditch efforts to avert a confrontation continued. On the day the timber crews went back to work (16 January), Internal Affairs staff tried to find someone to rescue the kokako before the logging reached them [28], and their Minister Allan Highet criticised Venn Young for not stopping it; Gwenny Davis, national president of NFAC, tried again to persuade Young to change his mind; nothing worked.

On 18 January Stephen King and his brother and several companions packed enough food for a month, climbed two of the giant totara in the Pikiariki Road area near where logging was about to start, and camped on platforms among the branches (Fig. 9.6). NFAC could not have people in every tree at risk, but the idea was to hide up there so well that the loggers could not see which trees were occupied, and therefore could not log any of them [9].

> The protest …proved irresistible to the media and ultimately brought New Zealand…to an awareness of the fragile and growing scarcity of these once-great forests. Stephen King… padded knowledgeably through the forest, imitating kokako and speaking both Maori and English… National attention was rivetted in 1978 by an act of desperate imagining and audacious theatre, scripted as if for television [34: 187].

Most NFAC supporters knew nothing of the protest before it took place, because success depended on total surprise, so none of the 80 members who visited the forest with the sitting group three days before the protest had any idea the tree top protest was being planned. Not even the whole NFAC committee knew, only those involved. They had tried every other avenue without success, and the final decision was made to proceed just three days in advance.

The tree-sitters whistled to each other as the crews began to start up their saws, but the loggers could not tell where the whistles were coming from. Dave Yanko looked hard, and swore that Stephen King was not there. The sitters revealed themselves only to reporters willing to interview them and pass their message on to the public.

The Prime Minister, Robert Muldoon, was flooded with telegrams from conservationists nation-wide, but, unwilling to allow a "young

Fig. 9.6 One of the tree platforms built high in an ancient totara by NFAC activists. They were based on pallets, and accessed by advanced abseiling gear. *Crown Copyright, Department of Conservation Te Papa Atawhai. Photographer unknown*

rabble" of protesters to get in the way of legitimate milling work, ordered them to leave. He could not enforce his order because King and his party had a permit for 14 people to camp in the forest for the month of January, so they had a legitimate reason to be there. They had been there for days already, and were refusing to leave until they got an assurance from the Minister of Forests Venn Young that the Pikiariki block would be saved. Young was out of touch, visiting the subantarctic Auckland Islands [11].

Undeterred by Muldoon's anger, a small group of protesters confronted a group of forestry workers in the forest. Police were informed, but did nothing. King declined an invitation to meet Jack Walker (the NZFS OC for Pureora), in case he took the opportunity to revoke the permit. The Timber Worker's Union sent Hamilton branch secretary Cliff Wall to make sure "these clowns" were not putting union members' lives at risk. Tempers frayed, and newspaper editorials in the main metropolitan papers added fuel to the uproar.

Muldoon's ultimatum was ignored, and cutting of the matai, rimu and kahikatea around the giant totara continued. His stance was significantly weakened by the public response when it emerged that NZFS knew very well that they were felling and burning an area that had been identified by NZWS as prime kokako habitat [1]. A door-to-door survey in Muldoon's own electorate the previous year had shown >90 % support for the Maruia Declaration petition.

Stephen King and his companions continued to sit in their perches. King's younger brother stayed steadfast in a tree just 10 m from two 30 m rimu being prepared for felling by Dave Yanko's logging gang without making himself known until after they had cut one tree. If he had shown himself before they cut the tree down, they would have simply moved and continued logging [29]. His courage shook the confidence of the logging gang and of NZFS.

OC Jack Walker tried to remain level-headed amid the fuss, but he remembered being well aware that "the silly bastards could have got killed… I guess we would have had to accept responsibility". He was often tempted to "lock them up in a bag and hang them on a fence" [31]. Local meetings went round in circles, and forestry staff felt they were being left to struggle through the crisis without much support. NFAC's presence was reinforced over the weekend by more supporters from Rotorua and Hamilton.

NZFS field staff realised that it wasn't possible to continue logging in the Pikiariki block without endangering human life [29]. So felling operations were transferred to the North Block, to ensure that NZFS's commitment to the Barryville mill's 15-year logging contract could be met. Dave Yanko said that they were planning to move there anyway and were just about ready to go, so from his point of view, the protests had done nothing at all to change NZFS' existing plans. For NFAC, that was not enough, so the tree-sitters sat tight.

In the absence of his Minister, Malcolm Conway, the increasingly embattled Director General of NZFS needed a semi-official dignified retreat.

On Sunday 22 January Conway announced a temporary pause in logging, out of concerns for the safety both of the protesters and of his own NZFS staff. Others were less sympathetic to the protesters. Wright [33: 106] quotes a comment made later by one timber man:

> When this bloke first climbed into the tree I thought he was a nutter. In view of what has happened since I wish I'd have cut the b—tree down with him in it.

The temporary halt did not stop the debate—it merely pushed the issue on to the next question, whether logging should be resumed. Martha Hepi and the Maori community living in Maraeroa supported the protesters arguing for long-term protection for their forest. With the help of the Tuwharetoa elders of the Mangakino marae, they collectively sent a telegram to Muldoon asking him to stop the logging for good [12]. They suggested that exotic timber be sent to the mill, a solution that had been used in the case of a Northland forest to protect kauri in 1974 [22]. Their plea was ignored.

The wives of the Pureora workers weighed in on the other side. They too sent a telegram to Muldoon:

> We, the wives of forest workers at Pureora Forest, object most strongly to the protesters unlawfully in our Forest. They are law-breakers and should be treated as such. We feel for our trees also, but to feed and clothe families are more important than 1000-year-old trees that are already rotten. We are 100 % behind [the request of] our staff and bushmen here to have the protesters removed and logging to continue as before. (Signed) J Walker, D Thackeray, J Hughes, J Reti, M. Anderson and 12 others [3].

Muldoon acknowledged the telegram, and assured the wives that their views had been noted.

Cabinet ministers were bombarded with telegrams, and at the Cabinet meeting on 24 January, the decision was made to confirm and extend Conway's ruling. Logging in the Pikiariki block was suspended until after the promised forestry seminar in Taupo in March.

King and his companions came down from the trees. They were allowed to return on February 11 to collect their gear, under escort by Woodsman Sneath [31].

NFAC's action was the first treetop protest publicized internationally, and it was over quickly because the surprise was complete, the story irresistible to the media, the case was very strong, and there was a good army of people from the grass roots to politicians, government department leaders and scientists who were ready and prepared to speak up to support the protesters. The story has become well known to many environmental groups ever since, although fewer know that it had an ironic twist

A month later, an NZFS burn-off destroyed 300 mature rimu trees in a riparian strip in the Pikiariki clearfelled area that had been protected by the tree-sitting action [30]. The fire spread into the crowns of some standing totara, and they had to be removed. "It didn't bring us any kudos…[but] we eventually did all the right things and got it out", commented Jack Walker. NZFS, embarrassed but not penitent, offered NFAC 25 ha to replant.

The Taupo Seminar

The seminar convened by NZFS at Taupo on 28–30 March 1978 was entitled *Management Proposals for State Forests of the Rangitoto and Hauhungaroa Ranges, Central North Island*. NZFS had previously held a similar seminar in Hokitika in 1974, to discuss the West Coast Beech Scheme (proposed in 1971: Chap. 6) so they had some idea of how to go about organising it, but this one was different.

NZFS staff had already been planning to hold a forestry seminar to consider the west Taupo forests to coincide with the completion of the massive King Country Land Use Study [20]. Nevertheless, the publicity surrounding the tree-sitting protests at Pureora forced the Minister to ask NZFS to bring forward that plan at short notice.

The seminar was attended by 52 mostly invited delegates representing a wide range of interests. They started with a one day field trip on March 28 (escorted by Dave Yanko) to inspect the disputed areas first hand. Comprehensive field trip notes were provided both by NZFS

[19], and by local interests [24]. Then the delegates listened to and discussed two days of detailed papers. The transcript of the seminar includes drafts of all the papers presented [18].

The structure of the programme and the affiliations of the speakers were affected, to a surprising degree, by history. Pureora Mountain is close to the geographic centre of the North Island (Fig. 0.1 in Preface), and has always been a natural boundary marker for different landholdings extending in all directions. Pureora Forest Park stands astride territorial boundaries dating back to Maori times, when the ancestral lands of three major tribal groupings met near there (Fig. 3.1), and it was natural that European administrative boundaries should follow suit. These local arrangements also simplified the differences between the Maori owners of the land in the way they chose to subdivide it for lease or sale, and between provinces in the geographical origins and development of road and rail links.

The same radial arrangement of boundaries persisted into the organisation of the timber contracts controlled by three different Forest Service Conservancies delivering timber to mills in six State Forests (Fig. 9.7). The three Conservancies had always managed their sectors of the west Taupo forests more or less independently. So, SF 96 Pureora (25,385 ha), SF 97 Hurakia (22,895 ha) and SF 92 Wharepuhunga (1245 ha) were in the Auckland Conservancy, and logs extracted from them went northwest to Auckland by rail from Mangapehi. SF 98 Tihoi (20,965 ha) was in the Rotorua Conservancy, and the logs went northeast to Rotorua and Putaruru by truck. SF 121 Taringamotu (6013 ha) and SF 112 Waituhi (3656 ha) were in the Wellington Conservancy, and the logs went south and southwest to Taranaki and Wellington by rail via Taumarunui [18, 21]. Hence, the 85,000 ha of State Forests under consideration at the seminar included land administered by all three NZFS Conservancies in the North Island.

The seminar was opened by the three NZFS Forest Conservators (Gavin Molloy, Auckland; John Rockell, Wellington; and John Ure, Rotorua). They started off by describing the areas for which they were responsible and summarising their past management record. Logging of podocarps in Pureora Forest and replanting with exotic species or release for farming had had a long history, as part of the long-term Pureora Working Circle plan (Chap. 7). Clearfelling and conversion to exotics had already ceased in North Pureora in the Okahukura Valley, giving way to partial logging in 1975. In the South Block the last clearfelling and conversion was done near Pikiariki Ecological Area in September 1977 [14], but selection logging was still going on.

Then two senior foresters (David Field, Rotorua and Erle Robinson, Te Kuiti) outlined the management proposals that were the cause of all the debate [6]. They described how the complex stands of remaining indigenous forests at Pureora had been zoned for different forms of management. They summarised five possible options for the future, and the consequences of each option for the existing logging contracts (Box 9.2).

Box 9.2 Management Options for the Remaining Unlogged Forest

These are the figures estimated by NZFS and given to the Taupo seminar on the future of Pureora in March 1978. The surplus/deficit figures refer to how much timber would be produced by each option in relation to the existing timber contracts [6].

Option	Strategy	Sawlog volume (m^3)	
		Available	Surplus/deficit
A	Clearfell all unlogged forest	882,000	+586,000
B	Selectively log all unlogged forest	302,000	+6000
C	Clearfell outside EAs	313,000	+17,000
D	Selectively log outside EAs	106,000	−190,000
E	Stop logging	0	−296,000

Excluding the exotic plantations, about two thirds of the 85,000 ha under discussion was already zoned for reservation in one form or

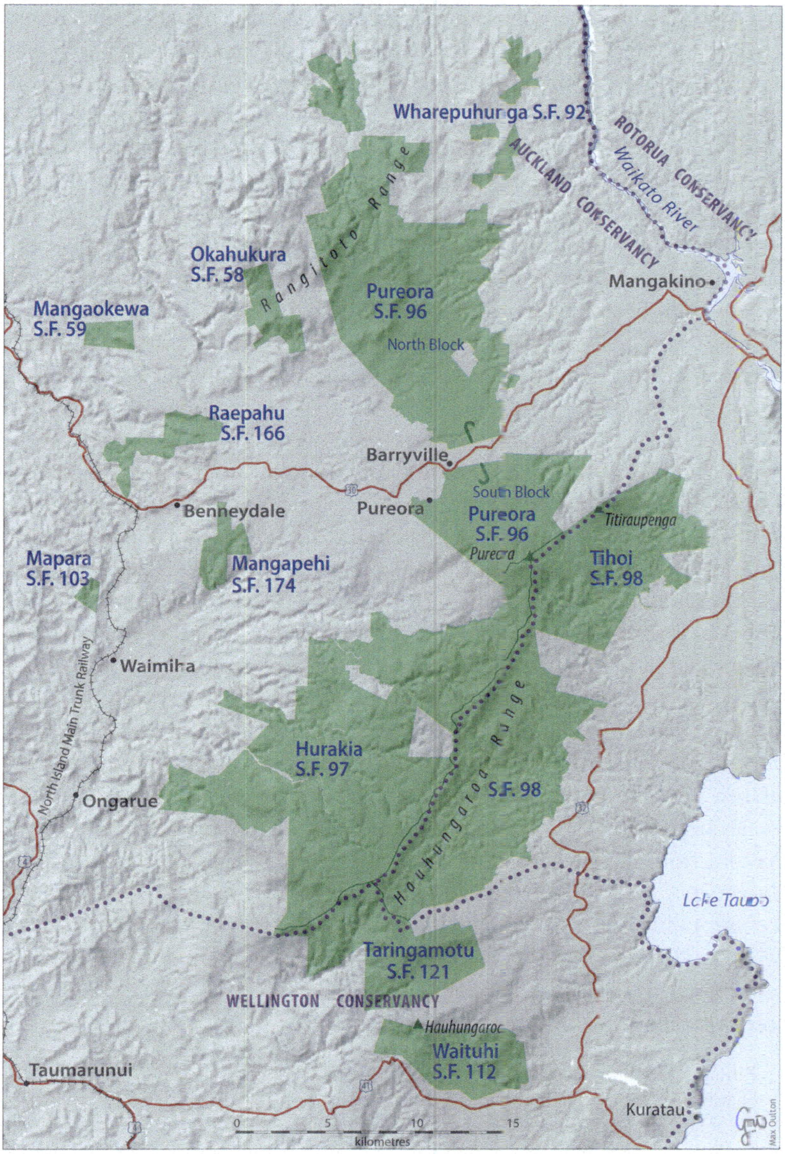

Fig. 9.7 Locations and numbers of the state forests under consideration at the Taupo Seminar, plus others nearby that were not considered. The plan for Pureora Forest Park included six State Forests formerly managed independently under three separate NZFS Conservancies. *Redrawn by Max Oulton from Molloy et al. (1978)*

another. Over the remaining third, the option favoured by NZFS would allow extraction of a vastly reduced selective harvest of timber over about 40 years.

NZFS managers were not used to having to justify their decisions to such a public and critical audience, but this time they had no option. They complied, providing figures to show that a decision to stop all logging immediately would leave NZFS short by 296,000 m^3 on delivering its timber commitments to the saw mills holding current contracts. No doubt they hoped that this patient and logical explanation would satisfy any reasonable critics.

The rest of the seminar programme was provided by a roll-call of virtually all the prominent scientists and foresters working in forest management and conservation at the time. Everyone was well aware of the intense public interest and scrutiny of their work, and of the amount of attention being focussed on the seminar by journalists and activists.

Most of those journalists and activists knew almost nothing about the technical business of forestry, or that NZFS had a long history of concern to slow down the rate of milling in native forests (Chap. 6), or that the Pureora project had been conceived by NZFS almost 40 years previously as a carefully managed plan to protect the future of the forests and the people who depended on them (Chap. 7).

A set of eight discussion papers followed, grouped under the general heading of "Forest Values". They included descriptions of forest composition and ecology (John Nicholls, John Herbert); hydrology (Colin O'Loughlin), wildlife (Christoph Imboden), recreation (Russell Dale), proposed reserves (John Nicholls, John Herbert), current and potential use of timber (John Vaney, Laurie Gibson), production forestry (David Field, Erle Robinson) and selective logging (Tony Beveridge, John Herbert) [17]. Then came an interim report on an unfinished study of the social impact of any reduction in logging at Pureora from the Business Development Centre, University of Otago, summarising a fuller account published a few months later [7].

Andy Kirkland (Assistant DG of Forests) contributed a 13-page summing up of the proceedings, ending with the memorable comment:

> I think…the differences in viewpoint that we have heard are essentially differences in philosophy… the mainspring of the conservationists' attitude to the forests is something that may be diminishing in other sectors of society – a reverence for the aged and the virgin [13].

Kirkland made very plain which was the option preferred by NZFS. It believed that the Government's recently revised management policy (Chap. 6) was working, and that there was room for both reservation and production in the west Taupo forests. He saw no reason to abort the policy prematurely; the juxtaposition of modified and unmodified forests could well be the best strategy, and it should be given a chance to prove itself.

Outside the reserved areas, he emphasised, clearfelling had largely ceased already; the issue was whether *selection* logging should continue as proposed in order to meet demands for high quality timber, such as rimu for furniture making and totara for Maori carvers, and to fulfill the existing logging contracts that guaranteed supply to timber companies in Barryville and Te Kuiti (Chap. 7). Replanting with native species was part of this policy, which was intended to maintain long-term sustainability of the indigenous timber resource.

The seminar ended with a final address by DG of Forests Malcolm Conway, emphasising his continuing support of selection logging, and then an invitation from the Minister of Forests Venn Young for public submissions on the NZFS proposals. Conway did not contribute a written record of his statement to the official transcript, but copies of the formal papers were lodged in NZFS offices and public libraries throughout the country. The debate surrounding them went on for years.

In Kirkland's last annual report for NZFS in 1985, he summarised the key question as "whether integration or separation of the various state forest functions is preferable". This was essentially the same issue that had been studied and rejected by at least two committees since 1969 [25: 381]. New answers to it were waiting just around the corner (Chap. 11).

References

1. Barrington, J. (1995). Profile: Stephen King. Forest and Bird, February 46–47.
2. Crook, I. G. (1973). Forest survey. Wildlife—A Review, 4, 37–41.
3. Department of Conservation. Archives. Hamilton: Department of Conservation.
4. Doak, D. (1989). Spotted owls and old growth logging in the Pacific northwest. Conservation Biology, 3, 389–396.
5. Evening Post, 11 April 1977.
6. Field, D. A., & Robinson, E. D. (1978). Management proposals for state forests of the Hauhungaroa-Rangitoto ranges [*Transcript of the Taupo seminar*]. Wellington: New Zealand Forest Service.
7. Higham, J. R. S., & Menzies, P. (1978). Social and economic impact of indigenous forestry in the Pureora area. Dunedin: Otago Business Development Centre.

8. Imboden, C. (1978). Wildlife values and wildlife conservation in the Hauhungaroa and Rangitoto ranges [*Transcript of the Taupo seminar*]. Wellington: New Zealand Forest Service
9. King, S. (2013). Personal communication.
10. King, M. (2007). The Penguin history of New Zealand illustrated. Auckland: Penguin Books.
11. King Country Chronicle, 10 January 1978.
12. King Country Chronicle, 24 January 1978.
13. Kirkland, A. (1978). Summary of information from seminar papers and discussion [*Transcript of the Taupo seminar*]. Welllington: New Zealand Forest Service.
14. Molloy, G. J., Rockell, J. D., & Ure, J. (1978). Introduction to the state forests of the Rangitoto and Hauhungaroa ranges [*Transcript of the Taupo seminar*]. Wellington: New Zealand Forest Service.
15. Murphy, B. (1984). Pikiariki ecological area, its pattern, processes and management. BForSc thesis, University of Canterbury.
16. Native Forests Action Council (1977). The Podocarp Rainforests of Pureora. Nelson: Native Forests Action Council (NZ).
17. New Zealand Forest Service (1977). Management policy for New Zealand's indigenous state forests. Wellington: New Zealand Forest Service.
18. New Zealand Forest Service (1978). Management proposals for state forests of the Rangitoto and Hauhungaroa ranges, Central North Island [*Transcript of the Taupo seminar*]. Wellington: New Zealand Forest Service.
19. New Zealand Forest Service (1978). West Taupo field trip notes. 28.3.78. Taupo: New Zealand Forest Service.
20. New Zealand Forest Service (1978). Submissions on the west Taupo state forests. Wellington: New Zealand Forest Service.
21. New Zealand Forest Service (1980). Pureora state forest park management plan Auckland: New Zealand Forest Service, Auckland Conservancy.
22. Orwin, J. (2004). Kauri: Witness to a nation's history. Auckland: New Holland.
23. Parker, S. (2000). Cutting edge: The Carter-Holt-Harvey story. Auckland: Viking Penguin.
24. Pureora Community Council (1981). Submission to King Country regional management plan. Pureora: Pureora Community Council.
25. Roche, M. (1990). History of forestry. Wellington: New Zealand Forestry Corporation Ltd
26. Savill, P. S., Perrins, C. M., Kirby, K. J., & Fisher, N. (Eds.) (2010). Wytham woods: Oxford's Ecological Laboratory. Oxford: Oxford University Press.
27. Waikato Times, 14 January 1978.
28. Waikato Times, 16 January 1978.
29. Waikato Times, 20 January 1978.
30. Waikato Times, 24 February 1978.
31. Wilkes, O., Gaukrodger, J., & Ritchie, N. (2011). Chronology of Pureora. Hamilton: Department of Conservation.
32. Wilson, R. (1982). From Manapouri to Aramoana: The battle for New Zealand's environment. Auckland: Earthworks Press.
33. Wright, L. W. (1980). Decision making and the logging industry: An example from New Zealand. Biological Conservation, 18, 101–115.
34. Young, D. (2004). Our islands, our selves. Dunedin: University of Otago Press.

The Moratorium and Afterwards, 1978–87

10

C.M. King and D.J. Gaukrodger

Abstract

This chapter describes the 1978–81 moratorium that halted logging of native forest to allow time for research on the kokako. Pureora State Forest Park was formally gazetted by NZFS in 1978. The temporary end of logging had serious consequences for the local community, and the last timber mills in the Pureora area were closed down. The NZ Government was forced to pay NZ$7.1 million in compensation for broken contracts.

Keywords

End of logging in the King Country · Pureora Forest Park · Kokako · Economic predictions · Economic value of logging · Business development models · Employment in the sawmilling industry · Compensation for broken contracts · Sawmill closures · Dismantling Barryville mill and village · Pureora village redesigned

Arguments and Submissions

Before and after the Taupo seminar (Chap. 8), public debate continued for months on the issues of logging and conservation, loss of employment, the future of Barryville and Pureora villages, and of logging of native forest generally. NFAC and RFBPS also organised a number of public meetings which focussed on the future of west Taupo and other native forests. The weekend before the Taupo seminar, NFAC held their 1978 annual Easter conference in Rotorua (a town with strong forestry interests) and invited the Leader of the opposition, Bill Rowling, to speak. The Minister of Forests Venn Young was overseas, but sent a statement re-emphasising Government policy.

The Environment and Conservation Organisations umbrella group (ECO), representing 57 environmental groups throughout the country,

C.M. King (✉)
School of Science, University of Waikato, Private Bag 3105, Hamilton 3240, New Zealand
e-mail: cmking@waikato.ac.nz

D.J. Gaukrodger
3a Te Angi Street, Turua, RD4, Thames 3574, New Zealand
e-mail: beejaygauk@xtra.co.nz

C.M. King et al. (eds.), *The Drama of Conservation*,
DOI 10.1007/978-3-319-18410-4_10

organised a joint protest meeting in Taupo, held shortly after the May 31 post-seminar closing date for submissions to NZFS. It was expected to be the largest in New Zealand's history—larger than the one that saved Lake Manapouri [33] (Chap. 9). It dispatched busloads of people to the forest (300 walked to the top of Pureora Mountain, and 500 visited South Block), as always escorted by NZFS staff. Included in the large crowd were Bill Birch (then Minister of Science and Technology in the National Government), and Labour Party environment spokesman Richard Prebble. They climbed to the top of Pureora Mountain and saw first hand the fragmentation of Pureora forests caused by the current clearfelling policies. They were astounded that the biggest political meeting they had ever been to was about the protection of native forests [16].

The following week, the King Country National MP Jim Bolger arrived to look at the selection logging being carried out in Pouakani SF by NZFS and the E&B's gang. Bolger had expressed himself "impressed" by what he saw [17], and the owners of the last four working sawmills (Fig. 10.1) saw that as a hopeful sign. Bolger returned to the NZFS HQ to talk to the assembled local people, including the farming sector (Bolger was himself a farmer).

Venn Young's office received a total of 1735 submissions on the NZFS proposals (Box 10.1). NZFS officers read them all, and dutifully compiled an analysis of all 3300 pages of typescript and handwriting [24].

Box 10.1. Submissions received by the Minister after the Taupo seminar

Submissions totalled 1735. They included 1658 individuals, families or informal groups, 7 service groups, 12 business firms and timber-processing interests, 7 local authorities, 11 professional research organisations or advisory bodies, 27 voluntary conservation groups and 13 recreation groups.

Only 15 submissions (0.86 %) supported the continuation of logging, of which more than half were from foresters who saw selection logging as compatible with conservation of both forests and wildlife. Among the rest, 72 % called for an immediate halt to all logging (led by the Native Forest Action Council and the Waihaha Forest Preservation Committee), 31 % argued for protection of wildlife values, especially the kokako; 18 % endorsed NZWS's call for biosphere reserve status for the forests; and 11 % reminded the Minister of the promises made by the present Government before it was elected in 1975 [13, 24]

They ranged from "brief, simple and obviously sincere statements, to detailed reports showing a depth of understanding of the proposals and lengthy association with the forests concerned". They came from all over the country, but 61 % came from the urban centres which at that time supported 53 % of the national population. Auckland (38 % of submissions from 25 % of the population) was clearly an important centre of concern, followed by Taranaki and Manawatu (12 %) and King Country, Rotorua and Taupo. The actual numbers and geographical distribution of the people behind the submissions were usually not known, but the anonymous NZFS compiler could not help commenting that "this suggests a slight tendency for those interested enough in the issues to forward submissions to be urban-dwellers rather than rural".

The NZFS summary of submissions described the main issues at stake under six broad headings.

1. *Wood production*. NZFS recognised the need to preserve two-thirds of the forest under dispute, but insisted that the remainder, some of which was already cutover, should be zoned for production. The opposing camp was adamant that all logging should stop immediately despite the two legal sale agreements binding NZFS to supply timber

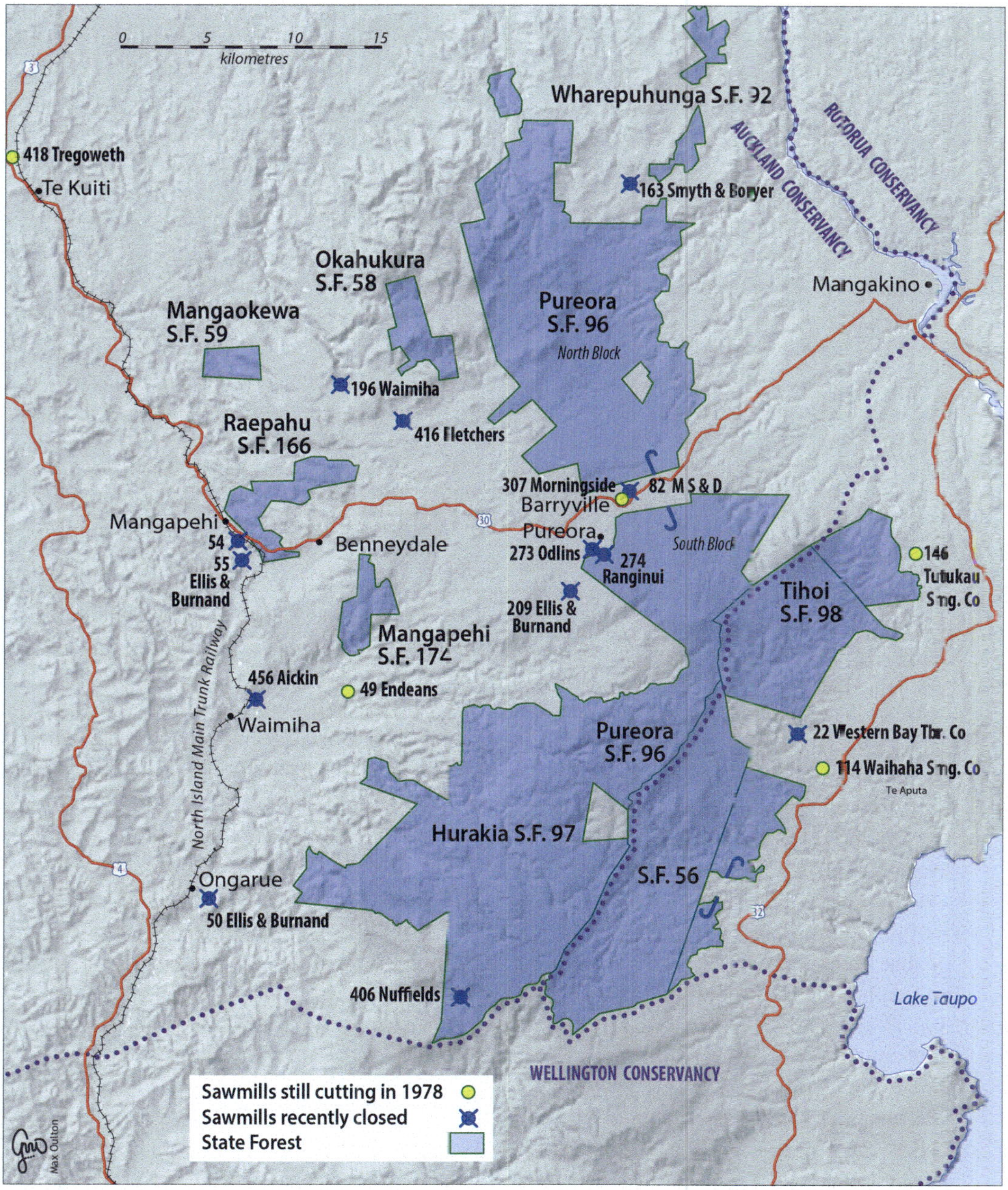

Fig. 10.1 By 1978, almost all the formerly numerous timber mills in the King Country were closed. The last four survivors drawing supplies from NZFS were mill numbers 114, 146, 307 and 418. *Redrawn by Max Oulton from NZFS file 6/96 with additions*

to two sawmills until 1983 and 1985, and to another two smaller mills until 1979.

2. *Kokako habitat*. NZWS strongly advocated extension of the reserves already established by NZFS, since biologists agreed that the forests at risk contain premium habitat for this bird, and protecting them may offer its last chance to avoid extinction. NZFS replied that kokako are widespread, although sparse, in State Forests throughout the North Island, and there was no conclusive evidence that careful selection logging would necessarily be disadvantageous. Both agreed on the urgent need for more information about the biology of the kokako.
3. *Waihaha*. Conservationists had long argued for special protection for the dense virgin podocarp forest in the Waihaha State Forest, with long, carefully researched and well reasoned submissions quoting both scientific studies and pre-election promises [36]. NZFS agreed in general, and undertook to set aside the Waitaia catchment as a longterm reserve where no logging would be done in the foreseeable future. At the same time they argued that part of that block could be selectively logged without causing serious damage.
4. *Social impact*. Implementation of the 1977 policy on indigenous forests had already made clear that winding-down of local sawmills was inevitable sooner rather than later, with consequent unemployment for local workers. Conservationists argued that, since this social impact was imminent and could not be avoided, the mills should be closed immediately, while there was still some forest left to preserve.
5. *Indigenous timber demand*. Foresters pointed to the high demand for decorative timber grades milled from the west Taupo forests. Conservationists replied that much of the native timber presently used could be replaced from other sources, including eucalypts planted elsewhere.
6. *Management status*. The suggestion put up by NZWS that the Pureora forests should be given UNESCO biosphere reserve status had collected a lot of support at and since the seminar, although there was little discussion of what criteria would need to be met. NZFS replied that State Forest Park status would provide sufficient recognition.

The main issues at stake were complex and distinct but closely inter-related. Naturally, no summary could do justice to the variety and depth of the arguments contained in the submissions, but NZFS did do its best to consider them all.

NZFS senior staff recognised that the conflicts could not be resolved for the long term unless they could find some way to hammer out a durable management policy for these important forests. Probably no-one at NZFS at that time could imagine the long-term policy solution that was to be imposed on them by a future Labour Government in 1987.

The Otago Business Development Report

The Otago Business Development Centre (OBDC) was commissioned by NZFS to summarise the current economic role of indigenous logging in the Pureora area, and assess the social and financial consequences of (a) business as usual, (b) a shift to selection logging only, or (c) complete cessation of all logging [14]. No other alternatives were considered, because OBDC assumed that these were the extreme options, and intermediate positions could be estimated as points on the range between them. A managed transition to supplies of exotic timber was recognised as an option discussed in previous literature, but not properly assessed during this exercise—a point emphasised by critics [2].

OBDC was directed to make use of all existing information and to conduct three surveys among local employers, employees' social organisations, clubs and social services. It was to report back with all speed, preferably with at least preliminary findings for the Taupo seminar. The published proceedings of the seminar included a progress report on the first of the surveys, and the full 55-page report appeared only three months later [14].

Considering that the three surveys were done only in March and April of 1978, the production of such a document so quickly was a remarkable bit of high-pressure research. The three surveys each supply some useful statistics of the state of the villages and their people at the time of the surveys, concentrating only on the four settlements of Pureora, Barryville, Benneydale and Te Kuiti (Box 10.2). The two surviving mill settlements at Tihoi and Arataki on the more distant eastern flanks of the park (Fig. 10.1), with contracts expiring in March 1979, were not included in the surveys. Endean's mill near Waimiha operated independently of NZFS, so was ignored.

Box 10.2 Employment statistics for the local logging industry in 1978

The Otago Business Development Centre was contracted by NZFS to survey the consequences of any reduction of logging in the Pureora State Forest [14].

In 1978, NZFS employed 66 men at Pureora and Te Kuiti in afforestation, felling and administration. The two commercial companies with binding contracts for sawlogs, valid until 1983 and 1985 (Chap. 7), were Pureora Sawmills Ltd employing 35 men at Barryville, and R. H. Tregoweth Ltd working under subcontract to Ellis & Burnand, employing 40 men at Te Kuiti.

A high proportion of the local population depended on these two employers and on jobs provided by associated concerns such as the Pureora school (teachers, cleaners, dental nurse, caretaker) and the transport operators. The survey had to use the out-of-date 1976 Census data (nothing more recent was available) to estimate the approximate total populations of the four settlements. Then it counted the numbers of people living in these places directly dependent on indigenous logging, that is workers and their families, and the numbers expected to be left behind a year after either the introduction of selection logging, or stopping it altogether. The effects on people living outside the immediate area were also considered, but were minor by comparison

These figures make an interesting contrast with those recorded during Margaret Smith's study of social life and working conditions in the village in 1953 [31], and the reassessment of J. A. Dawson [6] in

Location	Total population 1976 census	Proportion dependent on logging in march 1978 (%)	Number left after switching to selection logging	Number left after ceasing all logging
Barryville	134	100	0	0
Pureora	216	61	175	101
Benneydale	400	2.5	381	381
Te Kuiti	4832	4.3	No change	4800

1979, which concluded that most of the displaced workers were better off in their new jobs.

The Economic Model

The meat of the OBDC's report concerns economic assessments and forecasting, plus extensive documentation of the social consequences of adopting the alternative options. The OBDC economists were required to estimate the costs and benefits of managing the remaining resource under both selection and zero logging options. Much of the debate depended on an objective assessment of the capabilities and output of the four remaining local sawmills. Fortunately, the required data had been provided to the Taupo seminar in a paper by two NZFS forest utilisation specialists, John Vaney and Laurie Gibson [35].

The two largest mills, both holders of binding NZFS contracts and hence of greatest interest to the OBDC survey, were Pureora Sawmills' Barryville mill, and R.H. Tregoweth's mill at Te Kuiti. The technical (Box 7.2) and social differences between them played a significant part in the final decision, because they helped to determine which mill was most likely to survive and adapt through the inevitable transition in supply from huge and irregular native logs to smaller, more standardised exotic logs.

Vaney and Gibson concluded that, if logging were to stop immediately, there would be no locally available alternative State Forest exotics to keep the mills going, and the current exotic production forests in Bay of Plenty region were too far away to allow an immediate exotic changeover. If logging were to stop gradually, only Tregoweth's could readily adapt to cutting exotic logs in the future. The Barryville mill was primarily set up to handle large indigenous logs and would not be suitable or efficient for exotic milling.

OBDC acknowledged that the Barryville mill was vitally significant to the local economy, as shown by data previously commissioned by its joint owners, the Carter-Holt and Fletcher conglomerates [12]. The mill paid out NZ$189,000 on wages in 1977, much of which was spent locally on food, plus a rather higher amount on tobacco and alcohol than the national average (8 % of income compared with 5 %). It also spent NZ$24,000 on electricity, NZ$49,000 on engineering services, NZ$178,000 on road and rail freight and NZ$10,000 on fuel. Virtually all of this income would be lost to local people and businesses if the mill closed.

On the other hand, the Barryville mill used wasteful circular breakdown saws that could not be converted to handle exotic timber efficiently (Box 7.2); its plant and employee housing were well depreciated and with a limited future; and the Barryville settlement did not have its own shop or any other services. Tregoweth's mill at Te Kuiti was already equipped with more efficient band saws, it was a far less dominant local employer, and it had access to good worker accommodation in the town.

OBDC's main conclusions were as follows:

1. If logging continues at 1977 levels until the existing contracts are completed in 1983–85 and all of Pureora Forest is gone, the added value and savings created by the logging industry to local employers and workers would be NZ$2.9 million a year, totalling a net present value of NZ$12.7 million, and then it would close down.
2. If selection logging as proposed by NZFS (at no specified rate) is adopted, the value of the industry would be NZ$1.6 million a year, totalling a net present value of NZ$6.94 million for both mills for 2½ years, or NZ$7.53 million for Te Kuiti alone for 5 years, and then the logging would close down.
3. If logging is stopped immediately, the net present value of the industry, in incomes paid out to people managing the Forest Park (and to a few loggers still working on private land) would be NZ$2.73 million over 5 years, or c. NZ$0.6 million a year for the foreseeable future.

OBDC concluded that the logging of native timber had a very high net present economic and social value in the Pureora area, and should be protected as far as possible, so it was not in favour of the minimal returns available under option 3. On the other hand, no-one expected the timber resource to last until the end of the contracts, so option 1 was not fully viable either.

NZFS's stated preference was to continue operating under a regime of selection logging. OBDC stopped short of explicitly endorsing this option, but implied that a valuation of NZ $7.53 million for the net present value of the industry running a single mill to the ends of the contracts was the most reasonable one [14: 54].

Critiques of the OBDC

Not surprisingly, the OBDC report came in for some trenchant criticism by independent ecologists, sociologists and economists. South Auckland Conservation Association president Dr. Charlotte Wallace called it "incomplete and politically biased", and Waikato University researchers dissociated themselves from the report [37].

The Environmental Studies Unit of Waikato University organised a special seminar [11], as an immediate, academic and indignant response. The seminar was held on 29 June 1978, but the published proceedings, edited by N. J. Eriksen, did not appear until much later [11]. It produced some serious criticism of the economic model used by ODBC, on the grounds that it seemed to have been commissioned in order to influence the final decision in favour of NZFS's preferred outcome. It particularly disapproved of the way that the only question being considered was whether there should be some selection logging or no logging at all, to the exclusion of any other alternatives [10].

The seminar participants agreed that the vested interests of the agency that funded the surveys (NZFS), and the inadequately specified brief that OBDC was given, broke all the normal rules of contract research. The surveys did indeed document the serious social disruption to be expected, but ignored the consequences. The suspicion that the results were unduly influenced by the political needs and aspirations of the decision makers (rather than offering a fully objective contribution to the decision-making process) undermined all confidence in the conclusions [11].

An independent Victoria University economist, Geoff Bertram, added a lucid and withering critique of OBDC's methods and conclusions [2]. It was presented to ECO's annual conference in Wellington on 22 July 1978, and a transcript was included in Eriksen's publication.

Bertram pointed out that OBDC's calculations were quite invalid. The implication from the OBDC figures, that the difference to the local economy between logging as usual and a complete halt would be NZ$10 million (NZ $12.7 − 2.7 million), was wrong because it assumed that: (1) there is no monetary value whatever to the forest as a national amenity of wildlife, *i.e.*, that conservation and soil protection/erosion control have no financial benefit at all; (2) the commercial value of unlogged (standing) forest is zero; (3) none of the people who in 1977 were employed in the local logging industry would find any other jobs before 1985; (4) the capital equipment used in logging has no alternative use and no market value. All these assumptions are to various degrees false, so it is therefore equally false to suggest that the cost of stopping the logging (and therefore of any compensation for loss of net present value) is anywhere near so much.

A copy of the Waikato University seminar papers plus Bertram's critique was forwarded to the Otago Business Development Unit. It did not respond, although the Director-General of Forests rejected the criticisms, in a letter of 8 August 1978 included in the published version [11].

But time was short, and the pressure to reach a decision was intense. The OBDC report was accepted, and all critiques ignored.

High Tension in the Villages

The residents of Pureora and Barryville villages did not stand passively aside while their futures were being discussed by intellectual outsiders. They were very well aware of the risk of a serious population reduction that would certainly have a drastic effect on their working and social lives. Alerted by the questions being asked by OBDC's surveyors, the Country Womens' Institute (CWI) branch based at Pureora compiled a letter listing their apprehensions, and submitted it to Venn Young as part of the consultation process after the Taupo seminar.

> Dear Sir,
>
> The Pureora Forest CWI are concerned at the possible implications of the Indigenous Forest Policy as applied to our area.
>
> We accept that conservation of timber resources are in the "National Interest" and that a reduced level of logging is inevitable.
>
> We would like to point out just what a reduction in logging levels, or total closure, means to people living at Pureora Forest.
>
> 1. There will be no NZR bus to Te Kuiti.
> 2. There will be no mail, milk, bread or paper delivery.
> 3. There will be a reduced school roll, and possibly no school or pre-school unit.
> 4. The Post Office will close.
> 5. The local store will close.
>
> Community organisations will probably have to close, e.g., Play Centre, Football Club, Parent/Teacher and CWI, to name a few.
>
> A community working together is a dream not often realised. This has happened at Pureora, but, with a reduced population numbers would not be enough to keep any of our existing organisations going.
>
> J.E. Walker, V/President, Pureora Forest CWI [8].

Young's private secretary sent an acknowledgment on 31 May, and promised that "The Minister is giving consideration to your letter along with others received". When the OBDC report came out in the same month, it listed in confusing detail the data collected by the three surveys, all of which fully supported the CWI's predictions, but cost a lot more to collect.

The bushmen had their own opinions on what was the real issue.

> This forest controversy needs sorting out—and a little bit of sanity brought to the notice of those grossly selfish conservationists who would deprive 496 people of their living...we think of conservationists as mindless tree-sitter types buoyed up by emotion rather than facts...The future of the kokako would be determined not by the degree of Forest Service-controlled logging but by the existence of ground vermin [18].

They had a point there, and they were not the only ones to think of it (Chap. 13). But they followed it with a more ominous idea, that

> Any further "stirring" by conservationists would not be tolerated...They will not get into our forests again, and if they do they won't get out [19].

A Shock Announcement

Cabinet and the Government Caucus debated the issue at length throughout July of 1978. Finally, on 7 August Young announced a three-year moratorium on logging at Pureora, accompanied by strict implementation of the 1977 management policy (Chap. 6) in the native forests of the central North Island, which should reduce logging of native forest to a sustained yield. The policy declared that:

1. The supply of native timber from the west Taupo forests would be reduced from the then maximum of 60,000 m^3 a year to 6000 m^3, although up to 1500 ha of Hurakia State Forest would be made available for exotic afforestation if required to maintain the economic and social stability of local communities.
2. Logging would cease in Pureora (in an outlier on Ranginui Rd) and in Tihoi (in Waihaha north of Waihora Stream) State Forests by the end of the year. The Wharepuhunga, Pureora,

Tihoi, Hurakia, Taringamotu and Waituhi State Forests, a total of 71,870 ha (Fig. 9.7), would be immediately incorporated into a new entity, Pureora State Forest Park. The formal announcement was made on 7 September 1978 [43], and the decision was to be reviewed after three years, but that would not prevent local mill closures.
3. Research on the biology of the kokako would be undertaken over the next three years.

The decision had become additionally complicated because arguments over NZFS policy in the west Taupo forests had become mixed with similar arguments being conducted against the beech scheme on the South Island's West Coast (Chap. 6). The Ministerial decision on the West Coast project, imminent in mid 1977, was delayed when it became clear that it would be influenced by the outcome of the developing crisis at Pureora.

New Zealand's internal debates over indigenous logging of beeches on the West Coast and of podocarps at Pureora eventually reached a much wider audience. An independent analysis of both issues was published in the international journal *Biological Conservation* in 1980. It concluded that, in the end, environmental groups achieved more at Pureora, where timber companies and local communities were relatively small, than they could on the Coast where large timber companies were operating and larger local communities could pressure NZFS to protect their interests [45].

Even so, NFAC was not satisfied. "The 'battle' for the west Taupo forests is far from ended…conservationists could not afford to rest on their laurels", said Gwenny Davis [40].

The decision on Pureora was only a temporary reprieve, and for NFAC it was overshadowed by the "bombshell" announcement that the Government was considering a proposed 12-year logging programme for the Whirinaki Forest, adjoining Urewera National Park, to support the logging village at Minginui [25]. NFAC had almost finished a detailed submission recommending that Whirinaki be added to the Urewera National Park, said NFAC, but this Government decision, taken without a wildlife survey or the usual opportunity for public comment, appeared to be a deliberate attempt to circumvent the growing public movement favouring enlargement of the park. NFAC immediately declared its intention to continue fighting for Whirinaki, and to demand that normal procedures for taking decisions about management of state forests be reinstated [21].

For the residents of Pureora village, weary of all the high-powered arguments of the last few months, the switch in NFAC's priorities was very welcome. With the moratorium in place at Pureora, public attention, and its associated TV and newspaper interest, shifted to Whirinaki. Life in Pureora village quietened down a bit [43]. A new kind of work began in the forest, conducted by ecologists and ornithologists with clip boards and binoculars rather than chainsaws and tractors.

But for the older forest rangers, the relief from controversy was bitter-sweet. Years later, Jack Walker, OC throughout the upheavals (1974–80), summed up his reactions on tape.

> You try to discount your personal feelings in this sort of thing, and just carry on and manage your way through. That was my objective. I think closedown was inevitable once it all started, and the proof is that Minginui then closed down, and then everything except parts of Westland closed down. I think a change was inevitable, but at that stage our objective was to go ahead and do the bloody job we were paid to do. I think all the other stuff got a bit emotional, and they were involving people's lives, work and livelihood. It affected school roles and mail runs and groceries, shopping in Te Kuiti, all sorts of things in Mangakino and Tokoroa. In hindsight it was inevitable. But if they had waited another 10 years before they did it I would have been much happier [42].

Awaiting Developments

Although the NZFS's announcement of its intention to designate a State Forest Park in the Rangitoto-Hauhungaroa Ranges went back to 1975, uncertainty hung over the Pureora community during the three-year moratorium that started in 1979. The issues as they appeared at the time were well summarised by Tony Beveridge (from the Forest Research Institute in Rotorua) in his Sanderson Memorial address to RFBPS [3].

Beveridge reminded his listeners that NZFS's policy of clearfelling untouched stands of native forest and conversion to exotics had been already phased out after 1977, except where commercial contracts still had to be enforced (Chap. 7). So he could quote Andy Kirkland, Deputy DG of Forests, telling the ANZAS congress of January 1979 that any future reduction of the 6,200,000 ha of remaining native forest would mostly be done on private land, concentrating on clearing cutover or second-growth forest for farming.

Beveridge also pointed out that the Auckland Conservancy of NZFS had already started a programme of planting out 30,000–40,000 podocarp seedlings a year in Pureora Forest alone, and of tending the planting sites for long enough to ensure their survival. In the heat of the debate so well publicized during the moratorium, these important facts in NZFS's favour tended to be overlooked.

Certainly, the supporters of the protest movement achieved their aim of compelling NZFS to break certain contracts, but that was not the end of the matter. Key problems still to be resolved included, not only how to ensure the survival of the kokako, but also what might be the wisest future developments in native forest policy generally, and whether some form of logging could ever be resumed.

Beveridge drew a distinction between logging managed simply to maximise timber production, and logging managed to maximise the health of the residual forest. The second, harder option required calculating growth rates, yield, number and type of trees cut, replanting with nursery-raised seedlings, fostering of regeneration and minimising natural mortality, and he argued that this type of logging could safely continue in the new PFP. The people most affected had to wait to find out whether the Government would agree.

Later that year, J.A. Dawson of Victoria University produced a new and independent analysis of the economics of the Pureora logging industry. It put the net present benefit of the existing regime to the local community as NZ $0.69 million a year, not NZ$2.9 million a year as claimed by OBDC [7]. Dawson agreed with Bertram that net future values must be discounted by the industry standard of 10 %, not the 4 % used by OBDC (a discount of 10 % means that NZ$100 paid now would compensate for losses of NZ$110 next year). By using more realistic estimates of value added and savings created by the logging industry in its current form, and applying the 10 % discount, Dawson estimated its total net present value as NZ $2 million, not the NZ$12.76 million claimed by OBDC. He predicted that the future value of Pureora Forest Park as a recreational amenity would amount to much more than the NZ $2 million lost by stopping the logging. He was right (Chap. 14).

Arguments raged in Parliament between the Minister of Forests Venn Young ("I tire of the selfishness of the uncompromising environmentalists who have no regard for the employment needs of rural communities") and the Minister for the Environment Ian Shearer ("It continues to be my firm resolve to bring a halt to all logging in these forests as quickly as possible") [5].

The Labour Party promised to save both Pureora and Whirinaki if it won the next (November 1981) election [22], but the National Party was returned, with a reduced majority. On 6 May 1982, a new Minister of Forests, Jonathon Elworthy, announced that the moratorium at Pureora was to be made permanent. It was not until after the 1984 general election that a new Labour Government could stop the logging in Whirinaki.

Consequences

The immediate effect of the moratorium in 1978 meant the closure of three of the four remaining west Taupo sawmills supplied by NZFS (Box 7.2) and the loss of about 70 jobs—a lot, although fewer than had been predicted by the Otago Business Development report [44: 121]. Most other mills in the district had already closed (Box

Fig. 10.2 Nothing now remains of the Ranginui mill at the west end of Pureora village, but a huge pile of sawdust. *C.M. King (2013)*

7.2, Fig. 10.1), leaving only huge piles of sawdust to mark their sites. The sawdust pile accumulated by the old Ranginui mill just west of Pureora village still shows the rich dark colour typical of sawn matai timber (Fig. 10.2). Matai was the dominant timber species in the Pikiariki block (Box 13.6), near where the logging crews were working before the mill closed in 1976.

For the two mills still operating on the eastern flanks of the Hauhungaroas (the Waihaha Sawmilling Co Ltd at Tihoi, and the Tutukau Sawmilling Co Ltd at Arataki), supplies of timber from State Forest land (Fig. 7.4) ceased with minimal comment in 1979, since both had known for some time that their contracts were due to expire then anyway. The last remaining native timber mill in the King Country was Endean's, which continued its traditional operations, never upgraded or modernised, until it finally closed in 1996 (Fig. 5.16).

The owners of the Tutukau mill (Boreham Brothers of Rotorua) put the whole mill and village up for sale, including all the mill machinery, vehicles and five houses. Everything had to go, however run-down, and at knockdown prices (a Hough loader worth NZ$6500 went for NZ$4500, houses for NZ$170 each, two old trucks for NZ$500 each). The buyers salvaged what they could, and the rest was burned so that the site could be bulldozed clean, as NZFS required [32]. The Tihoi mill site was converted into the Tihoi Venture School, a successful outdoor education and leadership training centre [26]. Its website [15] shows a picture of the mill in its working days.

Tregoweth's mill in Te Kuiti survived, and is one of few King Country mills to have made a successful transition to exotics It was not easy, and the company received no compensation for the loss of 90 % of its business (Box 10.3), but it managed to adapt. The very last selection logging in native forest at Pureora was done at Waihora in 1983, and Tregoweth's purchased the timber at the skid. By 1996 the mill was fully converted to handle exotic logs, and it still operates in Te Kuiti.

The greatest bitterness was generated at Pureora Sawmills Ltd at Barryville, where the supply of timber from Pureora SF ceased on 15th December 1978 (Figs. 10.3 and 10.4), even though the two companies that jointly owned it (Odlins and Carter-Holt) still held a legal 15 year contract signed in 1970, and merchantable trees still stood (Chap. 7).

Fig. 10.3 The last log being taken out of Pureora State Forest to Carter's Barryville mill. *Crown Copyright, Department of Conservation Te Papa Atawhai (15 December 1978). Photographer unknown*

Fig. 10.4 Inside Carter's mill at Barryville, on its last day of operation, December 1978. *Graeme Reinhardt*

Compensation

The Government's decision to force the breaking of the two NZFS contracts required two payouts totalling NZ$7.1 million from the Crown in compensation to the three timber companies affected—NZ$2 million to E&B, and NZ $5.1 million to Odlin's and Carter-Holt as joint owners of Pureora Sawmill Ltd (NZ$30 million in today's money: see http://www.rbnz.govt.nz/monetary_policy/inflation_calculator/). All concerned refused to say how the payments were calculated, but they were probably not influenced by the OBDC's calculations, which turned out to be irrelevant. It is a mere coincidence that the NZ $7.1 million paid out was so close to the NZ $7.5 million estimated by the OBDC as the net present value of the contracts under NZFS's preferred option of continued selection logging.

The Government may never have seen Bertram's [2] demolition of the economic logic used by the Otago team to value the resource, or Dawson's [7] significant reduction of the estimated real valuation of the industry, but NZFS was not left having to accept the inflated figures given by the report they had themselves commissioned.

In fact, the compensation packages were calculated quite differently (Box 10.3). The full details are preserved in the Fletcher Trust Archive, in Penrose, Auckland. The Fletcher company under its various titles has a long history of taking over smaller enterprises, and maintains an immaculate archive of the histories of its component business operations, including both E&B and Odlins.

The itemised claim for compensation of NZ $6,467,803 was drawn up in August 1979 when the contract was terminated, as is documented in several very revealing PSL Board papers. Most of the claim was based on estimated loss of *future* profits, plus a small amount for two sets of real costs: lost assets (the mill machinery and buildings) plus a smaller amount budgeted for redundancies and relocations of staff, together totalling <NZ$1 million.

Box 10.3 Compensation details

The full details are held in the Fletcher Trust Archives, under catalogue numbers 1425/1/3 (PSL) and 0381/6/6,15 (E&B).

A Pureora Sawmills Ltd (PSL) Board document dated 30 August 1979 provides the itemised summary of PSL's first claim for NZ$6.47 million, compared with the equivalent figures for Ellis & Burnand's claim. When NZFS offered only NZ $4.7 million, PSL recalculated the schedule of lost profits to include their full contract annual entitlement (never yet taken) and sued NZFS for a total of NZ$8.4 million (itemised in an Odlins Ltd Internal Memorandum, 3 March 1980). NZFS reached an out of court settlement of NZ$5.1 million, and the formal discharge document from the Wellington High Court recording acceptance of that amount by both sides is in the file, dated 27 January 1981. The final PSL Board paper is dated 25/2/81.

The archive includes a letter dated 24 May 1982 to the PSL Board from its lawyer strongly advising it to "resist any attempt by the [Inland Revenue] Department to treat any portion of the settlement moneys as income …[it] was a lump sum payment by way of compensation for a breach of contract which has effectively terminated the Company's operation". PSL's Managing Director D.C. Hill then instructed the Company Secretary N.M. Manssen that "we should not in any circumstances provide a breakdown of the claim"…to Inland Revenue or anyone else. This suggests that the compensation was paid tax-free.

There has never been any public information on how the compensation payments

	Pureora Sawmills Ltd	Ellis & Burnand
Contract term	1970–85	1968–83
Permissible cut	33,985 m^3 per year	27,736 m^3 per year
Expiry date	31 May 1985	1 December 1983
Years left to run	6.6	4.9

(continued)

	Pureora Sawmills Ltd			Ellis & Burnand
Compensation (NZ$)	1st claim (30/8/79), NZ$	NZFS offer, NZ $	PSL lawsuit (3/3/80), NZ$	Claimed (NZ$)
Disposal of assets	590,400	475,150	543,950	0
Loss of profit on sawmill	1,746,360	1,000,078	1,600,830	2,663,097
Loss of profit on retail sales	2,932,162	2,026,655	4,123,818	
Loss of related sales	1,198,876	0	1,938,194	
Redundancies and relocations	105,000 (est)	Actual	115,468 (actual)	0
Other costs[1]	48,915 (est)	Actual	100,000 (est)	90,202
Total	6,467,803	4,700,759	8,422,260	2,753,299
Paid after negotiation	5,100,000[3]			1,962,871[2]

[1]"Other costs" are listed by PSL as "Conciliation and winding up", and by E&B as "Additional costs". E&B tried to claim NZ$90,000 compensation for Tregoweth's for breaking their subcontract, which at the time amounted to 90 % of Tregoweth's output, but NZFS did not accept the claim, and E&B denied liability (E&B Board Minutes, 29/11/78). Tregoweth's took E&B to court claiming NZ$653,780 for breach of contract (E&B Board Minutes 28/11/70), but E&B's Board was disbanded shortly afterwards so there is no record of what happened. The current CEO of the mill, Kevin Tregoweth, states that the court awarded the company about NZ$100,000 in compensation, about the same as it cost them to pursue the case. The loss of the contract put the mill under pressure to find new sources of native logs and slowly change to radiata pine, but it was not easy, as cutting native timber was based on lower volumes with bigger returns, whereas "with radiata everything depends on higher volume and lower pricing". It also cost "huge amounts" to convert the older style native mill to handle radiata [34]

[2]The breakdown figures for E&B's claim are quoted from the PSL Board document of 30 August 1979, but only the final payment of NZ$1,962,871 (without a breakdown) can be confirmed from E&B's own Board Minutes of 3 October 1979 (Fletcher Archive 0381/6/15). E&B had submitted a claim for loss of earnings of NZ$1,936,522 plus costs of NZ $32,038 (total NZ$1,968,560), but they considered NZFS's offer as so close to their own estimate that they were willing to accept it without further discussion. The Directors considered this an excellent result, and congratulated their team for negotiating it with NZFS in a "cordial manner". The same document records the E&B Directors' decision to accept the Fletcher takeover offer

[3]In the final agreement with PSL, NZFS undertook to pay the actual costs of redundancies and winding up incurred by the Odlin's/Fletchers consortium. The redundancies came to slightly more than the estimate, while the "other costs" were still not finalised at the date of this document. A later source [7] gives a figure of NZ$130,000 to cover redundancies and relocations, divided between 32 families.

were calculated. The archive material suggests that the final packages were worked out by the timber companies and their lawyers, without reference to the OBDC report or to any other outside economic analyses. By express permission of the Fletcher Trust, this information is published here for the first time.

NZFS disagreed with PSL's figures, and offered a mere NZ$4,700,759. There was a long period of fruitless negotiation attempting to close the gap, which ended when the owners of the company (led by Alwyn Carter) ran out of patience and recalculated their lost profit from the amount contracted for rather than the lesser amount taken. They filed a writ in the Wellington High Court for NZ$8.4 million against the Forest Service [9]. The Crown opted for an out-of-court settlement of NZ$5.1 million to PSL [28: 67], divided equally between the two parent companies, Carter-Holt and Fletcher. The final discharge document is in the Fletcher Trust archives.

E&B's claim was much smaller, because its remaining contract period was shorter, and it had always handled the logs through Tregoweth's mill in Te Kuiti, which was not being closed down even though the E&B contract comprised 90 % of its business. E&B did not have to dismantle a mill or pay out for redundancies or relocations of staff, but it still got almost NZ$2 million for loss of future profits. R.H. Tregoweth's claim against E&B for the breaking of the subcontract was not included in the settlement.

Many people said that no bird could possibly be worth so much money. Worse, the actual data confirm the speculation by NFAC, based on "independent economic studies" [23], that most of Pureora Sawmills' share of the payout was paid out to shareholders or retained by the two timber companies (Box 10.3). Only 2.6 % of the NZ$5.1 million went to the Barryville families affected. They shared NZ$130,000, divided between 32 laid-off employees in proportion to their length of service and family size. Payouts ranged from NZ$4000 to a man of 31 with seven years of service and three children, and NZ $1100 to a single man of 22 with five years of service, down to NZ$300 to a man with six children, of age unknown, with two years of service but on accident compensation for much of that. Yet all but one of the workers reported satisfaction with their severance pay [7].

In its final Board paper in February 1981 before it closed, PSL concluded modestly that "In retrospect…we could probably consider ourselves lucky to come out of it so well". NZFS had been thoroughly outfoxed, and their only small consolations were that, for the moment they did not have to guarantee that logging would never resume [23], and that the two mills at Tihoi and Arataki had no claim to be included in the compensation packages.

NZFS staff were stunned by the sudden change in their work and prospects, but at least most of them could be transferred to other jobs within the Service. The main effect was felt by the employees of the PSL mill at Barryville.

Goodbye Barryville

It was the end of the world as they knew it for the Barryville community. The village consisted only of the mill and one street housing 35 full time employees supporting 120 dependents, including 35 pre-school children, 58 school children and 27 "non-working" wives [12].

Until the upheavals of 1978, all the staff working there had assumed they had a secure future until 1985 when the mill's 1970 contract with NZFS expired. But suddenly, their jobs and homes were gone. There could be opportunities in Te Kuiti, but that was 53 km away, so daily commuting would be uneconomic.

Bruce Tricklebank, chair of the Pureora Community Council, pointed out that the villagers had supported the NZFS's changes from clear felling to selection logging, not only in the hope of saving their jobs but also because many were keen bushmen and hunters who knew and respected the forest. No-one had expected a reduction from 60,000 to 6000 m^3 a year from all the forests in the area—a total available cut that was much less than the 11,000 m^3 a year that a single mill needed to survive.

The workers were nearly all Maori, and many had lived in the area all their lives—some were second-generation residents [40]. They paid peppercorn rents for their houses (NZ$1.75 a week for 3 bedrooms, plus unlimited free firewood) [12]; their children went to school in Pureora or Benneydale; and a bus took them to Te Kuiti for shopping or to go to the cinema. Many had never been anywhere else.

For example, the Shepherds had lived in Barryville with their family all their married lives, and so had their parents [38]. They liked living there, where they had all the benefits of country living but also got daily deliveries of bread, milk and newspaper. They had a steady income, cheap accommodation handy to the job, and their children attended the Pureora school and pre-school. The community was supportive, the women played netball and tennis, or pool at the village hall, and the urban attractions of Te Kuiti were just a bus ride away. When the mill closed they would be moving to a house in Petone (Odlin's had a big mill there), where the rent was much higher and the community unknown. They accepted the need to move, but were not looking forward to it.

The local people started a petition to save the mill, which got 250 signatures in three days and could have got more from Te Kuiti businesses worried about the effects of local depopulation on their trade. When that failed, they argued for compensation for the mill workers, rather than for their employers. People think mill workers are transient, said Bruce Tricklebank, but he reckoned that was quite wrong—the only truly transient people at Pureora are the NZFS managers who stay for a year or two and then move on. In fact, only 12 of the 24 men interviewed a year later by Dawson [7] had been at the Barryville mill for more than five years.

The Government agreed that the mill workers were entitled to redundancy payments, but not to any special compensation. Their union maintained that their situation was unique, and that they were prepared to sit it out and refuse to move until the Government agreed, on the understanding that the company would hold their jobs meantime.

Barryville union delegate David Buchanan said that the Carter-Holt and Odlin's companies supported the men's claims for compensation, and any compensation awarded would be reclaimed from the Government by the company. Venn Young's response was that he would be willing to be flexible in negotiating redundancy agreements, but not in terms that would set a precedent for workers in other industries [38]. The Timber Worker's Union national secretary Ray Hamilton, visiting from Rotorua, saw that as evidence that the Government was putting birds before people, and was unwilling to meet its responsibilities. Hamilton promised to make a fight of it, and left "to make a statement to the media" [27]. Not surprisingly, the Barryville community was left with a lot of bitterness—at the conservationists, at the media, and at the Government [20].

Barryville people felt their resentments deeply, especially against those who had caused all the trouble. One teenager born and brought up in the village, JoAnn Shepherd (then aged 17), wrote a requiem for her home community, in part a poetic eulogy and in part a scathing protest against those "nose-blowing ecologists who forgot that we were there, and wrote for the nature-loving people in Auckland and Christchurch [who]…taught their children …that the kokako must be saved….Do people realize that damned bird isn't the only living thing here?" [27].

It was not as if the closing of timber mills was anything unusual—on the contrary, most were quite short-lived, and milling families had always expected they would have to move on when the forest was exhausted (Chap. 5). Nor was the number of jobs lost at all exceptional—by one estimate, more jobs were lost when the Te Kuiti telephone exchange was automated than by the protection of Pureora forests [21: 47]. The problem was that there was absolutely no alternative employment in Barryville, so the community had to be broken up while millable forests still stood, and for merely political rather than economic reasons. To them, that was a simple, unreasonable waste of opportunity and resources.

By December 1978, preparations for the closing of the mill were well advanced. Rusty Russell delivered the last log from Pureora Forest to the Barryville mill on Friday 15th December 1978 (Fig. 10.3), just as 31 years previously he had taken the first load to the now disused Ranginui mill at Pureora [38].

The Barryville mill's office manager Brian Harper reported that he had scoured the country, and outside it, on behalf of his men. By the time Dawson [7] tracked down and interviewed 24 ex-Barryville men, 22 had found other jobs (18 of them on higher wages), one had retired and only one was still unemployed. Harper and the mill manager Sid Russell, who had worked at the mill for more than 20 years, were transferred to their company's New Plymouth operation. Of the other employees, six or eight men remained to dismantle the mill and its old tramway, and the company houses. After they finished, Barryville

Fig. 10.5 Pureora Village after the moratorium. The road has been sealed, the school, the store and the Football Clubhouse remain, but many of the mill houses have already gone. *Crown Copyright, Department of Conservation Te Papa Atawhai (about 1986). Photographer John Mason*

no longer existed. The cleared land behind where the mill once stood passed into private ownership as Peacocke's farm.

It was not only the men who lost their livelihoods and identities; the companies that employed them did too. Pureora Sawmills Ltd. was wound up, and the partnership dissolved. Carter-Holt went on to become Carter-Holt-Harvey, still a major player in the building industry [28]. E&B and Odlin's, which had both been for years among the leading independent timber companies in New Zealand [1, 30: 323], were both taken over by the Fletcher empire.

After more than half a century of operation, the speed of E&B's demise was dramatic. E&B received on 24 September 1979 a Treasury cheque for NZ$1,962,871, and *on the same day*, Fletcher Holdings Ltd formally presented their notice of takeover of the E&B Company (Box 10.3). At their next meeting on 3 October, the E&B Directors advised their shareholders to accept it. E&B continued to trade under its own name until the early 1980s, when Fletchers itself changed the trading name of its business to Placemakers, and the Odlin's and E&B names vanished from the Companies Register.

Pureora Forest Village Redesigned

Pureora village did not close down along with the timber mills, but it became a lot smaller. In 1981 there were still 21 habitable houses, of which 19 were occupied by NZFS staff plus two set aside for hunters and visiting science staff (Fig. 10.5). Attention was switching to developing the park for recreation (Chap. 14), and to salvage logging—retrieving the logs that had been left behind in clear felled areas by the wasteful logging practices of earlier times. Some totara logs had been lying in bull-dozed windrows or old skids for 30 or 40 years, and had seasoned into prime carving wood.

Pureora village continued as a much smaller and differently focused community, determined to show that it had lost its logging but kept its heart. NZFS made available 1500 ha in the former Hurakia SF (previously cut-over) to be clear felled and planted in exotics in 1984–86. The rapidly growing plantations next to the Pikiariki protest site (Fig. 10.6) also needed tending. Between them these tasks should have provided 52 NZFS staff with work in pruning and thinning until 1985. In fact, only 500 ha in Hurakia SF

Fig. 10.6 Within a few years, the young pines planted in 1978 in the last clearfelled parts of the Pikiariki Road area overshadowed the burnt remains of the native forest. *C.M. King (1983)*

was suitable for planting exotics and free of regenerating forest, and only 302 ha was planted.

Only three men still worked on selection logging in the indigenous forest. Felling was still required occasionally, as for example, after lightning struck a hollow totara and Andrew Churcher was sent to bring down the burning trunk (Fig. 10.7). The school and the Post Office survived—for a while. Likewise, the workshop remained, but, as usual, desperately short of mechanics.

The closure of the Barryville mill meant that the Pureora primary school roll immediately fell from 60 to 39. Teaching staff were reduced from four to two (plus a teacher's aide), and the pre-school unit and play-centre closed. School principal George Keown told a curious reporter [40] that the community is now "just waiting to see what the Government would do next". Surprisingly, the OC's 1978/79 Annual Report commented that the immediate effect of the closure of the Barryville mill was not as severely felt in Pureora as was anticipated, but that was not to last.

After the mill was dismantled and the last families moved away, further decline in the school roll was inevitable: by 1986 it was down to 28 pupils, and less than a year later, only nine were left [41]. Four of these were expecting to move away with their families soon.

Reluctantly, the school committee and the then sole-charge teacher, Vic O'Rourke, accepted that they could no longer carry on. The school was closed on 8th May 1987, and the last nine primary-age children living at Pureora travelled the 22-km bus journey to school in Benneydale. The school buildings and equipment, all in good condition, were redeployed elsewhere.

Dave Yanko and his wife Mauven stayed on until November 1984. The remaining community farewelled them with great regret and a beautiful writing desk made of totara, which had been grown, felled and milled at Pureora. He and many others who had once lived in the village remained in NZFS but were transferred to other areas.

Pureora general store owners Rusty and Helen Russell were very disappointed that the Government had given into "pressure groups" and decided to cut logging rates so drastically [40]. They realised that Pureora would die when the Barryville mill closed, "all for the sake of a few birds"—but had hoped that, since NZFS itself survived at first, if only in a much smaller and very different form, they could carry on.

But lack of custom and the introduction of the new Goods and Services Tax (GST) forced the closing of the store in September 1986, and on 27 November 1987 the last postmistress Kui Watene closed the post office. From then until February 1988 the villagers had to drive 20 min down the road to Benneydale for mail and banking, until the Benneydale office was also closed. After that the nearest post office was Mangakino, 30 min away in the other direction [39].

Fig. 10.7 Although virtually all felling was now prohibited, occasionally it was necessary to remove individual trees, like this totara hit by a lighting strike. Andrew Churcher was sent to remove it while it was still burning. *Crown Copyright, Department of Conservation Te Papa Atawhai (April 1982). Photographer: John Mason*

The 1987 general election held on 15 August was seen by some of those interested in events at Pureora as a referendum on the 1984–87 Labour Government's economic reforms, especially the neo-liberal ideas that had driven further forward the previous (National) Government's 1975–78 policies on native forest management. Rusty Russell wrote a letter to the *Waikato Times* a few days before election day.

> Sir - In the late 1970s the then National Government, against all promises, bowed to a minority group and stopped all logging of native bush in the Pureora Forest. Barryville village disappeared; Benneydale was seriously affected and the heart of Pureora village was broken. With many good families having to leave, the school roll was reduced from 180 pupils down to 25. Now this year the Labour Government, the working man's party, with the corporatization of the Forest Service, have really driven home the final nail. Pureora, once a self-contained, thriving village, is now reduced to five families. Most of the facilities have gone. Bush United RFC was mana in the whole area, the teams were respected all around the central North Island and beyond. Pureora Store was the focal centre and supplied the many needs of the locals for many years. The lovely school and school house have been transported to other areas. All that remains is a few sheds and a heap of rubble. I proudly remember in 1949 casting my first ever vote at the polling booth in the Pureora School, and have done so, along with many others, at each election year since. But not this time! Saturday's election can come and go. It won't mean much to some of us who are left in Pureora Forest. Rusty Russell.

One does not need to have lived in Pureora village to recognize and understand Rusty's despair. He could not be expected to look past the ruin of his personal life to see that Pureora, while certainly no longer as thriving as it had been, was developing a new and different future as a centre of outdoor recreation that would bring in thousands of enthusiastic visitors (Chap. 14).

Over the next six years, some residents took up the option of buying their houses at Government valuation and removing them to Benneydale; others were sold to other Government departments or house removal companies.

"Then we started removing the large buildings —the store, the fuel shed, the old cook house, the line of garages. We had an infrastructure here that was set up for up to 100 people, and on the first day of DOC [1 April 1987] there were just seven of us here", remembered John Mason [42]. Most of the houses in Pureora village were removed, except the last six, now used by DOC [4]. In June

2003 DOC bought the empty shop and the Russell's old mill house behind it, to be retained as historic buildings. They are now in the care of the recently established Maraeroa A and B Trust.

The relict village became almost a ghost town, haunted by the spectres of the disused workshop, abandoned house sites, and overgrown tennis courts and footie field. It became just another timber settlement, much like Barryville, Maraeroa, Ngaroma, Piropiro, Mangapehi, Tihoi, Taringamotu and many others diminished or abandoned, except that at Pureora it was not because all the timber was cut out, but because of a political decision.

Old Mrs Hepi, a kuia (elder) of the busy Maori community formerly based around the E&B sawmill at Maraeroa (Fig. 5.13) that closed in 1967, lived on in an isolated house surrounded by rapidly growing pine trees. She stayed on as a kaitiaki, a guardian and caretaker of tribal land, because according to Maori tradition, tribal rights of take tupuna (Chap. 3) to ancestral land may be extinguished if it is left unoccupied for too long. Her son Ra Hepi (Fig. 13.17) continued to live in the village and work for DOC until he retired.

Just a decade previously, a district survey carried out by the senior pupils of the school on 17 March 1971 listed 89 houses, 44 single men's huts, and a total population of 434 (112 men, 96 women and 226 children) [8].

The sudden political changes caught out the carefully planned house upgrade programme which John Gaukrodger and his management team had been operating for the previous three years. The concrete roof tiles on twelve houses had been scheduled to be upgraded to Monier glazed tiles at a cost of NZ$4000 per house. Six had been done and the next three had been started when DOC sold them all. Their new roof tiles, which had to be removed to lighten the load for transporting, ended up as road surfacing.

When the King Country Regional Management Plan was announced, the depleted remnants of the Pureora community prepared an extensive and detailed submission, which summarised the changes suffered by the villagers since 1978, and their fears of further disruptions in the near future [29]. Such fears were perfectly reasonable, and anyone can empathise with them. On the other hand, an analysis based on field interviews with 24 of the 32 workers displaced from Barryville concluded that the majority of men and their families benefitted both financially and socially from relocation to other centres; and the social services and amenities at Pureora were not affected as much as the OBDC report and many residents had predicted [7].

The Grazing Co-op

The most enduring of the community initiatives of the 1980s, run jointly with NZFS, was the Waimiha Grazing Co-op. The tidy-up work associated with the closure of the mills had created a lot of open space in Pureora village. Maintaining these grassed areas with tractor and mower was time-consuming, costly and far from efficient. Damage to tractor and mower caused by falling into redundant septic tanks was also a danger and presented some risk to the operator.

With the help of a NZ$6000 budget from NZFS, stock fencing and water troughs linked to the existing water supply were established. With the addition of other grassed areas including the rugby field and the school grounds, a total area of 20 ha was prepared for tendering. The only tender received came from the Waimiha Grazing Co-op, run by the residents of the village, and in October 1982 the first load of sheep arrived.

The Co-op worked very successfully over 15 years. The benefits it contributed to the remaining community amounted to far more than simply achieving a well-cared for appearance for the village, and ensuring supplies of good quality and relatively cheap meat to the villagers, produced by their own efforts. The Co-op also made significant profits which were returned to the community as Youth Club trips, a new movie projector and copier for the school (until it closed in 1987), re-surfaced tennis courts, support to the Country Library Service, and donations of meat for social events and tangi (funerals). Most significantly for this book, the Waimiha Co-op gave DOC the first few thousand dollars needed to employ Owen Wilkes to start researching the history of Pureora.

Postscript

For a few years after all the upheavals, some visitors to Pureora village who knew something of its turbulent history and of the past confrontations with protest groups, noticed that one road on the edge of the village was named (by NZFS staff) Stephen King Place. Dave Yanko remembers a conservationist family who came up from Wellington to see the scene of the famous events of 1978, and discovered, to their astonishment, that this road led to the local rubbish dump (Fig. 10.8). (One un-named observer working at Pureora during the moratorium called this "the most intelligent insult" he had ever heard). The family stormed back to the NZFS office in high dudgeon. The staff all disappeared and left Dave to handle it.

The sign has been taken down now, but its previous existence raises the question: Does the naming of a road after the man who led the protest action show that history was, at least for a while, set aside in favour of recognition that reasonable debate needs to honour different points of view? Hardly, because the two sides to this dispute were, as always, the victims of their own crossed purposes, so were angry for different reasons.

On the one hand, NFAC admirers resented the offence to King's visionary leadership, although Stephen himself just laughed, calling it "a good humoured expression of local frustration" [16]. NFAC saw that the end of logging was inevitable, and continuation of it to the bitter end as madness, unjustifiable in both economic and ecological terms.

On the other hand, Dave Yanko and his loggers resented the perceived abuse of their hospitality to King during the three months before the showdown, when King had made friends with them, shared meals, and hitched rides in the forest with them. They felt that they had been

Fig. 10.8 Pureora residents expressed their reactions to the successful anti-logging campaign by naming the road to the local dump "Stephen King Place". Stephen thought it was a wonderful joke. *Bob Brockie (2013)*

taken in and done over by someone who was more interested in gaining the ear of politicians than in listening to them. Many people today admire King for what he achieved. Others might ask: Did the ends justify the means?

Without ignoring the passions generated by either side, John Morton et al. [21: 107] supplied a good answer to this difficult question, as part of their eloquent defence of Whirinaki Forest—the next conservation battleground after Pureora:

> If those with power over our native forests go on pressing for exploitation, even against the best technical advice close at hand, future generations will be sure to remember these things against us. They will be at a loss to find justification for actions that even today seem pointless. If they visit the cutover remains of the present forests, to find no giant trees remaining, they will think of us as improvident people. And our logging policies… they will call plunder

An aerial view of the clear felled area behind the village gives some idea of what that plunder looked like (Fig. 10.9).

Time heals many differences of opinion, even those profoundly held. Dawson's analysis [7] showed that most people who lost their jobs in 1978 were in fact better off later. Rusty and Helen Russell retired to Rotorua, and the DOC visitor centre at Pureora now has a photograph of

Fig. 10.9 Looking SW across Pureora Village towards the clearfelled Maraeroa C block, shortly before the end of logging. The village was already smaller than in its heyday in the 1950s and 60s. *Crown Copyright, Department of Conservation Te Papa Atawhai (May 1977). Photographer unknown*

them, taken at their home. Underneath it are these words:

> We were very angry back then…but looking back now, with the benefit of hindsight, the protestors should have been there 5 years earlier. And the Forest Service should have got into selection logging earlier instead of all that clear felling and the cutover should have been allowed to regenerate instead of being planted in pines. Rusty and Helen Russell, Rotorua, February 1998.

References

1. Anderson, K. (2008). Maoriland sawmillers. Ellis & Burnand Ltd, Sawmillers and timber merchants of Mangapehi. Manurewa: Ken Anderson family.
2. Bertram, I. G. (1980). Comments on "Social and economic impact of indigenous forestry in the Pureora Area", Report prepared by Otago University, Business Development Centre, for the New Zealand Forest Service. Hamilton: Waikato University, Environmental Studies Unit.
3. Beveridge, A. E. (1979). Management of the central North Island indigenous forests (Sanderson Memorial address). Forest and Bird, 214, 25–33.
4. Burns, R. (1998). Pureora conservation park strategic plan. Hamilton: Department of Conservation, Waikato Conservancy.
5. Bush Telegraph, August 1981.
6. Dawson, D. G. (1982). Critique of studies of the kokako, forest bird populations and vegetation in some central North Island indigenous forests. Lower Hutt, New Zealand: Ecology Division, DSIR.
7. Dawson, J. A. (1979). Pureora Forest Park: Costs and benefits. MA thesis, Victoria University of Wellington.

8. Department of Conservation (unpublished). Archives. Hamilton: Department of Conservation.
9. Dominion, 24 March 1980.
10. Edmonds, A. S. (1978). Comments on limited alternatives. In N. J. Eriksen (Ed.), Critique of the BDC report on the social and economic impacts of indigenous forestry in Pureora Hamilton: University of Waikato Environmental Studies Unit.
11. Ericksen, N. J. (Ed.) (1980). Critique of the BDC report on the social and economic impacts of indigenous forestry in Pureora. Hamilton: University of Waikato, Environmental Studies Unit.
12. Groome, J. G. (1978). The role of Pureora sawmill in the King Country economy. Taupo: J G Groome and Associates.
13. Hay, J. R. (1981). The kokako. Wellington: New Zealand Wildlife Service, Forest Bird Research Group.
14. Higham, J. R. S., & Menzies, P. (1978). Social and economic impact of indigenous forestry in the Pureora area. Dunedin: Otago Business Development Centre.
15. http://www.stpauls.school.nz/page/tihoi-history.
16. King, S. (2013). Personal communication.
17. King Country Chronicle, 15 August 1978.
18. King Country Chronicle, 2 June 1978.
19. King Country Chronicle, 4 July 1978.
20. King Country Chronicle, 19 December 1978.
21. Morton, J., Ogden, J., & Hughes, T. (1984). To save a forest: Whirinaki. Auckland: Bateman.
22. Native Forests Action Council (1981). Labour will save Whirinaki and Pureora. Bush Telegraph, 4, 1.
23. Native Forests Action Council (1981). $7 million for what? Bush Telegraph, 2, 1.
24. New Zealand Forest Service (1978). Submissions on the West Taupo state forests. Wellington: New Zealand Forest Service.
25. New Zealand Forest Service (1981). Whirinaki state forest park management plan. Rotorua: New Zealand Forest Service.
26. New Zealand Herald, 16 September 2012.
27. New Zealand Herald, 30 December 1978.
28. Parker, S. (2000). Cutting edge: The Carter-Holt-Harvey story. Auckland: Viking Penguin.
29. Pureora Community Council (1981). Submission to King Country regional management plan. Pureora: Pureora Community Council.
30. Simpson, T. E. (1973). Kauri to radiata. Auckland: Hodder & Stoughton.
31. Smith, M. W. M. (1953). Some aspects of family and community life in a New Zealand saw milling village. Diploma of Social Science, Victoria University College.
32. South Waikato News, 29 May 1979.
33. Sunday Times [NZ], 28 May 1978.
34. Tregoweth, K. (2013). Personal communication.
35. Vaney, J. C., & Gibson, L. J. (1978). Current and potential use of timber from the state forests of the Hauhungaroa and Rangitoto ranges [*Transcript of the Taupo seminar*]. Wellington: New Zealand Forest Service.
36. Waihaha Forest Preservation Committee (1978). Waihaha; A submission to the Minister of Forests in response to NZ forest service management proposal for the West Taupo forests, 29 May 1978.
37. Waikato Times, 11 July 1978.
38. Waikato Times, 16 December 1978.
39. Waikato Times, 28 November 1987.
40. Waikato Times, 9 August 1978.
41. Waitomo News, 5 May 1987.
42. Wilkes, O. (2005). Transcripts of oral history recordings. Hamilton: Department of Conservation.
43. Wilkes, O., Gaukrodger, J., & Ritchie, N. (2011). Chronology of Pureora. Hamilton: Department of Conservation.
44. Wilson, R. (1982). From Manapouri to Aramoana: The battle for New Zealand's environment. Auckland: Earthworks Press.
45. Wright, L. W. (1980). Decision making and the logging industry: An example from New Zealand. Biological Conservation, 18, 101–115.

11 Rethinking Forest Administration, 1978–87

D.J. Gaukrodger and C.M. King

Abstract

This chapter explains the statutes under which Pureora Forest Park (PFP) was established, its early management policies, and the creation of its Ecological Areas. It describes the traumatic disbanding of the New Zealand Forest Service (NZFS), once among the most powerful of NZ Government departments, and the reorganisation of all conservation activities formerly undertaken by NZFS, NZ Wildlife Service (NZWS), the Department of Lands and Survey, and the Historic Places Trust into a new Department of Conservation (DOC). The Ministerial Brief from NZFS to DOC accurately predicted many future problems.

Keywords

Institutional schizophrenia in NZFS · Economic restructuring · Rogernomics · End of NZ Forest Service · King Country Regional Management Plan · Establishment of Department of Conservation · Pureora Conservation Park · Ecological reserves · B9B dispute · Titiraupenga

Institutional Schizophrenia

NZFS staff have suffered a lot of criticism from conservation activists over the years, some of it well deserved, but it is also important to point out that some of it was uninformed and quite unfair.

In fact, long-term conservation policy was already being discussed from the very beginning of NZFS, long before the public battles of 1978 (Chap. 6). The Pureora project itself stemmed from that concern, as far back as the 1940s (Chap. 7). That did not stop things getting much worse between the start of the moratorium of 1978 and the end of the line for NZFS in 1987.

The irony of this story lies in the fact that neither of the two sides in the long-running controversy centred on Pureora used the word "*conservation*" with the same meaning. To NZFS it meant "*reservation for future wise use*": to NFAC and their allies it meant "*preservation to prevent any future destructive use whatever*".

D.J. Gaukrodger
3a Te Angi Street, Turua RD4, Thames 3574, New Zealand
e-mail: beejaygauk@xtra.co.nz

C.M. King (✉)
School of Science, University of Waikato, Private Bag 3105, Hamilton 3240, New Zealand
e-mail: cmking@waikato.ac.nz

C.M. King et al. (eds.), *The Drama of Conservation*,
DOI 10.1007/978-3-319-18410-4_11

Each understood their position to be the right one, occupying the moral high ground, which made the arguments doubly painful for both. It took a radical political decision to end them. In 1986, the politicians came up with (or agreed to) a more radical decision than anyone had anticipated.

It was always one of the oddities of an institution that had been developing over such a long time, that it had acquired responsibility for both the commercial use and the conservation of native forests simultaneously [26]. For example, during the postwar boom in house-building and industrial construction, native forest supplied almost all the huge national demand for timber, and logging companies pressured NZFS to release more and more areas for logging. The structural framework, flooring, doors and window frames of nearly all New Zealand houses built before about 1960 were made of hard native timbers, and are now valuable resources for demolition contractors.

With an eye to conserving stocks for the future, NZFS had resisted this accelerating demand long before any external conservation lobbies took an interest (Chap. 6). Stung by later criticism, NZFS pointed out to conservationists that it was because of this resistance that large areas of unlogged forest still remained in the west Taupo region as late as 1978 [14]. Colin Bassett, then head of FRI, attempted to put the record straight by listing NZFS' achievements: reducing wasteful logging practice, improving management, replacing native with exotic harvests, and reserving nearly 400,000 ha of ecological reserves up to April 1986 [1]. Activists and reporters who did not know that history (most of them) failed to take the point.

Being responsible for two largely contradictory functions sometimes caused conflicts within the ranks, especially after NZFS acknowledged the changing tenor of the times by creating in 1971 an Environmental Section within the Forest Management Division of Head Office (Chap. 6). "The schizophrenic nature of NZFS at this juncture was impressive" commented Wright [32]. Seldom mentioned was the advantage that, at the same time, some commercial revenues could also quietly be diverted to support conservation purposes, for example, by contributing to a five-year survey of small mammals at Pureora [12].

On the other hand, NZFS also sometimes made decisions which were very hard to defend. In 1984/85, two separate blocks of forest totalling around 200 ha in the old Waituhi SF, at the southern end of PFP off SH 41 (Fig. 9.7), had been identified as sites for conversion to exotics. The two blocks for potential conversion were linked by an access road through a high altitude area supporting significant numbers of Hall's totara. The plan was to convert the two blocks but to leave the intervening area, not only to save the Hall's totara but also because NZFS policy limited planting of radiata pines at high altitude. The plan was approved and the work was handed over to the NZFS office at Turangi.

The local office then made an unauthorised decision to extend the project area by including within it the high altitude area along the linking road. Some 50 ha of the supposedly excluded forest was cleared by bulldozer in 1985, Hall's totara and all. When environmentalist groups discovered this and asked why, they were told that the bulldozer driver had made a mistake.

But, as John Mason pointed out, it wasn't as if he was there for a day. He was there for weeks, even though it was contrary to the management plan and simply wasn't meant to happen. This incident is sometimes referred to as "the last nail in the NZFS coffin", although there were other incidents that would fit the title just as well. It and others like it were certainly among the factors, along with the associated intransigence of some senior managers of NZFS, that helped to swing the scales against NZFS when the big decisions were being made in the following year.

John Gaukrodger, one of the senior faces of NZFS but not responsible for the mess, inevitably shared in the consequent caning, and got the job of cleaning up. He organised a plan to recover the Hall's totara logs and disperse them to interested local Maori. The then Minister of both Maori Affairs and Forests, Koro Wetere, inspected the area, along with two master carvers [28]. Wetere approved the plan, so in July 1986 a group of carvers from the Kirikiriroa Marae in Hamilton, led by kaumatua Wikuki Kingi,

Fig. 11.1 A traditional-style carving of Te Kanawa, a chief of Ngati Maniapoto (Chap. 14), at the entrance to the Timber Trail near Pureora village. The art of carving remains a significant part of Maori culture. *Greg Martin (2013)*

arrived at Waituhi (including the son of Inia Te Wiata, the carver who created the pouihi in New Zealand House, London, from a totara log sent from Pureora: Fig. 8.6). The group spent a couple of months pulling out and testing selected logs, and trucking the best ones to five marae in the Tainui region for carving.

Carving has always been an important skill for Maori, and several carving schools have been established to train young carvers to supply the growing demand for new carvings executed in traditional style, like the one shown in Fig. 11.1. Since the end of large-scale logging in podocarp forests, supplies of prime carving timber had been limited, so the timing of this incident was fortunate in one respect. For a large-scale project like the new Waikato Museum, then being built in Hamilton (opened in 1987), the Waituhi salvage operation was an unexpected bonus. About 20 logs were needed to carve maihi (eaves), poupou (carved wall ribs), heke (roof ribs) and a 14.6 m long tahu (ridge pole) for the outside and inside of a special gallery designed to house Te Winika, a 200-year-old carved waka taua (Maori war canoe) of Ngati Tipa, Ngati Mahanga and Ngati Maru collaboration. It was gifted to Waikato Museum by the Maori Queen, Te Arikinui Dame Te Atairangikahu in 1973, as a gesture of fellowship and goodwill towards the city of Hamilton, so it needed very special housing. The carving work was shared between carvers at the marae of Tamaki, Taumarunui and Maramarua, and women from these marae wove the tukutuku panels for the gallery walls [27]. Totara salvaged from Waituhi supplied much of the timber needed.

The Waituhi incident was not the first example of strained relationships between foresters and environmentalists: NZFS staff had for years been trying to respond to public criticism, stating that "it is unlikely that NZFS will fell any more state-owned forest that could be inhabited by kokako" [10]. Many individual NZFS staff were intensely interested in the birds, and did their best to protect them.

Tony Beveridge relates how he observed flocks of kakariki feeding along the edge of a dense stand of podocarp forest then being clear-felled and planted by NZ Forest Products (a private company) in the Maraeroa block near Delaney's Corner (Fig. 14.3). It was also the site of the northernmost population then known of a threatened tree species, Turner's kohuhu, still common further south in PFP in Whenuakura Clearing. Tony wrote to NZFP asking them to save the small area of Maraeroa important to the kakariki, and got a reply saying that it would be left, as NZFP were "ecologically conscious". The area was later cleared by tractor

In fact the shifts within NZFS towards greater concern for forest wildlife and greater openness in decision-making go back to the mid 1960s (Chap. 6). But by the time the legislation was changed to allow more public input and debate, NZFS had lost a lot of credibility. One of the repeated criticisms of NZFS during the Pureora campaign and others, such as the West Coast Beech Scheme, was that NZFS was secretive and biased, denying public access to important information whilst sharing it with timber companies [32].

So, of necessity, the focus of attention for NZFS staff changed considerably through the 1980s, as emphasis shifted from management of native forest for wood production to the maintenance of healthy forest and wildlife. Tony Beveridge concluded his retirement address to a seminar at Pureora in November 1985 with the comment that:

> The objectives of ecological research must be to understand and explain natural processes in the forests and determine effective methods of managing these forests for all their values, probably with wood production as a minor one [2].

For NZFS staff not especially interested in wildlife, this change was not always made willingly. For those who still lived in Pureora village, after years of isolation and constant striving to increase production of native timber, the changes were hard to grasp. They disliked all the publicity and the threat to their livelihoods, so were understandably defensive and wary of newcomers.

The 1978 protest action was past, but the hard work for NZFS was picking up the people who had survived it, and who were now facing different challenges in their task of preparing Pureora village for its new role as the publicly accessible centre of a recreational and conservation park. John Gaukrodger and his team put their backs into it—the place and the people had always had spirit. The traumas may have got them down in 1978, but they were not out. They could not know that, after the 1984 change of Government, all previous debate on the future of NZFS and kindred organisations would be swept aside by a much deeper sea-change in the direction of national economic policy. Campaign promises (Chap. 9) were called in, and radical changes were on the horizon.

The New Economics

The new Labour Government was under pressure to make good on a number of promises made to the environmental lobby during the sudden 1984 election campaign, and one of them was to organise profound changes in the administration of New Zealand's remaining forest resources. The process was led by economic fundamentalist Labour MPs Roger Douglas, David Caygill and Richard Prebble. They proposed a structural adjustment programme requiring transfer of all the natural resources of the public estate, including land, forests and minerals to business corporations, known as State-owned Enterprises (SOEs). The idea (nicknamed "Rogernomics") was untried anywhere else in the world, and it was implemented without regard for social consequences in a process later described by Jane Kelsey as *The New Zealand Experiment* [11].

NFAC was at first delighted: "[Labour's] first Budget is based on a line of thinking that could be good news for the environment. The sweeping removal of subsidies for the use of natural resources…seems likely to take the pressure off many threatened habitats" [16]. But, as Kelsey explains, NFAC was deceived by a political masterstroke:

> The public service was widely perceived as inefficient, privileged and in need of a good shake-up, and no alternative models were being promoted. Few outside the state sector understood enough detail to challenge the logic of corporatisation or to foresee its impact. The restructuring of departments like lands and forests was linked to environmental reforms, turning potential critics into useful allies [11: 119].

Bob Brockie's cartoon showing Rob Muldoon dressed as a bushman, stomping defiantly away from a devastated forest, with his trademark wry grin and an axe in his hand (Fig. 11.2), illustrated the relief many conservationists felt at seeing the end of Muldoon's era. Brockie was well-informed on the subject, since he had been a staff scientist with DSIR Ecology Division during the height of the beech logging controversy. Amid the general rejoicing, the darker consequences of corporatisation remained well out of sight.

NZFS senior managers put up stiff resistance to the proposed corporatisation programme and its consequences for public service employees. "We are not severed at the stump yet", wrote Andy Kirkland to his staff [4]. But they could not win. Jane Kelsey explains why:

> Public service unions and workers mobilised against the State Sector Bill, but their protests made no difference. Their arguments and concerns were treated as special pleadings of a protected

Fig. 11.2 The departure of former Prime Minister Robert Muldoon was hailed with relief by environmentalists. *Bob Brockie (1984)*

> workforce who were afraid of being held to account for their performance [11: 177].

The proposals were even more damaging to Maori interests, and they put up real resistance, but in the end they could not change the outcome either.

> The government had promised to address long-standing Treaty of Waitangi grievances relating to control of natural resources such as land, forests and minerals. The government now proposed to transfer those resources to the SOEs. This gave rise to a stream of litigation aimed at stopping the corporatisation, and later privatisation, programme. Maori secured promises, set down in legislation, that any SOE lands over which the Treaty Tribunal upheld a claim would be returned to their tribal owners if the Tribunal so ordered. But the tribes still did not own the land, and there was no guarantee that they ever would. By 1995 the Tribunal had never exercised that power [11: 119] ... Affected tribes were furious that the sale [of State Forestry interests] had proceeded while their claims to forest land were outstanding [11: 374].

The End of NZFS

So, at the end of March 1987, amidst the anguished disbelief of its staff, sympathisers and Maori communities, the New Zealand Forest Service was dis-established with the loss of about 3,000 jobs. In its place, the NZ Forestry Corporation, the Ministry of Forestry and the Department of Conservation were established using some of those same staff.

John Gaukrodger had to manage the reactions of Pureora people, when they discovered that only two of the silvicultural group of about 30 staff employed by NZFS were offered jobs in the private sector (and those two were soon withdrawn). In effect, commented John Mason his astonishment still audible when interviewed in April 1998, the team went from 30 to absolutely nobody, all within the space of one day. That was largely because:

> Forestcorp eliminated almost all wage workers. Staff dropped from 7070 in 1987 to 2597 in 1989, and after tax profit more than doubled in the 2 years 1988–90, to NZ$138 million [11: 123].

By the second to last day, the crews had already had their farewell party, and there was nothing more to be said or done, so they were given that last day off. John Gaukrodger made sure that valuable equipment stored at Pureora village was located in various safe places, so not a single item was lost.

Graham Miller, the programme manager for tree crops, had invariably maintained a good output from his team, and he worked to the very end of the very last day. NZFS was blessed with many dedicated, loyal, "salt-of-the-earth" employees like him at all levels, and it was an organisation that John Gaukrodger was proud to have been part of. John wrote in his diary:

> In the words of Aldo Leopold, "the land is a community to which everything belongs". The people of Pureora Forest were an integral part of that land area. Their uprooting and removal was no different from the logging of the giant podocarp trees that had gone on for over 40 years previously. The breaking down of their spirit, pride and a sense of belonging easily matched that of a breaking-down saw dismembering logs extracted from the forest to which both people and trees belonged. Government policy had no favourites; first forests then people, both got dealt the same hand…it was their strength and character [both the men and the women] that ultimately enabled some families to move on and re-start their lives in a different world. Sadly, over the next 5 years there were several funerals of worthy, wonderful people that the world suddenly didn't have a need for. A way of life built around an industry and an organisation all died together.

Alan Familton, the last DG of NZFS, wrote a brave, stiff-upper-lip letter to every member of his staff individually, dated 27 March 1987. It is reproduced in facsimile, complete with NZFS crest over the Wellington Head Office address, at the end of a collection of NZFS memories edited by Dennis Harris [8: 187].

> This is the last official communication that will ever be written by a DG of NZFS…What past and present employees have achieved is truly remarkable…we have created a first class resource of commercial plantations that will endure as a renewable natural resource providing employment, revenue and…regional development and industrial expansion for generations to come…we have established a system of State Forest Parks for recreation and conservation…we have developed a magnificent FRI which has achieved world renown…we all now go our separate ways but I trust that, like myself, you each do so proud of the legacy we now bequeath to our successors.

Alas, Familton's confidence in NZFS's legacy of regional development and industrial expansion was overtaken by later political decisions that drove him and other former DGs of NZFS to despair [30].

It could be argued that the fate of NZFS represented the demise of the idea of multiple-use forest management, as NZFS had developed it over the previous 20 years. Disbanding NZFS represented a shift towards the assumption that the same range of benefits offered by multiple use organised by NZFS could be achieved more efficiently by a series of separate, mostly commercially-orientated organisations, each with more narrowly defined objectives.

Economic efficiency and accountability were certainly the favourite words of the 1984 Labour Government, especially the Minister of Finance Roger Douglas. So it may be that the successful critique of NZFS by the environmental movement was supplemented, more than is usually acknowledged, by independent shifts in political thinking that happened to favour single-goal organisations, including non-commercial agencies such as DOC [26: 430]. The local consequences for conservation of New Zealand's grand economic experiment [11] are working themselves out in Pureora Forest Park, in telling detail—and not always as expected.

The *Times* duly reported the critical moment on 31 March 1987, complete with a picture of John Gaukrodger looking sad but determined, on his last day in his uniform jersey bearing a familiar crest. NFAC welcomed this "historic decision" [5] and acknowledged the help of Labour politician Richard Prebble in achieving it. NFAC perhaps did not realise at the time that Prebble was one of the inner circle of economic fundamentalists responsible for pushing through Douglas' structural adjustment programme.

The Ministerial Brief

The extensive and expensive NZFS plan for the King Country could not be fully implemented before 1987, but it did at least provide the incoming new managers a useful historical overview of changing policies and events up to 1984.

The wheels of national government grind slowly and exceedingly small, and nowhere smaller than in the official documents surrounding the creation of a new Government Department. So in 1986 the Ministerial Brief, laid out for the new Minister of Conservation due to take over in April 1987, included many revealing details concerning the transfer of the powers and responsibilities from the old NZFS to the shiny new offices of DOC.

The document itself had a national focus, and many of the pressing issues it identified concerned places other than PFP, but it included some interesting details illustrating the modest place of PFP in the grand scheme of worries about to be inherited by the new Department.

The Preface started with the statistics, such as that, on 1 April 1987, NZFS was to transfer to DOC some 3 million hectares of State Forest land, which would comprise 38 % of the new DOC estate. Furthermore, NZFS would transfer the administration of the 1.8 million ha State Forest Parks network, plus responsibility for the control and management of native wildlife and game birds, recreational fishing, and wild animals in New Zealand.

DOC was to become responsible for most of the natural lands and forests of the Crown, the headwaters of all the major river systems, the principal areas of interest for mining, and important resources of scenic, recreational and tourist value including 600 huts, 17 visitor centres, 6000 km of walking tracks, 10 Recreational Hunting Areas totalling 290,000 ha, 14 Forest Sanctuaries (16,000 ha) and 48 Ecological Areas (160,000 ha).

Accompanying the transfer would be about 760 staff and wage workers, who were to comprise 40 % of the staff resources of the new Department. The author of the document added, with a massive understatement, that DOC will have to be, of necessity, a multiple-use land management agency.

NZFS had already been badly bruised by its experience over many years of managing the conflicts of interest involved in multiple-use land management, and 40 % of DOC staff on day one were going to remember that well, so one can almost hear the anonymous author add "and good luck to you". All the more credit, then, that the last three lines of this Preface read:

> The formation of this new department offers an exciting and unprecedented opportunity to develop a fresh approach to conservation management by the Crown. Forest Service staff are eager to take their part in this development [20].

It was a massive disruption to the normal work of such a long-established government agency, with its nine divisions at Head Office and its seven regional Conservancies plus the two campuses of the Forest Research Institute in Rotorua and Christchurch.

The document went on to list the staff, legislation, budget (NZ$13.4 million of protection, recreation and amenity in State Forests in 1985/86, excluding NZ$3 million for West Coast compensation packages) and the convoluted environmental policies of NZFS as they then stood.

It helpfully identified the key issues that DOC would have to tackle as soon as it got organised. Among them were the following.

1. As of 1986, national policies for Ecological Areas (EAs) were still being developed by NZFS, and it did not intend to release the final document for public consultation. Among them, of course, were several very important EAs in PFP for which the public had fought, so (read the subtext) DOC staff taking over control of PFP could expect some robust debate on that topic.
2. Equally significant for PFP was the warning that wild animal populations were spreading (especially feral goats escaping or being deliberately released from failed mohair farms: Chap. 12), and that Government attempts to control them were inadequate, resulting in accelerating damage to native vegetation. The termination of NZFS's long series of well-funded monitoring surveys would make official oversight of this process more difficult than ever, if not impossible.
3. The increasing problem of wise management of recreational hunting on public lands (including particularly the very popular Recreational Hunting Area (RHA) in the Pureora North Block: Chap 14), would be dogged with potential conflicts. The different aims of hunters demanding free and easy access to deer (a property of the Crown) as of right, and DOC managers who were suddenly expected to apply a user-pays concept in their management of Crown land, was predicted to be a continuing source of difficulties. As it turned out, DOC found it impractical to charge for permits that had formerly been free, but a compromise system of user-pays concessions was eventually established for commercial operators.
4. The briefing document gave the challenge of preserving the kokako a section to itself, because the large-scale possum poisoning operations in the forests of the King Country, which were needed to prevent the spread of bovine TB to the Waikato (Chap. 12), were certain to cause potential conflict with the equally urgent need to avoid accidentally poisoning kokako. Monitoring and control programmes against rats and possums to

protect kokako in the Pikiariki block had been going on since long before the end of NZFS (Chap. 13), and DOC would need to find ways to continue them.

5. The accelerating demand from school and community groups for educational facilities in forests were predicted to create an increasing workload for PFP staff responsible for managing Pureora Forest Lodge and supplying educational materials (maps, guides) to its users.

All these issues certainly did dog the early days of DOC's management of PFP, and most still do.

Pureora as a State Forest Park, 1978–87

Pureora Forest had originally been gazetted by NZFS as State Forest 96 in 1935, with the intention that almost all areas suitable for logging should eventually be clearfelled and converted to exotic plantations. At that time, Pureora SF 96 was much smaller than the area that later became known as Pureora Forest Park.

NZFS had begun to convert suitable State Forests into State Forest Parks in the late 1960s (Chap. 6), zoned so as to allow for protection areas and for public access for recreation into much of what native forest remained in them, but also to allow strictly controlled selection logging and production of exotic timber from the rest [26].

The idea of designating a Forest Park in the Rangitoto-Hauhungaroa Ranges had been suggested by the Otorohanga Rotary Club in 1973, and again in 1974 by the NZ Deerstalkers Association [17]. In July 1974, representatives of RFBPS visited the Pureora area, and later made a formal request to the Minister of Forests that all essential future logging should be carried out by NZFS (rather than private contractors), and that no further stands of native timber be transferred to private control [9]. Because NZFS was already inching towards the same idea, it was not difficult for NZFS to announce in 1975, in a submission to the Commissioner for the Environment [17], its intention to designate a State Forest Park in the Rangitoto-Hauhungaroa Ranges, by proclamation under Section 63A of the Forests Act 1949.

So Pureora Forest Park (PFP) was formally created in 1978. The remaining stands of native and exotic forests that had been previously managed by three separate NZFS Conservancies (Fig. 9.7) were cobbled together to form a single unit, administered from Te Kuiti by a single authority, the NZFS Auckland Conservancy [17]. PFP comprised six former State Forests (Pureora, Hurakia, Tihoi and Crown land, Wharepuhunga, Taringamotu and Waituhi) and was 80,313 ha in area, including all the last large untouched stands and the areas recovering from selection logging trials (Fig. 11.3). NZFS declined a recommendation to include a seventh area, Okahukura SF 58, but it was later designated as stewardship land, along with the adjacent Cowan block.

Calls for nominations for election to the new PFP Advisory Committee appeared in newspapers in August 1978, closing on 22 September. Sixty-two nominations were received for the nine positions other than chairman, and the successful candidates were announced by the Minister of Forests in March 1979. The inaugural meeting was held on 1 May 1979. The elected members represented a wide range of interests and skills: one company director, one university lecturer, one solicitor, one forester, one electrical fitter, two retired and two farmers (Box 11.1).

Box 11.1. Organisations represented on the PFP Advisory Committee (as at 20 March 1979) [17]. Membership has changed over subsequent years, but these organisations are usually represented.

Chair	Mr. G.J. Molloy, Conservator of Forests, NZFS Auckland	
Members	Nomination organisation	Location
Mr. M.W. Bellfield	Federated Mountain Clubs of NZ Inc	Tokoroa
Mr. J.A. Church	NZ Institute of Foresters	Tokoroa

(continued)

Chair	Mr. G.J. Molloy, Conservator of Forests, NZFS Auckland	
Members	Nomination organisation	Location
Dr. A.S. Edmonds	South Auckland Conservation Association	Hamilton
Mr. L. Goldsbury	Northern King Country Farm Forestry Association	Te Kuiti
Mr. J.V. Jerram	Royal Forest and Bird Protection Society	Taupo
Mr A.R. Meredith	Waitomo District Council	Te Kuiti
Mr. C. J. Peacocke	Waitomo District Council	Pureora
Mr. F.L. Phillips	Rotary Club	Otorohanga
Mr. R.A. Stuart	NZ Deerstalkers Association	Mangakino

Throughout the subsequent changes in staff structure and duties, from 1978/79 right up to the end of NZFS in 1987, the NZFS OCs based at the Pureora office faithfully continued keeping the regular station diary. John Gaukrodger had been the man in this hot seat since 1981, when he was sent there by NZFS Director-General, Mick O'Neill, with the terse instruction to: "clean the b—place up, it's meant to be a Forest Park, now get down there and make it one".

Gaukrodger set about tidying up the village and restoring its former self-respect, now that the protestors had gone and the mills were closed. He was supported in this daunting task by Dave Yanko (logging officer), Mike Diamond (tree crop officer), John Mason (environmental ranger), Bruce Tricklebank (mechanical overseer) and another 6–7 staff and around 55 wage workers. The workshop team numbered 5, the roading unit 5, the logging gang 3, builders 2, maintenance 2, and the rest in silviculture, tree crop and environmental work. They were a well-resourced and self-contained operation then, commented John.

Meanwhile, the Te Kuiti office of NZFS (led by Rob Guest) prepared a document for public consultation, addressing the issues raised during the debate following the moratorium. It outlined, amongst other policies, a range of options for the level of indigenous timber production in PFP, and also options for the degree of conversion to exotic plantation in 1500 ha of Hurakia SF. The document was published as The Draft King Country Regional Management Plan in late 1980, and sent out for public comment. It generated over 1500 submissions from various

Fig. 11.3 The 1961 selection harvesting trial area, 36 years after harvesting. Few of the emergent podocarps valuable to kokako (Fig. 13.9) remain. *M. Smale (1997)*

authorities, organisations and individuals, which Rob Guest then analysed into a summary report.

In late 1981, the Pureora Forest Park Advisory Committee considered Guest's summary, together with the report from the Forest Bird Advisory Group describing their three-year study on the kokako (Chap. 13). The Committee made its recommendations on future management to the Conservator in Auckland at the time, Gavin Molloy. All this information, together with the outcome of other research on silviculture, was then compiled into a policy for the future which was approved by the Minister in 1982 [18].

Unfortunately, the plan for PFP was written at a time of great uncertainty about the future direction of native forest policy, so although some work was done, many of its prescriptions soon became obsolete. It envisaged indigenous timber production only in one specific area, and only in exceptional circumstances. It outlined six criteria to be met before any felling of native trees would be permitted, but acknowledged that the criteria could not be met at that time in the King Country. It added that only 400–900 ha of Hurakia was suitable for conversion to exotic plantations.

Nevertheless, this policy was integrated into the King Country Regional Management Plan 1983–1993, which was published by the Auckland office of NZFS [19]. It was signed off by the Minister of Forests, Jonathon Elworthy in 1984.

Within days of signing off the Plan, Jonathon Elworthy was out of office, because a snap election, a gamble called and lost by the increasingly unpopular PM Robert Muldoon, swept the National party out of power, and a new Labour Government took over in July 1984.

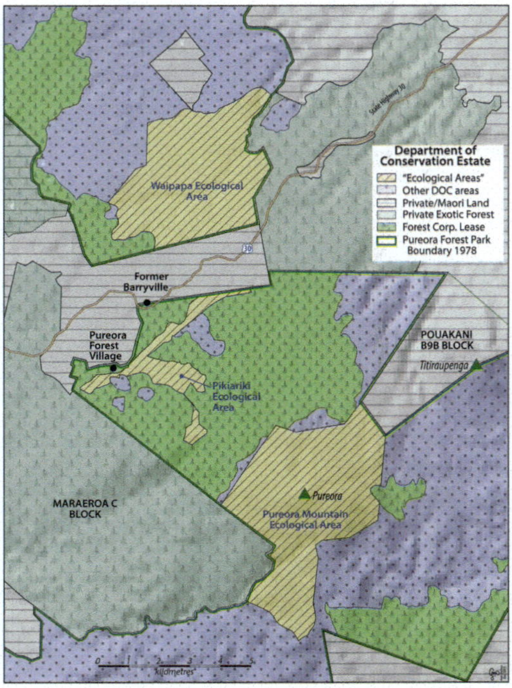

Fig. 11.4 During the transition from NZFS to DOC in 1987, Pureora State Forest Park as it had been in 1978 (bounded by heavy line) was broken up. Nearly all those areas already planted in pines, which had previously been within the park boundary, were passed to the Forestry Corporation. *Redrawn by Max Oulton from Waitangi Tribunal (1993: fig. 15.2)*

The Titiraupenga Track and the B9B Dispute

The new park could not avoid an uneven outline, because the separate bits didn't always join up very well—and some outlier blocks don't join the main area at all. It also enclosed several odd-shaped "holes"—lands which are not part of PFP because they are Maori blocks which for various reasons had never been sold to or taken by the Crown.

These anomalies look puzzling, until you know the history. One of the most puzzling of these is the irrational bite out of the boundary in the northeast corner of the park (Fig. 11.4). There is an old and sad reason for this.

The summit and the northern slope of Mt Titiraupenga lie just outside the present boundary of PFP, in the Pouakani Maori Land Block B9B (1076 ha). Both were expected to become part of PFP, as the block was considered to have been Crown land since the 1890s. At that time, the Crown had bought out a large number of individual Maori owners of land in this block (Fig. 5.2), including 140 acres transferred to the Crown in lieu of survey charges. But the surveys were not correctly done at the time, and the associated Native Land Court orders were later proved invalid.

The former owners of the B9B block researched the disputed boundaries, and laid a claim before the Waitangi Tribunal, which was heard at about the same time as Pureora Forest Park was being divided up into Production and Protection areas and handed over to DOC.

The session of the Waitangi Tribunal hearing the Pouakani claim concluded that the Crown should refund the reasonable expenses incurred by Maori in preparing the claim, and return the land previously taken in payment of survey charges [29: 86–8]. It recommended that title to the mountain should be returned to Maori ownership, vested in trustees to be appointed by the Maori Land Court, and that B9B be managed as part of Pureora Forest Park as a Maori Reservation under s439 of the Maori Affairs Act 1953. In due course, the Pouakani Claims Settlement Act of 2000 did indeed return the land to Maori [33].

DOC was keen to pursue the Tribunal's recommendation that the B9B block be managed as part of PFP, because it included a natural part of the park landscape, the northern slope of Titiraupenga. Since DOC could not afford to buy it back, the simplest solution would be to negotiate an agreement by which the Government would sell to the Titiraupenga Trust a nearby block of 942 ha of Crown land in exchange for adding B9B to PFP at an annual rent. Negotiators for the Titiraupenga Trust and the Minister for the Environment, Ian Shearer, agreed to this in principle.

The problem was that the two parties had very different estimates of how much the rent should be. So when they failed to reach an access agreement, the idea was cancelled, and the summit of Titiraupenga was declared Maori land off limits to the public. Modern maps still show the summit as outside the boundary, and some of the bush-clad northern slope has since been cleared for conversion to pasture.

DOC and its supporters were disappointed by this outcome. But, while regretting the loss of the forest, NFAC commented:

> There can be no suggestion that the Maori landowners are not free to do with their land what Pakeha owners have done with theirs. …Maori leaders are generally sympathetic to conservation ideals, which have deep roots in Maoridom But they are also deeply concerned by the lack of job opportunities in rural areas [4].

Like B9B, there are other formerly forested blocks that were not available to be included in PFP, and whose absence contributes to its apparently irrational boundaries. The forest has long gone from the swathe of farmland that separates the North Block from the rest, and from the straight-edged expanse of exotic forest on Maraeroa C block adjacent to PFP's western boundary. Both were once thick with dense podocarp forests held under private or Maori ownership, but their logging rights had been allocated to timber companies before 1939 (Chap. 5).

The old boundary disputes were reignited in December 1986 when NZFS staff working on a search and rescue mission for a missing hunter discovered that a logging operation on the B9B block was encroaching on to PFP land. A long series of discussions and protests, before and after the transfer of territorial responsibilities from NZFS to DOC in April 1987, and an official resurvey of the boundary line, had no effect until November 1987. Then a Maori Land Court hearing imposed a Preservation Order on both sides of the disputed area and halted all logging activity.

Peace was restored until 27 April 1988, when five more trees were removed. On the same day, the Pureora OC (John Gaukrodger) assembled a team including a DOC dozer driver and a police officer, and went to the site, where they found two bushmen in the process of loading another log from across the boundary. After the necessary formalities, the illegally operating dozer was confiscated and driven to the DOC HQ at Pureora. It remained there for almost a year, and was then transported back to the B9B block. No more trees were removed.

During this episode, DOC estimated that 152 trees (mostly mature rimu, measuring some 1400 m^3 and worth NZ$42,000, at NZ$30/m^3) were taken, but no compensation or prosecution followed. The problems arising from faulty boundary definitions and property ownership disputes in Cussen's time (Fig. 5.2) cast shadows a century long.

Pureora as a Conservation Park, Since 1987

Because Pureora Forest Park had been established by NZFS, it is still technically a Forest Park rather than a Conservation Park. The difference is that controlled logging is allowed in Forest Parks, but normally prohibited in Conservation Parks except for salvage under permit. In practice, PFP is a Conservation Park.

Under the new administration starting in 1987, PFP fell into the Waitomo District of DOC (later renamed the Maniapoto District) headed by the last NZFS OC, John Gaukrodger. His office was in Te Kuiti, because he was also responsible for Whareorino Forest, the Mokau River and Waitomo Reserves, and several smaller protected forest areas. To help him manage such a vast area he had 11 staff and 18 wage workers and hunters. Day-to-day management at Pureora was in the hands of John Mason, who also stayed on with the new title of conservation officer.

Almost all the exotic areas were excised from the State Forest Park and transferred to the newly created state-owned NZ Forestry Corporation to harvest under one or more Crown Forest Licences (Fig. 11.4). The removal of the exotic plantations shrank the forest park, and made its outline even more ragged than it was before. The new Department of Conservation took responsibility for protecting the remainder.

Licences for the exotic blocks at Tihoi and Waituhi were long-term lease agreements requiring replanting after harvest, the same system earlier established at Maraeroa C (Fig. 5.14). The licences were held by various owners to start with, but most were eventually sold on to private, often foreign-owned companies.

By contrast, some other licences were short-term "cut out and get out" harvesting, of the existing crop only, with no obligation to replant or to control weeds. The difference was due to another protest by Stephen King and his lobby group, the New Zealand Native Forests Restoration Trust.

King and the Trust argued strongly against the Government's massive sale of State forestry assets. In January 1988, King celebrated the success of his first tree-sitting campaign ten years earlier by returning to his platform, this time with a typewriter. He wrote a letter from there to Helen Clark, Minister of Conservation, and PM David Lange, pleading that some at least of the plantations should be withheld from sale, in the hope of preserving future options for replacing what had been destroyed. He advocated that all of the exotic plantings be managed to harvest the existing crop only, and thereafter be managed to ensure a gradual transition back to native forest.

The Government did not agree with this strategy, because at that time, the idea of converting "productive" pine plantations and farms back to native forest was considered heretical, and most of the sales proceeded. However, the Government did recognise the potential benefits of restoring native forest on some sites, especially the link between the North Block and the rest of the Park, so as a partial concession to the Trust, the North Block remained with DOC under harvest licences confined to the standing crop.

Now, under the Treaty settlement process, all of what used to be the South Block of the old Pureora SF 96, except a 1 km-wide strip of land linking Pikiariki and Pureora Mountain, is being transferred to Maori. After harvesting of the exotic crop, this area will be allowed to regenerate back to native forest as an ecological corridor.

Pureora Forest Park is administered under the Conservation Act 1987 section 61, and managed so "that its natural and historic resources are protected", and, subject to that, "to facilitate public recreation and enjoyment". Section 62 of the Act provides for State Forests and unoccupied Crown lands to be administered as Stewardship Areas. Ecological Areas classified as "Specially Protected Areas" are managed under section 21 of the Act.

The Reserves Act 1977 provides for the administration and classification of several different kinds of reserves, each with their own management objectives. Most are administered by DOC. Wildlife Refuges and Wildlife Management Reserves are managed under the Wildlife Act 1953 to protect the wildlife values for which they were set aside. Three Catchment Control Schemes (for Lake Taupo, Mokauiti, and Waitomo) also have interests in PFP [15].

John Mason was one of the former village residents who, in 1998, was asked by DOC researcher Owen Wilkes to record his memories of life in Pureora on tape (Chap. 8). He had had direct experience of the two quite different management regimes that had operated at Pureora within the previous 20 years. NZFS had controlled Pureora as a Forest Park (and previously as State Forest 96), for timber production and conversion to exotics: now DOC controlled some of the same land as a newly-reorganised conservation park, for protection of the native environment and for public recreation. The aims of the two were exact opposites, and yet, at least at first, the staff that had to carry them out included a large proportion of the same people.

Looking back on the transition period between 1978 and 1987, Mason agreed that the creation of DOC had been a "really positive thing" for Pureora. NZFS removed the former logging officers and appointed Mason as an environmental ranger, a change from his previous job focussed mainly on shooting goats throughout the King Country (Chap. 12).

> When you look at what we should have been doing, like managing really good pieces of forest and all the species within them, well we weren't doing it, we were sitting there watching the kokako disappear.. .. Yes, we killed possums and rats, but not effectively [because] the control techniques were still being developed and…the managers didn't have that sort of background…when [as a youngster] I said I wanted to go into environmental forestry I was told it wasn't a career option, a waste of time…I should get into pine trees promptly [31].

Luckily, Mason did not take that advice. He recognised that DOC understood Pureora's need for environmental management far better than had previous regimes run by staff trained mainly in silviculture and plantation forestry. He watched and envied the game-changing results from trials on protecting kokako from predators at Mapara (Chap. 13), where kokako were the only important wildlife, yet Waipapa had kokako *plus* many other species *plus* important remnants of rare lowland podocarp forest.

Mason reckoned that Mapara-style pest control applied at Waipapa would produce far more value per dollar spent than the mere declaration of reserve status. He lobbied hard for funding to support research in the Waipapa Ecological Area, and can now be pleased with what has been achieved there during and since his time.

Long-Term Consequences

At the time of the tree-sitting protests in 1978, a conservative government was in power, led by Robert Muldoon, an abrasive, aggressive and opinionated Prime Minister with a famously negative attitude to the conservation lobby (Chap. 9). Conservationists welcomed the change to a Labour Government in 1984, hoping for better things. But when six years of radical economic reform turned out to be far less favourable to conservation than expected, they were badly disappointed. The Labour Government had ignored all protests against the disposal of Crown forestry assets in 1988, pushing through a process which Alan Gibbs, the chair of ForestCorp, described as "the sale of the century" [8: 176].

Tenders inviting a second round of bids for plantations were being sent out in October 1990, shortly before the Labour Government faced a General Election. The NZ Native Forests Restoration Trust put in a competing bid covering some 6000 ha north of Pureora village, which had been logged in the 1940s and 50s by the first two Barryville mills (Chap. 5). The Trust pointed out that, if this could be restored to native forest, it would reunite the long-separated northern and southern sections of PFP [13]. This time, Cabinet agreed, and the Trust began to make plans to manage the logging and replant the area with totara and other native species [21]. But the headline, "Victory in Pureora Battle" was premature. In the face of a strong reaction from local Maori with Treaty claims over the same land, the Government backpedalled [22], and then lost the election.

The new conservative government was led by Rt. Hon. Jim Bolger, himself a King Country farmer and much more sympathetic to NFAC's argument than Muldoon had been. Bolger visited Pureora in December 1991, and was photographed striding along a gravel road in his boots, alongside Stephen King in his famous bare feet (Fig. 11.5). King credits Bolger with reviving the kokako recovery work at a time when the population had crashed to a critical state. Bolger supported the long-term hope of restoration, and has a superb young totara growing on his Te Kuiti farm [13].

Over time, successive New Zealand governments have continued or extended most of the same policies as those of the neo-liberal reformists of the 1980s. For economic rather than ecological reasons, DOC's programme of extending the national network of legally protected EAs had to come to a complete stop over

Fig. 11.5 Prime Minister Jim Bolger visited Pureora on 30 December 1991, and was escorted by Stephen King to inspect the Pikiariki protest area. *Rhys Palmer/New Zealand Herald*

the first five years of Bolger's term of office (1991–95) [25: Fig. 2]. During and since that time, further corporatisation and market liberalisation (including sales of State assets to overseas owners) have permanently damaged New Zealand's sovereignty over its own land and forest resources [11: 106]. The long-term example closest to Pureora Forest village is Peacocke's farm, the open space between the village and the North Block of PFP. It was bought by the Crafar group, whose 16 farms were eventually onsold to a Chinese consortium [23].

On the other hand, for some user groups there have been more positive long-term consequences of the upheavals set off in 1978. J. A. Dawson [7] predicted, from his independent economic analysis, that the value of Pureora as a conservation park would eventually be much greater than the cost of creating it. History has proved him right (Chap. 14).

Pureora as a Field Lab

The long cultural history, great biodiversity and protected status of PFP have made it an ideal open-air laboratory for field studies. It provides for all levels of interest, ranging from school groups to professional scientists, and all subjects from general observation to specialised research in history and the natural sciences.

The Forest Bird Research Group (funded by NZWS and RFBPS: Chap. 13) was the first of many teams of scientists to undertake new research programmes in the park since 1978. Dozens of other independent studies by university staff and students, research institutions or consultants have been done there at the time or since. Pureora is easily accessible from most parts of the North Island, so is an ideal field lab in many respects.

When visiting Pureora on field work, outsiders stayed in "the Blue House", set aside for visiting research teams. It was a former mill house at the western end of the village, similar to the one shown at the right of Fig. 8.2. The lives and sympathies of all concerned were caught up in the stormy re-organisation of forest research in 1987, but their work has mostly found safe harbour in the permanent literature

The Waipapa Ecological Area in particular has proved to be a fertile ground for important research on birds because it is easily accessible and supports a great diversity. Conservation authorities are required to make good management decisions about these and other iconic species, so they need reliable information about population density and trends. Some of these studies are summarized in their management contexts (Chaps. 2, 12 and 13).

Ecological Reserves

NZFS had originally proposed to reserve about 15,000 ha within PFP, but the Director-General asked John Nicholls and the PFP Scientific Co-ordinating Committee to examine the NZFS proposals on the ground. The Committee visited the forests in March 1977, and subsequently presented back to the DG an enlarged and modified set of proposals for 11 reserves covering 22,000 ha [24]. Only one of the Committee's proposals, to be called Oruangungu EA south of Benneydale, was not eventually accepted (Fig. 11.6).

The first management plan for PFP [17] showed that the ecological areas in the park contained all the remaining areas of virgin forest, despite some significant gaps created by recent logging. The plan described these proposed ecological areas in detail, and the Scientific Co-ordinating Committee contributed an appendix (pp. 93–97) describing the 'uses acceptable for ecological areas'. Inevitably, the final arrangement is not quite the same as the one first proposed at the 1978 Taupo seminar by John Nicholls.

In Norton and Overmars' review of 2012, the ten EAs in PFP and their gazetted areas are given as: Mangatutu (2600 ha), Waipapa (1839 ha), Pikiariki (426 ha), Waimanoa (790 ha: Fig. 11.7), Pureora Mountain (2074 ha), Rata-nu-nui (946 ha), Waihaha 12093 ha), Maramataha (7165 ha), Nga Morehu (216 ha) and Whenuakura

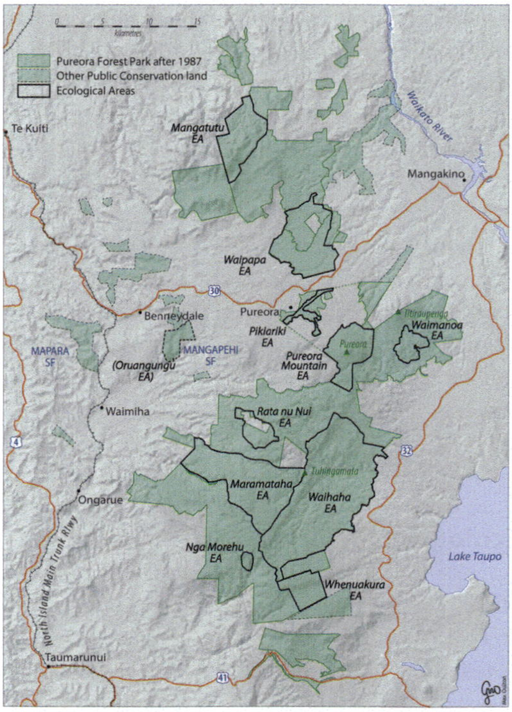

Fig. 11.6 Pureora Forest Park after it was handed over to DOC on 1 April 1987, showing the changed outline and the proposed Ecological Areas. The pine plantations next to the protest site (the grey area within the dotted outline between the Pikiariki and Pureora Mountain EAs) were removed and allocated to the new Forestry Corporation. Mapara SF was always outside the park but, with the extended Waipapa and Pikiariki EAs, later became an important area for research on kokako. *Updated and redrawn by Max Oulton from Nicholls (1978)*

(1766 ha) [25], constituting about 40 % of the area of the Park [3]. Dates for the legal gazettal notices for each area continue to 1986 [25: 118], and one very large extension doubling the size of Waipapa EA does not appear on the list.

All ten EAs in PFP (of a national total of 140) are important conservation areas that were rated as having outstanding wildlife values by NZWS (Fig. 9.1). The main consideration in choosing them was "the preservation of intact ecosystems, for their intrinsic interest and for their future value in contributing to land-management decisions" [24: 1].

With hindsight we can see that history has gifted the Pureora area so varied a landscape that the ten ecological areas do indeed represent the unique range of unmodified altitudinal vegetation sequences, dense stands of lowland podocarps, and natural post-Taupo eruption vegetation within PFP. Each of the ten EAs has particular distinctive qualities, and all were first established in the face of fierce pressures to continue logging (Chap. 9).

The Waipapa EA is the only one that contains extensive unmodified podocarp stands, a nationally scarce forest type. It also includes some important wetlands which are especially sensitive to tracking and browsing by ungulates. Along with Pureora Mountain EA, Waipapa EA and Mangatutu EA represent most of the biodiversity native to those habitats [6].

Fig. 11.7 Part of Waimanoa Ecological Area, with Titiraupenga in the background. *SCION image 8043736. Photographer: John Barran*

The peat bogs on the crest of the Hauhungaroas are the best examples of montane wetlands in the North Island (Chap. 2). Two small remnants of silver beech survive in the Waihaha and Taringamotu valleys, but beech has not recolonised the rest of the park since the Taupo eruption. Pikiariki EA has become a significant area for forest research (Chap. 13).

References

1. Bassett, C. (1987). The role of the forest service in conservation of native state forests. NZ Forestry, August issue, 15–18.
2. Beveridge, A. E. (1985). Ecological research of the Indigenous Forest Management Group at the Forest Research Institute, Rotorua. In B. Veale & J. G. Innes (Ed.), Ecological research in the central North Island volcanic plateau region: Proceedings of a NZ Forest workshop (p. 18–26). Rotorua: Forest Research Institute.
3. Burns, R. (1998). Pureora Conservation Park Strategic plan. Hamilton: Department of Conservation.
4. Bush Telegraph, August 1985.
5. Bush Telegraph, November 1985.
6. Clegg, S. (1986). Resurvey of Pureora North Block 1981–1982. Auckland: New Zealand Forest Service.
7. Dawson, J. A. (1979). Pureora Forest Park: Costs and benefits. MA thesis, Victoria University of Wellington.
8. Harris, D. (Ed.) (2009). Timber cruising and other forestry stories. Wellington: FirstEditionPublishers.com.
9. Hay, J. R. (1981). The kokako. Wellington: New Zealand Wildlife Service, Forest Bird Research Group.
10. Innes, J. G. (1983). Personal communication.
11. Kelsey, J. (1997). The New Zealand experiment: A world model for structural adjustment? Auckland: Bridget Williams Books and Auckland University Press.
12. King, C. M., Innes, J. G., Flux, M., Kimberley, M. O., Leathwick, J. R., & Williams, D. S. (1996). Distribution and abundance of mammals in relation to habitat in Pureora Forest Park. New Zealand Journal of Ecology, 20, 215–240.
13. King, S. (2013). Personal communication.
14. Kirkland, A. (1978). Summary of information from seminar papers and discussion [*Transcript of the Taupo seminar*]. Wellington: New Zealand Forest Service.
15. Leigh, A. H., & Clegg, S. E. (1989). Maniapoto District wild animal management plan, 1989 to 1999. Hamilton: Department of Conservation.
16. Native Forests Action Council (1985). Environment could benefit from Budget logic. Bush Telegraph, 19, 6–7.
17. New Zealand Forest Service (1980). Pureora State Forest Park management plan. Auckland: New Zealand Forest Service.
18. New Zealand Forest Service (1982). Report on Pureora Forest. Auckland: New Zealand Forest Service.
19. New Zealand Forest Service (1984). King Country regional management plan: 1983–1993. Auckland: New Zealand Forest Service.
20. New Zealand Forest Service (1986). Ministerial brief for Minister of Conservation. Wellington: New Zealand Forest Service.
21. New Zealand Herald, 3 October 1990.
22. New Zealand Herald, 11 October 1990.
23. New Zealand Herald, 1 December 2012.
24. Nicholls, J. (1978). Scientific reserves recommended by the Scientific Co-ordinating Committee [*Transcript of the Taupo seminar*]. Wellington: New Zealand Forest Service.
25. Norton, D. A., & Overmars, F. B. (2012). Ecological areas—premier protected natural areas. New Zealand Journal of Ecology, 36, 108–120.
26. Roche, M. (1990). History of forestry. Wellington: New Zealand Forestry Corporation Ltd.
27. Waikato Times, 14 July 1986.
28. Waikato Times, 5 June 1986.
29. Waitangi Tribunal (1993). The Pouakani report. Wellington: Waitangi Tribunal.
30. Whyte, A. G. D., & Poole, A. L. (2001). New Zealand forestry politically compromised. Fremantle, Australia: Commonwealth Forestry Association.
31. Wilkes, O. (2005). Transcripts of oral history recordings. Hamilton: Department of Conservation.
32. Wright, L. W. (1980). Decision making and the logging industry: An example from New Zealand. Biological Conservation, 18, 101–115.
33. www.legislation.govt.nz.

Protecting the Forest from Introduced Herbivores

12

C.M. King, J.G. Innes, M.C. Smale and G. Nugent

Abstract

This chapter describes the arrival of introduced browsing mammals (red deer, feral goats and cattle, and possums) in Pureora Forest Park (PFP), the long-term damage to native vegetation they cause, and the persistent dilemmas surrounding the use of 1080 poison to protect the forest. PFP has developed a new role as a significant open-air laboratory for field research, so this chapter summarises some of the important studies conducted on the fauna and flora of the park, especially on the development of ways to protect them.

Keywords

Introduced herbivores · Feral goats and cattle, red deer, brushtail possums · Browsing damage · Pest control operations · Possum fur industry · Possums and bovine TB · Aerial 1080 · 1080 and bird populations

Browsing herbivores feed on the leaves of shrubs and trees rather than on grass. The original native forests in New Zealand included dense stands of trees that had co-evolved with moa, and so their leaves were adapted to being pulled off by beaks rather than cut off by teeth. Saplings of some species, such as the lancewood, have evolved a different "juvenile" form for leaves less than 3 m above the ground, which are less vulnerable to clamping and tugging (long, thin, hard leaves edged with spines). Their adult canopy develops only when it can be held out the reach of a moa. Some 20 % of the native woody flora of New Zealand have thin, wide-angled branches, divaricating and intertwining to form a tangled hedge, strong and elastic enough to resist being torn off [2].

C.M. King (✉)
School of Science, University of Waikato, Private Bag 3105, Hamilton 3240, New Zealand
e-mail: cmking@waikato.ac.nz

J.G. Innes · M.C. Smale
Landcare Research NZ Ltd., Private Bag 3127, Hamilton 3240, New Zealand
e-mail: innesj@landcareresearch.co.nz

M.C. Smale
e-mail: smalem@landcareresearch.co.nz

G. Nugent
Landcare Research NZ Ltd., Box 69040 Lincoln 7640, New Zealand
e-mail: nugentg@landcareresearch.co.nz

C.M. King et al. (eds.), *The Drama of Conservation*,
DOI 10.1007/978-3-319-18410-4_12

By contrast, forest trees in countries with co-evolved mammalian browsers tend to evolve spines or chemical defences against damage by mammals that chew twigs, leaves and fruit with teeth. For example, many Australian forest trees pump toxins into their leaves as they harden [13]. Fewer New Zealand trees (except, perhaps, horopito or pepperwood) have any defences effective against the very different challenge posed by mammalian browsers.

Introduced Browsing Animals

Four species of introduced browsers have appeared in the area that is now Pureora Forest Park since the arrival of the first European farmers and settlers in the early twentieth century. All have modified the native vegetation to at least some extent, with consequences that are becoming clearer the more we find out about them [26, 29].

Two species are the feral (i.e., gone wild) descendants of domestic cattle and goats; the other two (deer and possums) were deliberately introduced, for different reasons which, however regrettable in retrospect, did seem good at the time. Some users of the park (at least the hunters) see them as a welcome wildlife resource, some see (or have seen) them as providers of meat, natural fur or fibre; and others see them as simply animal pests with no redeeming values at all. Horses do not browse the forest, but semi-feral horses were once a nuisance round the village (Chap. 6) and occasionally grazed the sensitive native frost-flat vegetation.

Escaped domestic stock founded populations of feral cattle, which were once common but have now disappeared, and feral goats, which have never been widespread in the park. Red deer have been abundant in some parts of New Zealand since the 1890s, but reached Pureora only in the mid 1950s. The late arrival of possums in the 1960s was unplanned but lucky for PFP, because "it explains why we still had kokako in the 1980s" [22] and see Chap. 13.

Feral Cattle

Fencing is expensive, and the earliest pioneer farmers were short of money, so many cattle escaped into the forest from the new farms established along the line of the Main Trunk Railway. They retained the physical characters of their domestic (mostly Shorthorn) ancestors, but in behaviour they quickly became shy and aggressive. Forestry staff working on the 1946–55 National Forest Survey [31], and early PFP field staff including Colin Sutherland and Dave Yanko, reported hair-raising encounters with angry wild bulls, and they often only just managed to get up a tree in time (Chap. 14). Some local populations of feral cattle reached high densities in the broad forests of the Hauhungaroa and Rangitoto Ranges until the 1960s [23]. Farmers also valued the time-honoured tradition of wintering their stock in the shelter of the bush.

Feral cattle are very large and compulsively sociable animals. Their substantial appetites, high browsing reach and intensity, and their concentrated trampling created a maze of tracks and clearings and spectacular destruction in the undergrowth. That in turn could have helped to open up these forests for the later-arriving deer and possums more than the much smaller and longer-resident pigs had done.

Peter McKelvey [30] compiled the first comprehensive and seminal ecological account of the west Taupo forests as they were documented by the National Forest Survey of 1946–1955. He reckoned that, at least in favoured patches, feral cattle were the most significant cause of forest modification at that time, although he told Graham Nugent that the understorey was then still so thick in places that it could take an hour to get through 100 m of it [42].

The abundant bush cattle were attractive not only to hunters with rifles but also to resourceful local farmers armed with a different technique. One, Arthur Pittams, would catch a cattle beast, and then leave it for a couple of days tied to a tree by a rope wound tightly around the base of the horns. By the time he returned the subdued animal would be willing to be led behind

Arthur's horse, and then penned until he could drive it with the rest of his stock to the sale yards in Waimiha [60: 153].

After the early 1960s this trick was no longer an option, because the numbers of wild cattle had been greatly reduced—although complete extermination took a while, as NZFS records were still commenting on cattle damage in 1969/70 [63]. Today the only evidence of their former presence is an occasional bone or skull, and a contemporary forest structure fitting McKelvey's description of how cattle had modified the forest [29].

Feral Goats

Goats have been domesticated for at least 9000 years, mainly because they can provide meat, milk and fibre, are small enough to be easily handled and transported, and not too expensive to feed. Unlike sheep, goats are primarily browsers, but famously unfussy ones.

Goats were usually among the first livestock on poorly-fenced pioneer farms, from which escapes were inevitable. The descendants of the escapees rapidly became feral, and eventually established large and independent populations throughout New Zealand.

Goats were first brought to the King Country in about 1910 as ideal rough country stock. Woody weeds such as blackberry and gorse rapidly invade newly cleared forest, and goats have long been the cheapest and most effective weed-control agents. That was the reason why goats were brought to the western flanks of the Rangitoto Range in the 1930s [9], and, further south, to the Maramataha and Waione valleys in the 1960s [29].

Box 12.1 Goats and deer

Goats have very high natural fecundity (when forage is abundant, most adult females and half of the females under a year old breed twice a year, twins comprise up to a third of all births, and productivity can reach 1.7 embryos per female per year) [45: 386]. Nevertheless, goats colonised the future PFP slowly. They were hardly mentioned by McKelvey [30], and were still absent from the Pikiariki kokako area in 1978–81 [28] and the Waipapa Ecological Area in 1980–81 [11].

In 1975, Russell Dale [9] reported the highest concentration of goats in the Okahukura State Forest and along the Waipa gorge, Rangitoto, Owawenga and Tunawaea valleys, and made specific recommendations for control of both goats and deer in this area.

The only part of the North Block where goats were numerous in 1980–81 was the north-western edge, closest to their first liberation site. The general conclusion of NZFS surveyors at that time was that goats were much less significant a problem for managers than were deer and possums.

By contrast, the eastern forests of the Waihaha, Tihoi, Pureora and Waipapa areas were then still relatively free of goats [29], and still are. Only fenced exclosure plots can demonstrate the diversity and abundance of preferred plant species that once grew in the absence of deer and goats.

The Auckland and Wellington Acclimatisation Societies enthusiastically organised translocations of red deer for liberation in the King Country from the early twentieth century onwards. Most of these animals came from the Paraparaumu Game Park. Their bloodlines went back to nineteenth-century imports from Windsor Great Park and Warnham Park in England, and from Invermark Forest in Scotland, among other sources [26]. The first known releases were near Taumarunui (8 liberations totalling 17 deer between 1913 and 1922), Tokaanu (7 liberations totalling 15 deer between 1911 and 1917), and Te Kuiti (5 deer in 1920).

The most spectacular forest damage is produced by a combination of high numbers of deer on the ground and possums in the canopy. Less visually shocking effects can be achieved by the selective browsing of relatively few deer concentrating on ecologically valuable and sensitive vegetation, destroying whole populations of the preferred woody species within browse range (15–135 cm height) [4].

The controlled re-measurements and pellet counts done routinely in various parts of PFP until the demise of NZFS in 1987 [3, 7, 9, 11, 23, 27] have consistently found evidence of increased deer use with each survey over that time.

The problem in managing any mainland forest area supporting both goats and deer is that the damage done by these two invaders can seldom be separated (Box 12.1). But on their own, goats like to concentrate on favoured sites where they can browse a wide variety of plant species, remove nearly all seedlings, kill mature trees and shrubs by ring-barking, and over the long term, suppress regeneration so completely as to turn tall forest into open grassland and threaten soil conservation. So there is every reason to keep goats out of any areas of PFP that they have not already reached. Culling by DOC keeps their numbers low and confined to the west and south of PFP.

Deer

Deer colonised the Hauhungaroa and Rangitoto Ranges from the early 1950s, earlier in the south. NZFS records list a large stag seen in September 1952, and increasing numbers in the forest behind Pukemako Camp (Fig. 7.3); then two further records in 1953/54, and a group of 16 in the following year [63]. Thanks to Te Rohe Potae (Chap. 4), this was rather late by comparison with the arrival of deer in the neighbouring forests of Tongariro National Park and the Kaimanawa Ranges (Fig. 0.1 in Preface).

The early phase of forest modification attributable to red deer cannot be distinguished from that of any other ungulate browsers. Repeatable surveys began only in the 1970s, by which time browsing by deer had replaced that by feral cattle, and goats and possums were also contributing to massive alteration of forest structure. By 1984, deer and goat density was highest in the Mangatutu EA, though lower than in 1981, while the densities of deer and possums had increased in the Waipapa EA despite the issuing of many hundreds of recreational hunting permits every year [29].

Where red deer invade a previously deer-free area, their numbers typically increase to a peak, when they can completely eliminate all of the palatable vegetation in the undergrowth, in places often leaving a bare forest floor of litter and moss, and in other places a sometimes dense understory comprised of a few unpalatable plants such as horopito (pepperwood) and the soft tree fern. As the peak numbers of deer decline, these unpalatable species may gradually increase and fill in the understorey. Horopito can form dense groves and live for up to 200 years [52], so the effects of deer browsing in producing shifts in understorey composition on the forest ecosystem as a whole can last a long time.

In the central Hauhungaroa Ranges, commercial hunting appears to have started before deer numbers peaked, so the forest there was possibly not quite as depleted as in other areas. Fewer deer are usually correlated with better condition, both of the vegetation and of the deer themselves.

Possums

The brushtail possum (Figs. 12.1 and 13.5), an arboreal marsupial native to Australia, was introduced into New Zealand to provide a fur harvest from standing forest. Possums colonise any new area in a wave like a furry tsunami, which follows a predictable pattern, repeatedly observed and analysed in other New Zealand forests.

For some 15–25 years after their first arrival in any new area, possums have access to rich food supplies well out of the reach of any competing browsers. Possum numbers increase rapidly to a peak, at which immense densities (30 to 45

Fig. 12.1 The Australian brushtail possum. There are three colour phases of possums: black and brown, illustrated here, and grey, shown in Fig. 13.5. Their relative proportions vary locally, but all are the same species. *Painting by P. Barrett, from King (ed) (2005)*

possums/ha) can be achieved—but not for long [19].

Inevitably, the most favoured species of trees and shrubs vanish, other food supplies are harder to utilise, and the possum population crashes to a much lower but more sustainable post-peak density of up to 10–12 possums/ha, but as low as only 1–2/ha at higher altitudes or in simple beech forest. The post-peak densities seem to have been intermediate in PFP, e.g., 4 possums/ha in Waihaha in the 1990s).

Pre-peak, peak and post-peak populations are always quite different in age structure, reproductive output, body weight and fat reserves. Forests that have supported post-peak populations for a long time show a significant shift in tree species distribution and diversity, towards species at the bottom of the possums' preference list (Box 12.2; Fig. 12.2).

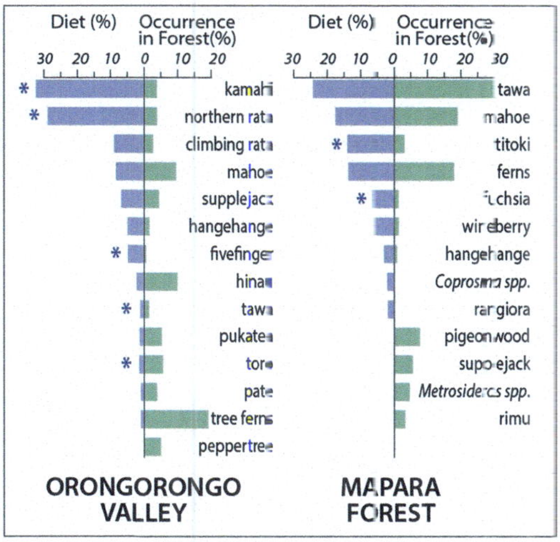

Fig. 12.2 The composition of forest vegetation is changed over time by persistent possum browsing on some tree species more than others. The Orongorongo Valley near Wellington has been occupied by possums for much longer than has Mapara: their preferred species (kamahi, rata, fivefinger) are still selected out of all proportion to availability, and will eventually be eliminated: the same process is under way at Mapara. Asterisks mark species at risk or already reduced by browsing. *Redrawn by Max Oulton from Green (1984: fig. 4)*

Box 12.2 Possums

Significant modification of the forest by peak populations of possums in PFP has been observed only since the mid-1970s [29]. Possums are not, of course, the only causes of these changes, except in the canopy which deer and goats cannot reach. Surveys in 1974/75 found significant forest depletion in the North Block on the upper slopes of the Ranginui and Rangitoto Ranges [9], and especially along the crest of the Hauhungaroa Range in 1978/79, where many palatable species were reduced. In most (not all) gullies the collapse of fuchsia, one of the possum's first targets, left only a dense understorey of unpalatable pepperwood [23].

Repeated surveys of permanent plots and exclosures first established in 1975 have documented the dismal story. In 1993 Keith Broome compared the re-measurements he made that year with 12 years of previous data. Among the species preferred by possums, whole trees were found dead, and the crown density and basal area of the survivors were declining [4]. Possums destroyed most of the pioneering species recorded during trials of selective logging, even at sites established before possums arrived. After one 1961 Pureora trial, wineberry, fuchsia, pate, fivefinger, kamahi and ferns colonised the sites, but the broad-leaved species mostly succumbed to possums after their arrival nearly a decade later, and by 1998 those sites were dominated by tree ferns and ground ferns [53].

Analyses of possum diet from pellets collected in three forests including the Pikiariki Ecological Area in 1978–81, around the time that possums were at peak numbers in that part of PFP [27], found that possum diet comprised mainly fivefinger, raukawa, mahoe and supplejack, which overlapped with that of kokako (fruit of fivefinger, raukawa, kaikomako, putaputaweta, matai and supplejack, and leaves of mahoe). In all, some 22 plant species each contributing more than 5 % of kokako diet in at least one season were also eaten by possums (Box 13.3).

In the Waihaha catchment in 1990–93, where possums have declined to post-peak numbers, possums at up to 3/ha were eating only about 3.3 % of total annual foliage production—figures described as "low" and unlikely to induce a catastrophic collapse, although a continued major shift in forest composition seems unavoidable [39].

The activities of possums do not affect only indigenous forest vegetation, but also the endemic animals that evolved with it. Possums and rats browse and destroy most unprotected inflorescences of the root parasite *Dactylanthus taylorii,* which has an evolved mutual dependency with its pollinator, the short-tailed bat (Fig. 2.14), hence threatening two endemic species simultaneously [12]. The distribution in PFP and survival of one of the native mistletoes, *Tupeia antarctica*, is also linked closely to the success or otherwise of possum control [54].

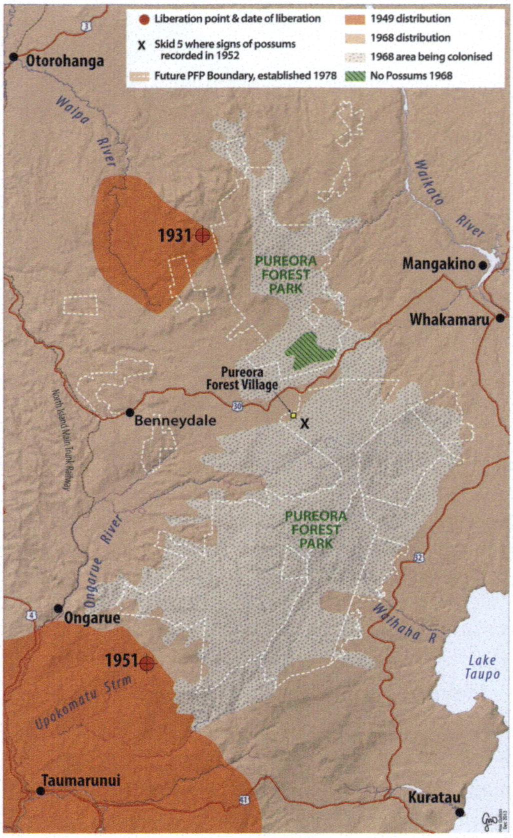

Fig. 12.3 Spread of possums into the King Country, 1931–68. The ranges still being newly colonised in 1968 included some important areas that became part of Pureora Forest Park after 1978. The most significant of these, in part because it was the latest colonised, is the Waipapa Ecological Area (*centre*). *X* marks the site of Skid 5, where a very early (1952) observation of possum sign was recorded. Possums are now present throughout. *Redrawn by Max Oulton from Leigh and Clegg (1989: map 4)*

The first possums to reach the central North Island arrived relatively late compared with those that colonised more accessible country. Liberations in coastal Maniapoto and Pirongia date from 1925–29, but the two nearest releases to the boundary of the present PFP [51] were not made until 1931 and 1951 (Fig. 12.3). There were probably many other unrecorded releases.

The NZFS Pureora Forest station diary recorded the first signs of possums in a cutover area on 21 July 1949, along with a prophetic comment: "They may at some future date cause damage to young [Douglas fir] which are to be planted this winter". No-one then realised that damage to Douglas fir would soon be seen as the least of the effects of possums on the forest.

Signs of possums were noted in July 1952, for only the second time in 3 years, "behind skid 5" [63]. Skid 5 was only 1.5 km from Pureora Forest Park HQ (Fig. 12.3). At the time of the first survey of native tree seed crops and seed dispersers in Pureora over the years 1958–64, possums were only just starting to build up in numbers, and still had no obvious effect on the forest [1].

A 1968 survey found possums present in much of the country between and west of the known liberation points, and also approaching the eastern flanks of the ranges from the Lake Taupo side. Only the Waipapa Ecological Area, one of the last patches of unlogged forest in the whole district, was still more or less free of possums in 1968 (Fig. 12.3), although the NZFS annual report for 1962/63 remarked on "increasing evidence" that they were moving in [63].

Possums were still spreading down the eastern side of the North Block when G. T. Jane [23] mapped the distribution and density of deer, possums and goats by regions of PFP in 1979. By the summer of 1980–81, possums had completed their colonisation of the North Block [11], and by 1983–84, they were at their highest density there [27].

Monitoring Browsing Damage

Browsing mammals have been for decades at the centre of management conflicts between parties with opposite views on them, as causes of undesirable browsing damage to vegetation versus as animals with significant economic or recreational values. By contrast, ubiquitous and destructive small mammal pests such as rats rarely got more than cursory attention until the mid 1980s, and their important influences on ecological communities were ignored.

Under the Wild Animal Control Act 1977, a "wild animal" was an introduced grazing or browsing herbivore whose effects on the forest directly damaged NZFS interests. The original definition as used in the early surveys included only deer, feral goats and pigs, although possums were added later.

On a national scale, the economic value derivable from wild animals was once considerable: NZFS estimated in 1986 that, for the year ending 31 July 1985, the combined export value of venison, deer skins, live deer, possum skins, wild pork, goat meat, live goats and tourist hunting was NZ$48.71 million [35]—a massive NZ$125 million in today's money. Nowadays however, the value of wild animals is much lower in places such as Pureora, because pigs are no longer commercially harvested, commercial deer harvesting is rarely viable, and possum numbers are often reduced to densities well below commercially viable levels.

In huge areas of forest country, the most important browsing animals now are deer. Because deer on open ground are very vulnerable to helicopter hunting, most of them now live within the roughly 59,000 km^2 of tall forest scattered throughout New Zealand where the canopy provides adequate cover [17].

The changes induced by introduced herbivores in the structure and composition of the native vegetation, with direct or indirect consequences for the endemic fauna, are serious conservation costs that must be minimised. So herbivore populations and their activities have been systematically monitored for decades, throughout the conservation estate formerly controlled by NZFS and now by DOC.

During the 1960s and 70s, NZFS developed a long tradition of employing an army of rugged bushmen every summer, willing to follow compass directions across any terrain, and to live rough in mountain huts for long periods [5]. They systematically counted faecal pellets along transect lines (capable of detecting differences of at least 30% in the usage by animals of the sample areas, since pellet counts are supposed to be related to the approximate number of animals); built exclosures (paired 20 m × 20 m plots, one to fence animals out and one for comparison); and produced standardised vegetation assessments at permanent plots inside and outside the fences

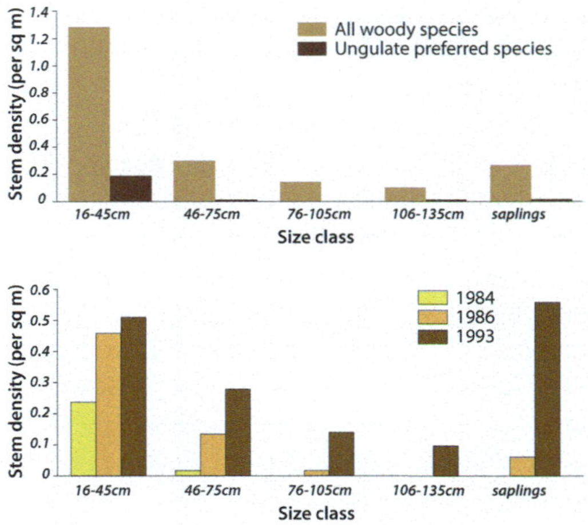

Fig. 12.4 *Above* By 1993, selective browsing damage by ungulates (mainly deer and goats) in unprotected forest had virtually eliminated regeneration of their preferred understory plants in the North Block. *Below* Understory species preferred by ungulates continued to thrive when, as here, protected within fenced exclosures since 1983. Note different scales. *Redrawn by Max Oulton from Broome (1995: figs. 3.1.1 and 3.1.9)*

(tagging and measuring stems of >2.5 cm dbh and counting smaller ones in size classes).

They aimed to repeat these methods at intervals, and compare the results to detect trends in numbers and damage indices, whenever scheduled, whatever the weather. Those working during the autumn "Roar" season (Fig. 14.6) liked to make themselves obvious to hunters by wearing high-visibility vests and carrying a radio playing loud music.

In Pureora North Block, a baseline array of 256 constant-count plots and 32 permanent 20 m × 20 m plots was established in 1974/75 [9]. Another 26 permanent plots were added after the 1981/82 re-measurement [7]. After the plots were re-measured for the third time in 1993, Keith Broome assessed 12 years of vegetation trends (Box 12.3; Fig. 12.4), and reported that the species preferred by possums, deer and goats had declined steadily over that time; seedlings palatable to deer and goats had become scarce above a height of 45 cm, while unpalatable species increased [4].

Box 12.3 Surveys of wild animal damage in the North Block

The two main ecological questions to ask about herbivores in PFP are: first, does possum damage in the canopy actually kill trees of the species that possums prefer to browse? Paired comparisons of forest sites with or without possum control over 10 years shows that extensive and sustained possum control significantly reduces mortality of susceptible trees [18].

Second, can tree populations replace themselves despite deer and goats browsing in the subcanopy? That is the subject of ongoing research, in PFP and elsewhere.

A 1990–93 survey assessed the condition of 25 km^2 of podocarp-dominated forest in the headwaters of the Waihaha and Waitaia Streams [39]. The results showed the effects of red deer and possums in that area before the start of possum control in 1994. The density of deer at the time was c. 6/km^2, and of possums up to 3/ha.

In the Waihaha forest, a transect first surveyed in 1958 was dominated by matai, totara (2 species), miro and tanekaha, and had a basal area then of 85.1 m^3/ha. By 1983 the basal area had decreased by 3 % and two of the large totara had died. Broadleaf and lancewood were the two main foods for deer, which ate mostly adult tree foliage, much of it as litterfall. Possums relied heavily on foliage of Hall's totara, and on fruits in good seasons. Kamahi was the only species with foliage important to both possums and deer. Many large Hall's totara trees were dead or dying, but other conifers—rimu, matai, miro, and toatoa—were seldom browsed.

A 1995 report by Keith Broome [4] reviews past and current work to monitor the impact of deer, goats and possums on the vegetation in and near the North Block of Pureora Forest Park. Surveys have been done since 1974, using vegetation assessment in 20 × 20 m permanent plots, animal population assessment of faecal pellet counts on transects and other methods. Vegetation trends are assessed over 12 years, following remeasurement of plots in 1993. Species preferred by possums, deer and goats have continued to decline. There has been substantial mortality of some possum-preferred species and an overall decrease in basal area of preferred canopy species.

Plots were first established in 1975 after substantial change induced by browsing, but forest condition has not improved since, despite frequent animal control measures. Graphs show stem density in understorey size classes for the main ungulate-preferred and non-preferred species. All species show a normal regeneration pattern, but species preferred by ungulates have almost disappeared above 45 cm height, while unpalatable species have increased. Within fenced exclosure plots, preferred plant species consistently increase in diversity and abundance in every survey. Mean crown densities for possum-preferred species show severe depletion of some species.

The report concludes that deer populations must be strongly reduced to allow adequate regeneration, but the level of control measures needed is in conflict with the objectives of a Recreational Hunting Area. There has been no funding for deer control and barely adequate funding for possum and goat control. Regular monitoring of indicator animal species as well as plant species is required for adequate protection of the forest ecosystem. Appendices to Broome's report present data and maps on wild animal distribution and location of bird species recorded during surveys.

Some possum-preferred canopy species had died, and the survivors had smaller crown densities and basal areas. By contrast, the fenced exclosure plots continued to support the former diversity and abundance of those preferred plant species that could survive only when protected from deer and goats. These observations showed that regular monitoring of indicator plant species, and management of pest animals, will be required for years to come.

There was some respite from possum and deer browsing during the 1970s and 1980s, when large numbers of both were removed, especially during the boom in commercial deer hunting (Box 12.4). However, the removals were not enough to improve forest condition. Broome concluded that adequate regeneration depended on serious and co-ordinated reduction of deer, goat and possum populations, yet DOC had not been provided with any funding specifically for deer control, and barely adequate funding for possum and goat control. Worse, the level of control deemed necessary was hard to decide, according to who was asking the question.

This dilemma still dogs all attempts to achieve multiple use of PFP by different groups of people with conflicting interests. It is important to survey a problem, describe the values at risk, and consult widely, because the double leap from inventory to policy to decisions is one that requires balancing the different opinions of all stakeholders. Deciding what to do with that information is decidedly more tricky. DOC managers continue to invest in possum control research, and in models predicting and testing whether deer control superimposed on possum control could change the seedling growth rates of selected trees. "To me, Pureora is a forest where possums, rats and goats are determinedly managed, and where questions about deer management are actively investigated" says DOC scientist Clare Veltman [59].

In the late 1950s and early 1960s, some 50 long-term ecological transects for monitoring vegetation change were established by NZFS in central North Island indigenous forests. Six of them, in podocarp and podocarp/hardwood forest within the present boundaries of Pureora Forest Park, are amongst the oldest permanent monitoring lines in that forest type. After 25 years, the six transects were re-surveyed, and measurements or counts were made of all trees, saplings and seedlings, plus calculations of basal area and stem density for all species.

Williams and Leathwick [64] concluded that the composition of the forest in the Pureora transects had changed markedly over the 25 years. Deer and possum populations were low at the time when the transects were established, but had reached moderate to high densities 25 years later. Williams and Leathwick suggested that browsing animals are the 'primary agent' responsible for the changes in forest composition, as most palatable plant species (e.g., fuchsia, broadleaf and *Pseudopanax* species) had become significantly less abundant. As usual, unpalatable plants, particularly horopito, had generally increased. Unless browsing animals can be controlled, these trends are likely to continue.

On the other hand, the montane forests in PFP (above 850–900 m elevation on Pureora Mountain) are still largely intact [36]. They are adapted to a harsher climate, grow on different soils (shallower Taupo pumice), and have little fire history (they were too cool for plantations and too damp to burn easily), and possums and deer were late arrivals.

An early 1990s study of the browsing habits of possums and deer in the headwaters of the Waihaha and Waitaia Streams estimated that possums and deer consumed about 88 and 30 kg of forage/ha/year, respectively, or 3.3 % of total annual foliage production for possum and 1.1 % for deer [39]. Most of the food eaten by deer was obtained as fallen leaves from a few preferred species such as broadleaf. The species most strongly sought by deer were reduced to so few plants surviving on the forest floor, that only a few deer could still prevent their regeneration [40].

Of the canopy trees that were still common about 30 years after possums arrived, only adult Hall's totara seemed to be browsed heavily enough to still be under immediate threat—somewhat curiously, however, seedlings and small saplings of Hall's totara remained common in the understorey, because possums rarely browsed shaded foliage. For other common species eaten by possums (such as kamahi and toro), defoliation by possums causing somewhat increased mortality was likely to result in a slow long-term decline only in some places. At the other end of the scale, highly palatable species such as mistletoe, and (in some places but not others) fuchsia, can be completely suppressed by only a few possums (Box 12.2). The conclusion seems to be that Hall's totara is the only species at risk of severe canopy dieback in the future, and no further changes in abundance of the remaining common species seem likely.

Nevertheless, deer and possums have a dominant influence on patterns of regeneration, and their numbers should be reduced. The main downside of targeting possums to protect the forest is the potential benefit to ship rats, and thence to stoats that damage the birds on which the forest depends (Chap. 13).

Continued monitoring is important, but since 1987, NZFS-style professional surveys have been far too expensive for DOC to run. So there is no organised reporting of official records or collation of survey responses. Instead, Wayne Fraser [15] collected data on red deer from diaries of recreational hunters from 1989-1993 and from a separate postal survey, and gathered data on the sex, age and condition of deer killed from his own study of jawbones. Fraser found that diary data on sightings and kills correlated well with data from faecal pellet surveys, suggesting that information from hunters could be used for monitoring deer densities at Pureora at considerably less cost than traditional faecal pellet surveys [15]. His data suggested that 2–3000 deer a year were being taken out of the entire park (including the commercial hunting zone: Fig. 12.5; Box 14.1).

The net result of these and other studies up to the early 2000s has been a quantitative, detailed picture of a dynamic process of change.

In forests, deer prefer most of the broad-leaved hardwood tree species (typically in the subcanopy tier) such as various *Pseudopanax* spp., pate, and broadleaf, and some ferns such as hen and chickens fern. The beech and podocarp canopy species, and the remaining subcanopy trees, shrubs, ferns, herbs, and grasses are generally edible but less preferred. A few species, such as pepperwood and crown fern, are almost never browsed, or only in very small quantities.

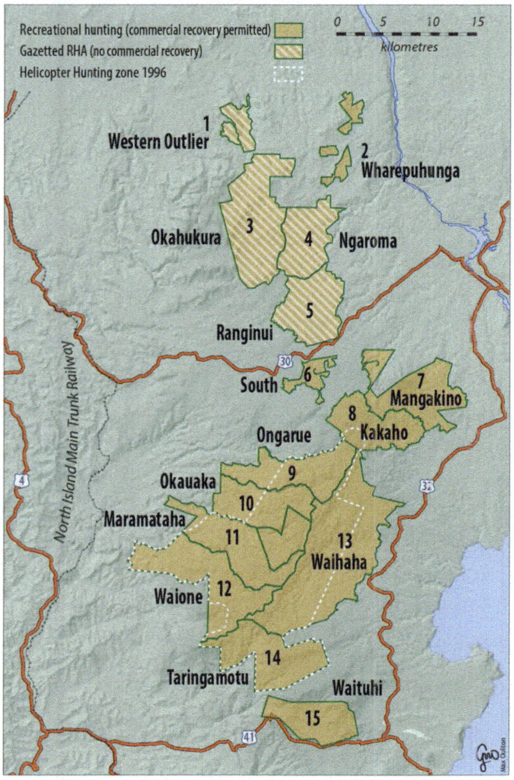

Fig. 12.5 Location and boundaries of the commercial and recreational hunting blocks in Pureora Forest Park in 1996. For tally data, see Box 14.1 *Redrawn by Max Oulton from Fraser (1996)*

> Deer-preferred species were typically abundant in the understorey before colonisation but are virtually eliminated from this tier as deer numbers increase. Subsequently, they are replaced to some extent by a smaller number of less palatable species. This sometimes produced localised forest dieback in areas where the canopy was dominated by short-lived deer-preferred species. In areas where possums are also present, canopy dieback can be far more immediate and severe because possums can kill established trees and deer may then prevent any replacement. In tussock grasslands, the overall pattern of deer impacts has been similar to that in forests, with selective removal of preferred foods such as the large-leaved herbaceous species and some tussock grasses.
>
> The level of control required to protect a particular plant species depends largely on its vulnerability to deer. In the absence of hunting, deer numbers increase to and remain near the ecological carrying capacity of the habitat; the biomass of plants which are eaten by deer is reduced and maintained at low levels. Harvesting the population reduces deer density and browsing. . [allowing] some increase in the biomass of the plant species which are eaten by deer. The scale of that increase or recovery depends on the extent of the reduction in deer density. Furthermore, unless this reduction is sustained, the deer population tends to return to carrying capacity [16].

This conclusion was updated and reinforced yet again by a new analysis of data stretching from 1974 to 2002. S. W. Husheer [21] used data from repeated measurements of forest monitoring plots (20 × 20 m) (n = 32) and nine ungulate exclosures (paired fenced and unfenced plots; 20 × 20 m) to show the effects of introduced ungulates on tree regeneration in PFP. The results confirmed that introduced ungulates, particularly red deer, have suppressed the regeneration of at least six major hardwood (angiosperm) species that can now remain common as saplings and small trees only in the complete absence of ungulate browsing. However the overall number of trees actually increased by 15 %, as the least preferred species were able to 'grow into' the canopy gaps left by trees killed by possums. So the forest canopy and the basal area are not declining, but the species composition is changing substantially.

So if it is agreed that management should aim to enable regeneration of palatable tree species in Pureora Forest, then possums, goats and red deer will need to be culled to low densities indefinitely. Next question: how?

Deer: Commercial Hunting

The introduced game animals were originally seen as a valuable asset, often brought in with official help, and in the nineteenth century they could be hunted only under a system of restrictive permits, seasons and bag limits. But by the late 1920s the effect of their browsing on native vegetation could no longer be ignored, and all management of game as a resource was abandoned when responsibility for control of wild animals was transferred from the acclimatisation societies to the Department of Internal Affairs (DIA) in 1932 [5].

DIA employed ground-based deer cullers in an effort to reduce deer numbers and slow their spread into new areas, although not in the future PFP, where deer had not yet arrived. By 1956 the national management priority had changed from reducing animal densities to protecting water and soil values [6], and the job of controlling wild animals was transferred from DIA to NZFS.

Commercial hunting of deer from the air started in the South Island in the mid 1960s. It rapidly achieved a dramatic reduction in deer populations by at least 75–90% over much of their range in open un-forested country, where deer were extremely vulnerable to aerial hunting. Reductions were far smaller in areas where a continuous forest cover provided deer with good cover, as at Pureora [37]. Helicopter hunters reached the Hauhungaroas in about 1975/76 [29].

Most helicopter crews were based in Taupo, and they usually worked in the more accessible open country east of the lake. When they did visit the Hauhungaroas, they took the carcases back to Taupo airport. Another very consistent and successful operator was Ben Dellow from Kopaki. From June 1980 to May 1981 inclusive, 11 helicopter operators took 1509 deer from the range. The official statistics for 1979/80 to 1987/88 (Box 12.4) include some goats and pigs taken incidentally, but many kills were not reported, so Leigh & Clegg reckoned that the tallies were probably underestimated by some 50 %.

Box 12.4 Commercial recoveries of deer, goats and pigs from PFP, 1979/88 [29: 18]

Year	Number of operators	Deer	Goats	Pigs
1979/80	9	155	–	–
1980/81	11	1509	–	–
1981/82	5	470	28	13
1982/83	4	930	54	65
1983/84	5	300[a]	3	19
1984/85	3	202	27	7
1985/86	6	252	143	5
1986/87	4	366	67	7
1987/88	4	437	6	20

[a]Live capture largely replaced meat recovery from 1983/84 until 1986/87

A survey of the conservation role of commercial deer hunting in New Zealand generally was compiled in 1992 [38], but its recommendations have been overtaken by events. DOC has no resources to report recreational kills or to collate responses, so trends in deer numbers since then are not known. Commercial hunting has all but ceased in PFP, so deer numbers may be increasing. Continued use of 1080 without deer repellent may occasionally curb the increase [42].

The area zoned for helicopter hunting (the Wild Animal Recovery Operations or WARO zone) was about 29,000 ha south of Pureora Mountain (Fig. 12.5). It was chosen on the basis of the survey by G. T. Jane [23] in 1978/79, who found high numbers of deer in that area. Unfortunately, only a fraction of it was huntable from the air, because much of it was under a thick forest canopy that offered shelter to the deer. There were also many ground based commercial hunters active throughout the Hauhungaroa Ranges, except during the "Roar", when competition with other hunters was highest (Chap. 14). Some used horses to carry out carcasses, along marked tracks leading all the way down from the crest of the range.

The helicopter crews had to rely on searching the scattered clearings, slips, creek heads and mires where the vegetation was lower. Success demanded an intimate knowledge of these sites and of the times of day when deer were likely to venture out from cover. Every visit cost hundreds of dollars an hour, so they had to be quick.

For all these reasons and others, the effect of helicopter hunting in the Hauhungaroas was much less spectacular than on the open tops of the South Island. The 1982/83 forest survey in the recovery area showed that deer numbers remained constant between 1978 and 1983 [3]. More recent comparisons of paired study areas of 3600 ha (one pair was at Waihaha), one with and one without helicopter and/or ground hunting and all monitored over eight years, found no consistent difference in abundance of deer attributable to the level of control applied [14]. The authors of that study concluded that "Reducing the abundances of deer in forests may require substantially more control effort than is currently believed".

So why do the commercial deer kill figures for PFP record a sudden drop in the numbers of deer killed in 1983–87? Because the helicopter crews had temporarily switched to live deer recovery, a competing and more lucrative strategy.

Deer: The Short-Lived Live Capture Boom

Legislation allowing the establishment of deer farming after 1969 prompted a change in strategy: wild deer no longer had to be shot for meat or left to rot as pests, because they were now much more valuable alive. The development of new techniques for catching live deer, not necessarily from helicopters (very difficult in thick forest) but in nets or pens, produced a new boom in deer hunting from 1979/80, this time concentrating on live capture of animals for sale to deer farmers.

Prices were huge: at the first auction of live deer in 1977, mature stags fetched NZ$750, and six-month-old weaner hinds NZ$550. Interest in the deer of Pureora increased likewise, especially as by then the easier hunting grounds were becoming depleted [29]. In 1981/82, prices for live deer skyrocketed: hinds fetched NZ$1000–2000 straight off the helicopter. NZFS records show that live capture operations continued until 1987, despite attempts to discourage them by the Ministry of Agriculture, concerned about exporting bovine TB from the park to deer farms.

In the late 1970s, eleven deer capture pens were built in the Hauhungaroas and serviced by helicopter or vehicle. The pens were privately owned and operated, but licenced under bond by NZFS. The efficiency of the pens varied with site and construction, but some were very successful: ten of them caught 145 deer in 16 months. On the other hand, they were expensive to construct, and worthwhile only when the catch was valuable. Eventually they were dismantled on the instructions of the PFP Advisory Committee in 1981, but deer traps on private land adjacent to the Park continued for some time.

Within a few years, by June of 1987 there were 25,429 farmed deer in Otorohanga and Waitomo Districts and Taumarunui County [29]. Deer farms became self-sufficient in breeding stock, and the market for wild-caught deer of unknown pedigree and disease status simply collapsed.

Helicopters returned to meat recovery operations, but by then the economics were against them. In 1980 the retail price for venison had been about NZ$3/kg, and was still only about NZ $3.55/kg in 1988. But then the average wholesale price for all carcase weights and all shot sites (hunters earned more if they shot in the head or neck) fell to NZ$2.60/kg for an average gutted carcase weight of 40 kg. The costs of operating a helicopter tripled in the same period.

Two independent operators were asked what revenue they had to earn to maintain a helicopter operation with a Hughes 300 over a year. They agreed on at least NZ$180,000, more than could be earned by a helicopter crew who recovered and sold more than 1700 deer a year. The yield of a few hundred a year from PFP, divided between several operators, was clearly not going to keep the commercial guys in the park for long. So perhaps it is not surprising that the helicopter crews killed many fewer deer in Pureora than did the recreational hunters (Chap. 14). Nonetheless, for a few more years many operators (both in PFP and nationwide) were able to continue hunting part-time by hunting only at the times of year when deer are most vulnerable—typically the late spring and autumn—and on days when the weather was particularly likely to encourage deer into the open.

In 1986, NZFS warned the incoming Minister of Conservation that DOC would no longer be able to count on free help from commercial helicopters in the battle to reduce the numbers of deer in forests [35]. On the other hand, within a few years this shortfall would be met by intensive aerial 1080 operations against possums, which also killed many deer (Box 12.6).

Feral Goats and Mohair Farming

Feral goats have long been regarded as noxious pests by foresters, and Government-funded control operations have accounted for tens of thousands of goats in and around the King Country (Fig. 12.6). In the forests between the Rangitoto Range and the Main Trunk Railway, NZFS specialist goat hunting teams killed over 14,000 goats between November 1975 and December 1980. Contracted helicopter-based hunters working under a "kill to waste" policy, over lands of all tenure, started in 1976 and had killed 15,500 goats by 1980, once including 900 in one day. But, impressive though these figures might seem, they achieved little lasting benefits for the forests [29].

John Mason remembers why: at that time, NZFS was interested only in tallies, not in ecological benefits. "Once your tallies were down to 3–4 goats a day, it was time to move to another area, when in reality that was the very time when you should have persevered, to get those last few, most valuable goats. But we didn't", he said. Adding to that, of course, deer were also present (but untargeted) in many of the goat-control areas.

There is, however, another consequence to the pest status of goats. It means, among other

Fig. 12.6 Feral goats (here photographed north of Mangapehi SF) are sociable, fertile, and abundant, and they eat almost anything. *Graeme Reinhardt*

things, that they do not belong to anyone. They are free to all comers interested in capturing the values they still have for various purposes, including meat, weed control and (dramatically but briefly) fibre.

Mohair and cashmere are fine fibres produced by certain breeds of goats. The best fibres come from domestic herds of specialized angora and mohair breeds, but even feral goats can produce 50 g of cashmere per goat per year. Selective breeding from these can improve the yield to 200 g per goat per year. In the early 1980s, the fledgling NZ mohair and cashmere industries were widely and skillfully promoted as another diversification strategy to revitalize national agriculture, one with a huge potential to make profits from the world's undersupplied goat fibre markets [29].

Unsurprisingly, the establishment of hundreds of goat farms set off a new wave of live capture operations in the back country. Feral does were fetching NZ$15 a head in 1983, rising to between NZ$100 and NZ$250 each by 1985/86. NZFS staff at Te Kuiti were issuing dozens of capture permits a day in 1985 and 1986, the busiest years, to provide foundation stock.

At least 10,000 feral goats were removed from the Maniapoto Crown lands during a three-year boom (1984/86). The official records refer only to those operators who did things legally and returned capture permit data, so the real tally was probably more than double that. Leigh and Clegg [29] quote Ray Scrimgeour as remembering one yarding of 10,000 feral goats brought from all over the King Country to the Te Kuiti sale yards in 1985. The two principal stock agent companies in Te Kuiti traded at least 59,000 goats in the period 1984–88. The figures do not show what proportion of these animals came from PFP, but at least some probably did.

In January 1988, the demand for live feral goats collapsed as suddenly as it had begun. The unrealistic high prices being paid for feral does evaporated as farm crossbreeding and embryo transplant programmes were completed. In February 1988, feral does fetched NZ$3–NZ$5 per head at auction. Applications for capture permits dropped from 47 in 1987 to one in 1988.

The immediate benefits for the forests of this massive exodus of feral goats from conservation land were not measured at the time, and anyway they would have been short-lived. By 1987 there were about 78,000 goats held on farms in Otorohanga and Waitomo Districts and Taumarunui County, but there was no effective legislation to ensure safe containment of the captured goats, no compensation for their sudden decline in value, and nothing to prevent them from being simply abandoned. Many goats made their own way back to the forests, and truck loads of goats were driven to the far reaches of dead-end roads and released into the bush. After about 1985, ear-tagged feral goats were often shot by DOC hunters inside conservation areas [29].

Worse, goats have a very long history of domestication, and they demonstrate superbly the high fecundity that was deliberately favoured by selective breeding of domestic goats over the millennia. Exploited populations of goats have a staggering potential recovery rate: in the last stages of the eradication campaign on Raoul Island (in the Kermadec Group, north of New Zealand), the well-fed survivors could have doubled their numbers in only 20 months [45: 386], if the eradication had not succeeded. In PFP, the forest populations depleted by the live-capture operations would have been back to normal within five years [29].

The most serious effects of the 1984/86 bubble have been a long-term, unintended redistribution of feral goats escaping disturbance, including into areas that had previously been goat-free. Live-capture teams targeting easily accessible forest-edge populations may have prompted goats to disperse deeper into the forest [29]. Hunters have always been willing to shoot goats for dog tucker, or occasionally for the pot, but goats do not compare with deer as game meat or trophies, especially in remoter areas.

This, plus the temporary shortage of goats to shoot, may explain why recreational hunters did not immediately step in as the live-capture teams left. The recreational tallies of goats from PFP are poor (Box 14.1). Parties of official hunters working in the north-west Rangitoto Range did a little better, but not much: they killed 689 goats over 928 hunter-days in 1985/86 and 356 goats over 297 hunter-days in 1986/87 [29].

Possums: Bounties and Fur

In PFP, the first signs of possums causing damage to forest vegetation were recognised in 1949, and later their role as a potential competitor for the foods of native birds including the kokako became clear during the FBRG programme of 1978–81 (Chap. 13). So conservation authorities have long had a policy to permit, and in places to encourage, the harvesting of possums wherever it can be done to the mutual benefit of the fur industry and of conservation management.

Until 1947, possums were regarded as a resource to be harvested only under licence. By 1951 (before possums reached PFP), this system had switched to a bounty of two shillings and sixpence (about NZ$7.90 in 2015 money) per head, which ran until 1961. It failed because, as usual with bounty schemes, most animals submitted for payment were taken from roads, farms or other easily accessible places, while in the forests where removing them was most important, the possum populations continued to grow, sometimes aided by illegal, deliberate introductions ensuring future bounty income.

Once possums reached harvestable numbers, and the necessary infrastructure of pelt buyers and processing plants was established, people began to make a living from them. Hunters were encouraged to harvest possums under a free permit system. Large forests were divided into possum blocks and allocated to applicants claiming the right to harvest skins for sale by trapping or poison baiting. Throughout the 1970s the pelt prices made possum-trapping a viable trade, subject always to variations in the market for fur. National exports ranged from 200,000 pelts in 1960 [29] to 3.4 million in 1981 [24, 44].

Naturally, the effort put in to commercial possum hunting is driven by the fur-buyers. The entire area of the DOC estate within the Maniapoto District administered for possum hunting in 1988 was about 132,000 ha, over which the harvest collected by 302 hunters for the 1987/88 season was about 149,188 possums. PFP contributed almost half of them (Box 12.5), mostly greys (Fig. 13.5), with a few blacks and browns (Fig. 12.1). In the 1980s the revenue from the entire Maniapoto District averaged about NZ$6 per skin, or NZ$895,128 a year, none of which went back to DOC [29].

Box 12.5 Commercial possum harvest from PFP [29]

Year	Permits issued	Possums caught	Average export price per skin[a]
1979/80	?	38,892	?
1981/82	268	74,033	?
1982/83	186	43,975	?
1983/84	232	51,462	NZ$5.45
1984/85	209	53,166	NZ$6.27
1985/86	194	42,072	NZ$6.62
1986/87	229	81,932	NZ$7.03
1987/88	256	68,106	NZ$7.87

[a]Export, free-on-board prices estimated from the total annual exports of unprocessed possum skins after payment of all charges

Allowing for around 4–5 possums per hectare, the hunters removed only about one possum per hectare, or a quarter of the population overall (less than the annual recruitment of young). However, trappers do not spread their efforts evenly, but focus on the most-easily accessible areas where they can maximise the number of possums captured each day. In those places possum densities can be substantially reduced, but there is little incentive for the hunter to return to the same place until numbers have built up again—so over time there is only a relatively small reduction in average density, not nearly enough to make any difference to the forest or the birds, even if possums had been taken equally from hard-to-access areas as from road-sides, which assuredly they were not.

Rather, possums were a nice little side earner for local people including Pureora residents—there were always traps along Perham Avenue (Fig. 7.7), and many people pulled in 4–5 possums on every morning trap round. DOC's policy, commented Leigh and Clegg [29: 80], was to maintain rural employment and the benefit from some cropping of animal populations, whilst receiving no significant revenue from its wild animal resources. Neither aim was acceptable to the economists running the 1984–90 Government [25], and their successors. Could this system be improved?

Part of the dilemma is that possum hunters and DOC managers do not want the same things [24]. Hunters need to make a living from their work, but the most economically sustainable harvest strategy still cannot reduce possum numbers down to the level consistent with a good conservation outcome unless fur prices are very high. Managers do not have accurate information on possum densities in the different hunting blocks, or on which individual hunters would get the best returns from given blocks, because such information is too expensive to obtain except as part of special research projects funded from outside.

The international fur business was devastated by the global anti-fur campaign of the 1980s (most effectively in a famous 1985 TV ad "It takes 100 dumb animals to make a fur coat, but only one to wear it"). But that had only limited effect in New Zealand, because catching common pest animals that happen to carry a marketable fur was never a moral issue in the same category as the indefensible trade in pelts of rare spotted cats.

More significantly, the New Zealand fur industry has adapted to trade, not in whole bloody pelts but in loose fibres plucked cleanly from the fresh skin, and sold for spinning with merino wool into a soft and luxurious fabric. Prices for loose fur vary a lot of course, but can be surprisingly high. In November 2011, possum fur buyers were paying NZ$140/kg, for which a trapper would need to catch about 12 possums [24].

The luxury-garment industry based on merino/possum fibre grew at about 10 % a year between 2005 and 2012, to reach a value of about NZ$100 million a year [24]. Plucked possum fur now accounts for about 1 % of the worldwide fur market by value, but not much of that now comes from PFP. Commercial possum hunting in PFP has virtually stopped, thanks to a new dilemma completely independent of the fur business.

Out of the blue, all the old issues that had long dogged the economics of possum control via fur trapping, bad as they were and are, paled beside a new one, which suddenly appeared in the mid 1970s.

Possums: Bovine TB and the Threat to Pastoral Farming

Bovine tuberculosis (TB, caused by a bacterium, *Mycobacterium bovis*), an old-world livestock disease that can also affect humans, has been widespread in New Zealand cattle since the earliest days of settler farming. In 1967, it was discovered in Westland possums. It probably also became established in possums in the King Country at around the same time, because surveys in the mid 1970s showed that bovine TB was already endemic in populations of wild possums on both sides of the ranges, in the western bays of Lake Taupo and also in the Waipa Valley. Agricultural authorities scrambled to investigate, and came to the shocking conclusion that TB was being transmitted from possums to cattle, and also between possums.

The discovery that possums had become a self-sustaining, independent wild reservoir of the disease changed everything, because it meant that diseased possums (Fig. 12.7) could continue to re-infect cattle indefinitely. TB also often infects pigs, deer and ferrets (and, less commonly, many other species), but in places like PFP, only possums are true wild reservoir hosts.

Although the risk has been much reduced over the last 20 years, transmission of TB between possums and cattle remains a severe hazard for New Zealand's dairy industry. The status and importance of possums in the central North Island therefore makes them, not only a serious cause of damage to native vegetation and a minor

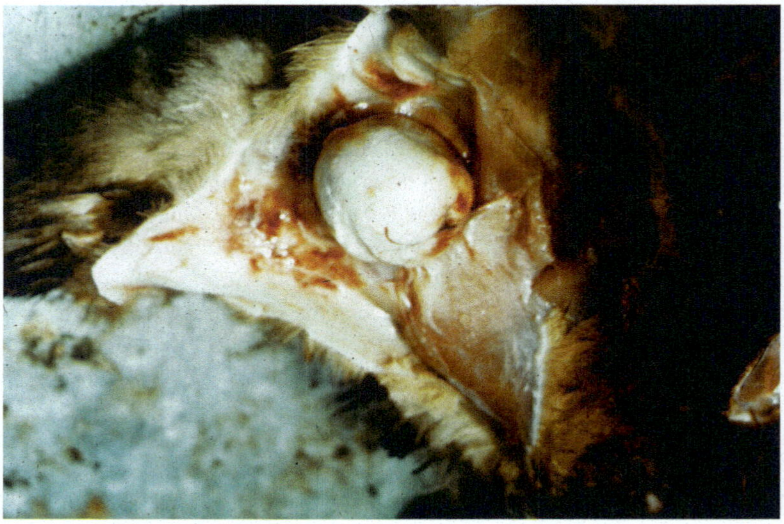

Fig. 12.7 A possum with an advanced infection of bovine TB, showing a large lesion in the auxiliary lymph node. Moribund possums staggering about on open pastures attract the interest of curious cattle, which nuzzle and lick them, so picking up the infection directly. *John Bathgate*

source of income for fur trappers, but also a pest of national economic significance, and a very real threat to New Zealand's pastoral exports.

Aerial Control Operations

NZFS and the Animal Health Board (AHB, then called the Animal Health Division of the Ministry of Agriculture, and now TBfreeNZ) took charge of an intensive, nationally funded TB-control campaign. Their weapon of choice for forested areas was and is aerial distribution of sodium monofluoracetate (1080) poison, at first in chopped carrot baits and later, in purpose-manufactured cereal baits. Cereal-based pellet baits containing 0.15 % 1080 are attractive to possums and rats, also to some deer, and a few pigs and goats. 1080 is by far the most effective method of reducing possum numbers at landscape scale, but is also very controversial. In the central North Island the use of 1080 started in Waimahora in 1976, and has continued in and around the park more or less ever since.

It is possible that the apparently rapid establishment of TB in possums in the PFP area was aided by the fact that the possum population was increasing to high levels at about the time that TB made the jump from livestock to possums. Possums were already very abundant in the central Hauhungaroa Range by the 1970s. In 1978 Graham Nugent killed 250 possums along a single 1.5 km cyanide line close to the crest of the Hauhungaroa Range [43].

From 1976 onwards, 1080 bait was spread from the air, initially up to 2–3 km into the forest from the bush–pasture margin. Populations of possums, pigs and deer were reduced, the TB reactor rate in cattle fell dramatically, and the problem was considered fixed [29]. Not so. By 1982 TB in cattle had returned to higher levels on both sides of the park.

The first formal estimate of TB prevalence in possums in 1982–83 examined 6083 possums from the Hauhungaroas, of which 5.4 % of males and 3.9 % of females were infected [47]. From then to 1988 large-scale aerial and ground operations were done every year in various places around the periphery of PFP. In total some 24,000 ha of Crown lands were treated in the 1983–88 campaigns, mainly with 1080 cereal bait. Parallel and consequent studies have carefully documented the effects of 1080 operations on birds (Chap. 13).

The early start-then-stop attempt to control TB in PFP possums was repeated elsewhere, resulting in a rapid expansion during the 1980s in the number of infected herds nationally, and in the total distribution of infected possums around New Zealand. By the 1990s, the cattle and deer industries and government had begun dramatically to increase funding for TB-possum control, amounting to about NZ$50 M a year since 2002. The consequence for PFP was a serious renewal and expansion of possum control efforts from the early 1990s onwards, mostly with aerial 1080.

To check the effectiveness of these operations, a standardised monitoring technique was developed, analysing the Trap-Catch Index (TCI) (number of possums caught in a fixed trapping regime) after each operation. The operations were sometimes extremely effective—an operation covering the whole of the eastern Hauhungaroas in 1994–95 achieved a very low residual trap catch immediately, and possum surveys found no evidence of TB persisting in that area more than two years later. In contrast, the prevalence of TB was close to 6 % in possums, 47 % in deer, and 80 % in pigs within the uncontrolled central-western area during 1997–2000 [43].

In 1998, the TB-possum control programme was formalised as a National Pest Management Strategy (NPMS) for bovine TB, under the Biosecurity Act 1993. The Strategy has been revised twice since. To start with, it aimed simply to reduce TB levels in cattle and prevent expansion of the area occupied by TB possums, but as funding increased and as the programme became increasingly successful, the key objectives of the NPMS were refined. From 2004, it next aimed to reduce the national herd infection rate to no more than 0.2 % by 2013 and to halt the geographic spread of TB infection in possums beyond the 10.6 M ha (c. 40% of NZ) then affected by wildlife TB. This programme was to be achieved by controlling TB both in livestock and in infected wildlife populations.

The NPMS objectives were reviewed again in 2011, and the new key goal is to eradicate TB from wildlife over 2.5 M ha of New Zealand by 2026. The goal includes demonstrating that TB can be eradicated from wildlife in two large areas of difficult operational terrain (i.e., heavy forest) before 2026, and PFP has been chosen as one of those two demonstration areas. The TB-possum control programme in PFP has made great progress: at the time of writing, the last known case of TB in a possum was recorded in the Maramataha catchment in 2005 [41].

TB continues to be found occasionally in deer (and in pigs that scavenge on deer carcasses) because deer can remain alive long after becoming infected from possums. It is therefore still too early to stop possum control, as the disease could re-establish, but at present it is expected that the last aerial 1080 operations for TB-possum control will be completed by about 2018 [43].

The success of the 1994–95 TB-possum control operations in reducing TB levels, not only in possums but in deer and pigs as well, was of course largely restricted to the areas in which control was applied early in the programme. Although that area did include a large part of the total, there was a so-called 'hole in the middle' of the western-central Hauhungaroas that was not treated until about 2000, and even then not very well. In that area, pre-control trapping surveys in 2005 identified around 5000 ha in which all but one of the 17 transects recorded TCIs of >15 %, much more than the acceptable threshold value of 5 % [41]. The implication was that TB could be persisting in some places.

Hence, in 2005 a huge aerial 1080 operation covered the whole Hauhungaroa Range (82,976 ha: Fig. 12.8) for the first time, with dual prefeeding at what would now be regarded as very high sowing rates [8] applied to most of the area. That operation reduced possums to extremely low densities—just seven possums were captured from 514 traplines of 10 traps set for three nights (a TCI of 0.04% compared with the 20–30% likely before 1994) – suggesting that possums had been virtually eliminated from the Hauhungaroa ranges. Inevitably, however, numbers soon began to recover, requiring continued efforts to prevent TB from persisting.

In the Rangitoto Range, possum control dates back to 1987. The first attempt to cover the whole 40,000 ha area in 1996 achieved an estimated kill of 99.3 % [57]. Part of the c. 4000-ha Waipapa Ecological Area has been protected by long periods of annual ground-based possum and

Fig. 12.8 Aerial distribution of 1080 bait over the Hauhungaroas during an extensive possum control operation in 2005. *Kane Stafford*, Epro Ltd., Taupo

ship rat control, to minimise predation on kokako nests, and contractors are confident TB will have been eliminated from possums there. Taken together, possum control work in the Hauhungaroa and Rangitoto Ranges has played an enormously important role in TB research—much of what we know about TB in possums and deer comes from Pureora Forest Park.

Can Trappers Help?

The need to conserve native forests in general, and to protect the food supply of the kokako in particular, are both ongoing problems for conservation management. Many people assume that possum trapping can still make a contribution to this effort, at least in flagship reserves like PFP.

Most possum trappers firmly believe that widespread poisoning of potentially valuable fur-bearing animals and leaving them to rot is a waste of resources, hence possum control should be left to them. Against that is a powerful argument from simple economics: trappers cannot continue to work profitably on a population after they have reduced the number of possums available by up to 40 %, more often 20 %, whereas possum numbers in the Hauhungaroas after the 2005 poison operation (Fig. 12.8) could increase at up to 59 % a year in the first few years [55]. That inescapable mismatch between the rates of removal versus replacement ruled out trapping alone as a practicable strategy decades ago [33].

There is a possible solution to this apparent impasse. Chris Jones et al. [24] estimated that a subsidy of NZ$13.60/ha could enable trappers to go on removing possums to lower-than-normal economic densities. Moreover, this subsidy would be cheaper than a full-cost ground control operation (NZ$45–80/ha) in the kind of steep forested country worked by the trappers interviewed by Jones and his team. At the moment, the word "subsidy" is politically unacceptable, but that could change, for two reasons.

First, aerial distribution of 1080 can at present remove possums from huge areas for NZ$20 or less (including overheads) in unforested semi-arid country, or <NZ$30–40/ha in places such as PFP [43]. The massive anti-TB campaign driven by TBfreeNZ has so uniformly and massively reduced the numbers of possums in PFP, to well below the densities at which any harvest is economically viable, that there is now no commercial hunting of possums anywhere in the park. But if the 1080 option is ever ruled out, an integrated strategy including trapper subsidies might begin to look more feasible, especially considering the added benefit in employment for rural communities.

Second, TBfreeNZ's TB-control operations and surveys are unlikely to continue beyond about 2026 once they have achieved their objective of declaring the wildlife (possums, pigs, and deer) in PFP free of TB. When that happens, DOC will have to run the whole possum control programme in PFP independently.

There is now more than enough evidence for DOC to justify the expense, at least in the most critical areas, and it will be needed indefinitely, because low-level modification of the forest by possums is on-going, and, more importantly, the survival of the kokako is now known to depend on how far it can be protected from predation by possums and ship rats (Chap. 13). The benefits of possum control for the forest are well known—witness, for example, the recovery of rata in the Waipapa Ecological Area (Chap. 13). The problem will be how to fund it.

But the complex interactions within and between animal communities mean that environmental management always affects far more than the target species, and can produce an outcome more deadly than the original problem. The most significant example of this Gordian knot is the relationship between possums and ship rats. Possums and rats compete for the same habitats and foods, but possums are dominant. Both are killed by poison campaigns, but rats breed faster, and recover more quickly (Chap. 13). For a few years until possums catch up, rats can reach higher numbers than usual, with all-too easily predictable consequences for birds.

The 1080 Dilemma

Years of repeated and reliable research show that 1080 is the single most effective method of controlling the numbers of introduced vertebrate herbivores in New Zealand. Carcases or moribund animals that have taken 1080 are also attractive to predators (stoats and cats, and, regrettably, uncontrolled dogs). This combined effect gives well-managed 1080 operations the best chance of removing most introduced pests without affecting native animals, and yet it is subject to ongoing and passionate protests by several well-organised lobby groups aiming to have it banned.

The arguments against use of 1080 have gathered a lot of public support despite many careful scientific studies and multiple critical reviews, all concluding that the objections to the use of 1080 do not outweigh the benefits. The most authoritative and thorough of these, by the Commissioner for the Environment [46], did not end the debate. Why not?

First, hunters deplore the effects on their sport of accidental poisoning of valuable game (Box 12.6). Surveys confirm that 1080 not only reduces their hunting success, but also risks leaving sublethal residues in surviving animals, making the meat unsafe to eat The hunters' objection to the apparently needless slaughter and contamination of game animals is a significant driver of the anti-1080 lobby, and is strongly supported by those for whom hunting is their main priority, but again, hunters cannot remove deer, pigs or goats faster than they can be replaced. In response, an effective deer repellent has been developed, and although expensive, it can be used in areas where hunters' concerns might otherwise prevent 1080 use.

Box 12.6 Possum control in the Hauhungaroa and Rangitoto Ranges

Table 12.1 Summary of recent monitoring outcomes after possum control in the Hauhungaroa Range

Study block	CCI (%) Autumn 2011 (Pre 1080)	CCI (%) Spring 2011 (immediately post-1080)	CCI (%) Summer 2012 (6–8 months post-1080)	RTCI (%) Summer 2012 (6–8 months post-1080)	CCI (%) Autumn 2013 (pre-1080)	CCI (%) Spring 2013 (post-1080)
AS1				0.34 (±0.50)		
AS2 E	34.4	28	22	2.48 (±1.42)	24	3
AS2 W	20.6	1	0	1.29 (±1.14)	9	0
AS3	47.5	8	8	1.72 (±1.17)		
AS4	23.3	9	1	0.0	6	0
AS5				0.14 (±0.29)		
AS6 N	5.6	3	1	0.21 (±0.45)	1	0
AS6 S	2.5		0	0.0	0	0
AS7 W	38.2	4	7	0.86 (±1.06)		
Tihoi 3B	17.6	1	2	0.34 (±0.49)	2	2
Tihoi 4				0.34 (±0.49)		

Study blocks are TBFreeNZ management units. Data collected before and after aerial 1080 poisoning of most of the area in 2011, and additional control in some areas in 2013. *CCI* 7-day Chewcard Index. *RTCI* Residual Trap-Catch Index. For explanation of CCI and other indices, see [56]. *Unpublished data courtesy G. Nugent*

Table 12.2 Red deer kills monitored during 1990s aerial-1080 poisoning operations in PFP, mainly targeting possum populations

Area	Date	Bait type	Toxic loading, % 1080	1080 Sowing rate (kg/ha)[a]	Deer kill (%) of est. pop.
North Pureora	1 July 1988	Pollard	0.08	10	43
Pureora, 7-km buffer	2 June–August 1994	Carrot	0.15	15[a]	30
Pureora, 3-km buffer	2 June–August 1994	Carrot	0.15	15[a]	31
Pureora, no pre-feed	2 June–August 1994	Carrot	0.15	15[a]	42
North Pureora	3 May 1996	Carrot	0.09	15[a]	57
Titiraupenga	4 August 1997	Carrot	0.08	15[a]	93
Titiraupenga	4 August 1997	Carrot	0.15	15[a]	92

[a]These operations were pre-fed with non toxic bait. Note that these sowing rates are 5–10 times higher than the 1.5–2.0 kg/ha sowing rates used in recent operations. Data sourced from multiple unpublished research reports by W. Fraser and P. Sweetapple, copyright Landcare Research and the former Forest Research Institute

Second, as the Prime Minister's Science Advisor, Sir Peter Gluckman, has pointed out in a different but comparable context (the anti-fluoridation debate), the two sides are arguing from different positions. The science of 1080 use is settled, but the objections are more personal. A philosopher would recognise the important distinction between facts (how things are) and values (which things matter) as a means of untangling such differences (Chap. 3). Managers have to find ways of acknowledging different viewpoints while getting on with the business of efficient protection of native species. There is no end to the row in sight, but the thorough review assembled by the Parliamentary Commissioner for the Environment did at least support the continued use of 1080 by statutory authorities for the foreseeable future.

Questions surrounding the legitimate use of 1080 poison continue to rank high among the most intractable dilemmas facing managers of large forests in New Zealand. No-one regards 1080 as ideal—indeed, all parties, including its current supporters, hope that better methods will be found eventually—but for now, 1080 is the best option available. In the meantime, a lot of work is being done to address the legitimate concerns of conservationists about the side effects of 1080 operations for bird populations, and much of that work is being done in PFP.

1080 and Birds

In the winter of 1983 an aerial 1080 poison drop was planned against possums, using cereal pellets, to cover some 1500 ha on the southeastern side of PFP, between the Waihaha River and the Waihora Stream. The effects were monitored by the five-minute bird count method in both poisoned and control blocks, before and after the drop [61]. Bird counts were made at 30 stations on circuits in each of the treated and untreated (control) blocks of the trial area. Most of the counting stations were in podocarp forest dominated by rimu. They found no significant reduction in populations of twelve native and four introduced bird species. A search revealed no dead birds.

Of course, this and every similar study has limitations: the five minute bird count method does not sample equally well the rarer, nocturnal or aquatic species in the area (kaka, parakeet, NZ falcon, morepork and blue duck) or the strictly territorial species with limited distributions (North Island robin); and there were differences between the forest blocks compared because the treated block was mostly in virgin forest, and the untreated block had been selectively logged.

So this was followed by many other studies, some using the much more demanding methods of territorial mapping of known colour-banded individuals. PFP is proving to be one of the best available areas to do that difficult work. For example, it was imperative to work out whether kokako were at risk from operations to remove rats and possums by aerial distribution of 1080 bait. Non-toxic pellets were placed on tree platforms near ripe natural fruit, to see if kokako would confuse them with their normal foods. They didn't. Subsequent detailed monitoring of kokako survival in real 1080 operations confirmed this conclusion [58].

Kaka and Blue Duck

Kaka are potentially at risk from 1080 carrot baits because captive kaka are known to eat carrots. Blue duck are unlikely to take carrot baits, but they are easy to find by hungry stoats and cats deprived of rodents after a 1080 operation. So in 1994 Terry Greene led a programme to monitor marked individuals of both these vulnerable species during a planned 1080 operation over 24,600 ha in the Waihaha EA [20].

None of 20 kaka and 18 blue duck that were radio-tagged and followed throughout the 1994 operation were directly affected. That does not mean that no unmarked birds died, but there could not have been many of them. So why did they escape? Greene pointed out that kaka prefer larger food items than the average carrot bait (6 g), and one toxic bait of that size would not carry enough 1080 to kill a kaka. More importantly, kaka feed in the canopy, so are unlikely to encounter the baits laid on or falling to the ground.

The possible effects of prey switching on both kaka and blue duck during the nesting season was suggested by the shortage of adult females (only six female blue duck and three female kaka were caught), but a broader study comparing kaka nesting sites at six locations (three with and three without predator control) was more encouraging. The team concluded that, given protection from predators, nesting adult female kaka suffered significantly lower mortality (5 % vs. 65 %), and the numbers of young female kaka surviving to sexual maturity more than compensated for the losses of adults [32].

Over time, the improvements in techniques used by DOC's intensive management programme in Waipapa EA have produced some spectacular improvements in forest health and increasing numbers of native wildlife. As of 2014, Waipapa is home to New Zealand's largest kaka population, numbering >800 birds [10].

Robins and Tomtits

North Island robins and tomtits are small, friendly birds that are active on the ground or in the sub-canopy, so are easily captured for banding. At Pureora these birds have lent themselves to some important experimental observations, several led by Ralph Powlesland [48]. His team monitored colour-banded robins (Fig. 12.9) in treatment and non-treatment areas to determine the costs and benefits of 1080 operations to robin populations.

A disastrous operation in August 1996 had resulted in 43 % mortality of known territorial birds (banded and unbanded) or 55 % mortality of colour-banded birds. During the same period there was no robin mortality in the non-treatment area. This episode caused great concern, and has been widely quoted by opponents of 1080 operations as proof that use of 1080 should be abandoned. But, as John Gaukrodger pointed out, there was a simple reason for this unfortunate outcome, and it was nothing to do with whether or not 1080 is dangerous to birds.

When chopped carrot is to be used as bait, it always has to be passed over a screen attached to the cutter, which removes the fine particles of bait ("chaff"). That ensures that the toxin is spread only on large pieces of carrot, acceptable

Fig. 12.9 Catching North Island robins for banding during research at Pureora in 1996, to determine the effects of 1080 baiting operations on native birds. Robins were trained to collect mealworms from familiar sites, where they could be harmlessly caught in a clap trap. Inset, a banded robin with a mealworm. *Crown Copyright, Department of Conservation Te Papa Atawhai. Photographer: Ralph Powlesland*

to possums but too big for birds to pick up. The speed of this screening process determines the volume of fines removed. Normally, around 25 % by volume of the carrot is screened to waste.

On this occasion, the operating team decided to break the record for the most bait distributed in a day. To keep up the supply of prepared bait to the helicopter, they had to speed up the rate of cutting the carrot and passing it across the screen. That meant that the screening process was less effective, and when the operation was checked by John Mason, only 12 % of the total volume of carrot was in the waste heap, less than half the expected 25 %. Hence, the bait distributed included too much chaff loaded with toxin and available to the birds.

For the next operation in September 1997, the team ensured that all the carrot chaff was removed. The result was much better—the 8.6 and 9.7 % robin mortality (as determined by territory mapping and of banded birds, respectively), was about the same as in the non-treatment area. Possums and rats were reduced to very low densities during the robin nesting seasons (September–February) following both operations.

In response, the birds did their part, laying two or three clutches averaging 2.6 eggs in the season from September to March [49]. Robin breeding was particularly successful in the treatment areas following the two 1997–98 poison operations (producing a mean of 3.7 and 3.8 fledglings per pair) compared with that in the non-treatment area in 1996–97 (0.4 fledglings per pair) and in the post-treatment area in 1997–98 (1.5 fledglings per pair).

Through recruitment of the fledglings, both populations had more robins, and a greater proportion of females, one year after the possum control operations than immediately before. The team concluded that, as long as carrot baits are properly screened to remove chaff, and the baits are distributed over large blocks of forest so that pests remain uncommon during the next robin nesting season, the robin *populations* will benefit from 1080 possum control, producing many more young birds than are needed to replace any *individuals* that die during the operations. When 1080 opponents choose selective quotes from these studies as evidence that 1080 damages bird populations long term, they are ignoring the subsequent increased breeding rates—as if a bank balance could be predicted without listing deposits as well as withdrawals.

The importance of minimising the sowing rate was illustrated by Powlesland's team again when they colour-banded resident tomtits to monitor the costs and benefits of two successive aerial 1080 operations to tomtit populations. In August 1997 they used carrot baits with very little chaff, sown with 0.08 % w/w 1080 at 10 kg/ha, and were alarmed to find that 11 of 14 banded tomtits disappeared, but none of nine from the non-treatment area [50].

In August 1998 they used cereal baits, sown with only half the amount of 1080, at 5 kg/ha. The cereal baits sown at that rate were just as effective as the carrot baits were in driving down the numbers of possums and rats, but much kinder to the tomtits, because no tomtits in either the treatment or non-treatment areas disappeared.

This encouraging result was confirmed a few years later by a study comparing cereal and carrot baits spread at an even lower 1080 sowing rate, 3–5 kg/ha. Both baits reduced possum numbers below the target level, but cereal bait operations with low sowing rates and large bait size had little, if any, immediate impact on tomtit populations [62]. Studies like these are crucial to determining the least toxin needed to obtain the greatest benefit for wildlife, such as in the 20,000 ha Tongariro Forest Kiwi Sanctuary, where kiwi chick survival usually doubles in the breeding season following a 1080 drop at the standard sowing rate of 2kg/ha of 12 g cereal baits (Fig. 0.1 in Preface and Fig. 12.10).

1080 and Non-target Pest Mammals

Conservation agencies in other parts of the world are often astonished that 1080 can be aerially distributed over New Zealand forests without permanent harm to native fauna. The reason is, of course, that in most places where it is used

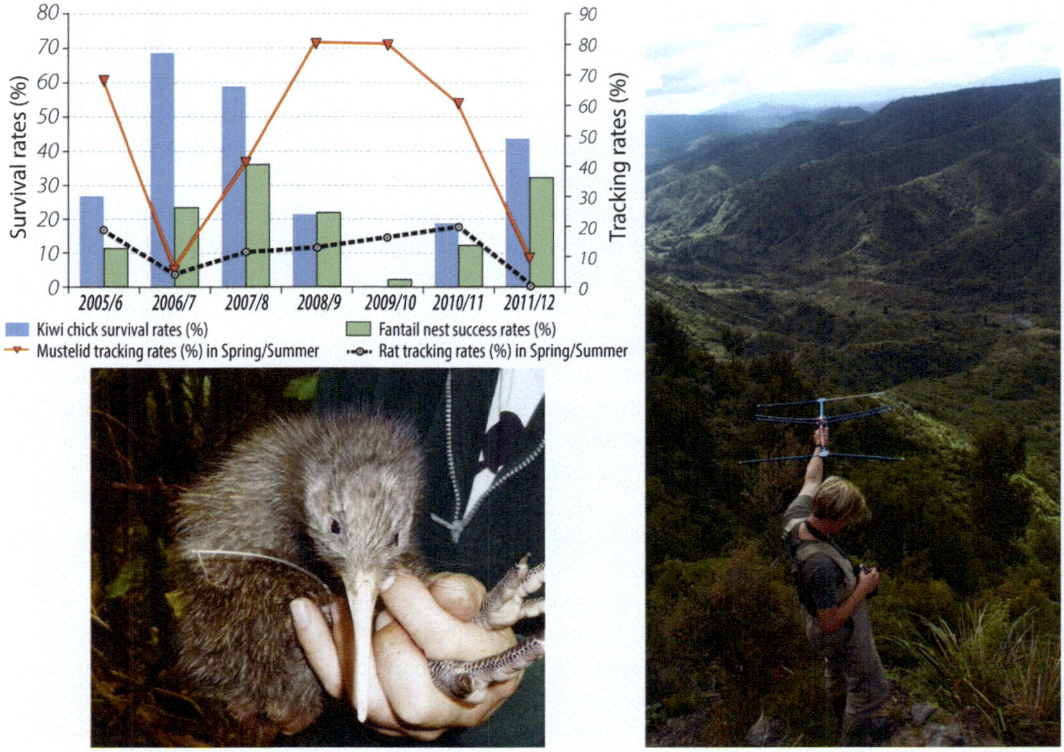

Fig. 12.10 Monitoring of radio-tagged kiwi chicks at Tongariro Forest Kiwi Sanctuary shows that aerial 1080 operations in 2006 and 2011 against rats and possums had a temporary but predictably positive effect on kiwi chick survival, mainly via secondary poisoning of stoats. *TFKS data courtesy Renee Potae, Crown Copyright, Department of Conservation Te Papa Atawhai. Photographs from Ross Martin (telemetry) and C.M. King (kiwi chick with radio antenna)*

correctly it reliably causes significant population declines among the unwanted pest mammals that are very vulnerable to it, but very rarely does it have that effect on the native terrestrial fauna of greatest value.

Better still, 1080 affects not only the primary target pests, usually possums and rats, but also anything that eats a poisoned carcase. So one of the important arguments for the continued use of 1080 is that it is not only remarkably efficient in reducing the numbers of possums and rats: it is also much more efficient than are traps in removing stoats by secondary poisoning.

A moribund rat or possum staggering about carrying a lethal dose of toxin is defenceless against a stoat. In turn, stoats find such an easy prey completely irresistible, and they do not reject fresh carcases either. In August 1997 an area of forest at Waimanoa EA in PFP was about to be treated with an aerial 1080 operation, and 13 stoats living there were marked with radio-collars beforehand (Fig. 12.11). Over the next 2–18 days, 11 of the collared stoats were monitored. Every one of them died, and two-thirds of them had remains of rats in their guts. No mustelids were tracked or trapped in that area for the next 3 months [34]. So a 1080 drop has exactly the opposite effect on stoats (Fig. 12.12) compared with kiwi (Fig. 12.10)—just what is needed.

On the other hand, rats recover from a 1080 operation more quickly than do possums, and for the next year or two, rats have access to a lot of food that had previously been monopolised by possums. So removal of possums may permit increases in abundance of rats, thereby reducing

Fig. 12.11 An anaesthetised stoat being equipped with a radio collar by DOC scientist Craig Gillies, before a planned 1080 operation in Waimanoa Ecological Area. *Crown Copyright, Department of Conservation Te Papa Atawhai. Photographer: Nic Gorman*

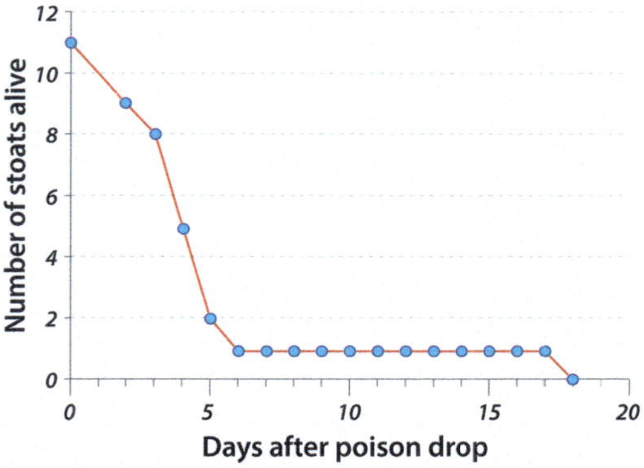

Fig. 12.12 Survival of the 11 radiotagged stoats known to be in the area of an aerial 0.08 % 1080 operation in Waimanoa Ecological Area in August 1997. All died within 18 days of the drop, and the area remained free of mustelids for another 3 months. Two thirds of the dead stoats had recently eaten a rat, and all but one had residues of 1080 in the tissues. Stoats may also scavenge dead possums. Hence, a single 1080 drop simultaneously removes the three most significant predators of forest birds. *From Murphy et al. (1999: fig. 3)*

the overall net benefit from pest control. And, of course, not everyone applauds the wider effects of 1080 on forest mammals, even if they are unwelcome invasives. Working through this ongoing debate will be part of the work of PFP pest management for many years to come.

References

1. Beveridge, A. E. (1964). Dispersal and destruction of seed in central North Island podocarp forest. Proceedings of the New Zealand Ecological Society, 11, 48–55.
2. Bond, W. J., Lee, W. G., & Craine, J. M. (2004). Plant structural defences against browsing birds: A legacy of New Zealand's extinct moas. Oikos, 104, 500–508.
3. Broome, K., & Krzystyniak, S. (1985). Wild animals of the Hauhungaroa Ranges. Auckland: New Zealand Forest Service.
4. Broome, K. (1995). The impact of wild animals in Pureora North Block 1982–1993: A remeasurement of permanent plots and exclosures. Hamilton: Department of Conservation, Waikato Conservancy.
5. Caughley, G. (1983) The deer wars. Auckland: Heinemann Publishers
6. Caughley, G. (1989). New Zealand plant-herbivore systems: Past and present. New Zealand Journal of Ecology, 12 (supplement), 3–10.
7. Clegg, S. (1986). Resurvey of Pureora North Block 1981–1982. Auckland: New Zealand Forest Service.
8. Coleman, J. D., Fraser, K. W., & Nugent, G. (2007). Costs and benefits of pre-feeding for possum control. New Zealand Journal of Zoology, 34, 185–193.
9. Dale, R. W. (1975). An animal and vegetation assessment of the Rangitoto Range. Auckland: New Zealand Forest Service.
10. Department of Conservation (unpublished). Archives. Hamilton: Department of Conservation.
11. Deuss, F. (1981). Animal survey 1980: Pureora State Forest Park. Auckland New Zealand Forest Service.

12. Ecroyd, C. E. (1996). The ecology of *Dactylanthus taylorii* and threats to its survival. New Zealand Journal of Ecology, 20, 81–100.
13. Fitzgerald, A. E. (1984). Diet of the possum (*Trichosurus vulpecula*) in three Tasmanian forest types and its relevance to the diet of possums in New Zealand forests. In A. Smith & I. Hume (Ed.), Possums and gliders (p. 137–143). Chipping Norton, New South Wales: Surrey Beatty and Sons.
14. Forsyth, D. M., Ramsey, D. S. L., Veltman, C. J., Allen, R. B., Allen, W. J., Barker, R. J., Jacobson, C. L., Nicol, S. J., Richardson, S. J., & Todd, C. R. (2013). When deer must die: Large uncertainty surrounds changes in deer abundance achieved by helicopter- and ground-based hunting in New Zealand forests. Wildlife Research, 40, 447–458.
15. Fraser, K. W. (1996). The effect of recreational hunters on deer populations in Pureora Conservation Park. Science for Conservation, 31, 1–39.
16. Fraser, K. W. (2000). Status and conservation role of recreational hunting on conservation land. Science for Conservation, 140, 1–46.
17. Fraser, W., Cone, J. M., & Whitford, E. J. (2000). A revision of the established ranges and new populations of 11 introduced ungulate species in New Zealand. Journal of the Royal Society of New Zealand, 30, 419–437.
18. Gormley, A. M., Holland, E. P., Pech, R. P., Thomson, C., & Reddiex, B. (2012). Impacts of an invasive herbivore on indigenous forests. Journal of Applied Ecology, 49, 1296–1305.
19. Green, W. Q. (1984). A review of ecological studies relevant to management of the common brushtail possum. In A. Smith & I. Hume (Ed.), Possums and gliders (p. 483–499). Sydney: Surrey Beatty & Sons.
20. Greene, T. (1998). The effects of compound 1080 on populations of specific non-target species, Waihaha ecological area, Pureora Forest Park, winter 1994. Science for Conservation, 69, 1–54.
21. Husheer, S. W. (2007). Introduced red deer reduce tree regeneration in Pureora Forest, central North Island, New Zealand. New Zealand Journal of Ecology, 31, 79–87.
22. Innes, J. G. (1983). Personal communication.
23. Jane, G. T. (1979). The wild animal populations and forest conditions in the Hauhungaroa Ranges. Rotorua: New Zealand Forest Service, Rotorua Conservancy.
24. Jones, C., Barron, M., Warburton, B., Coleman, M., Lyver, P. O., & Nugent, G. (2012). Serving two masters: Reconciling economic and biodiversity outcomes of brushtail possum (*Trichosurus vulpecula*) fur harvest in an indigenous New Zealand forest. Biological Conservation, 153, 143–152.
25. Kelsey, J. (1997). The New Zealand experiment: A world model for structural adjustment? Auckland: Bridget Williams Books and Auckland University Press.
26. King, C. M. (Ed.) (2005). The handbook of New Zealand mammals, 2nd edition. Melbourne: Oxford University Press.
27. Krzystyniak, S. (1984). Pureora North Block animal survey, summer 1983. Auckland: New Zealand Forest Service.
28. Leathwick, J. R., Hay, J. R., & Fitzgerald, A. E. (1983). The influence of browsing by introduced mammals on the decline of the North Island kokako. New Zealand Journal of Ecology, 6, 55–70.
29. Leigh, A. H., & Clegg, S. E. (1989). Maniapoto District wild animal management plan, 1989 to 1999. Hamilton: Department of Conservation.
30. McKelvey, P. J. (1963). The synecology of the West Taupo indigenous forest. New Zealand Forest Service Bulletin, 14, 1–126.
31. Millen, J. (1999). Through Trackless Bush: The story of the New Zealand national forest survey 1946–55. Rotorua: P. F Olsen & Co.
32. Moorhouse, R., Greene, T., Dilks, P., Powlesland, R., Moran, L., Taylor, G., Jones, A., Knegtmans, J., Wills, D., Pryde, M., Fraser, I., August, A., & August, C. (2003). Control of introduced mammalian predators improves kaka *Nestor meridionalis* breeding success: Reversing the decline of a threatened New Zealand parrot. Biological Conservation, 110, 33–44.
33. Morgan, D. R., & Warburton, B. (1987). Comparison of the effects of hunting and aerial 1080 poisoning for reducing a possum population. Christchurch: New Zealand Forest Service.
34. Murphy, E. C., Robbins, L., Young, J. B., & Dowding, J. E. (1999). Secondary poisoning of stoats after an aerial 1080 poison operation in Pureora Forest, New Zealand. New Zealand Journal of Ecology, 23, 175–182.
35. New Zealand Forest Service (1986). Ministerial brief for Minister of Conservation. Wellington: New Zealand Forest Service.
36. Nicholls, J. (1986). A descriptive overview of the central North Island Volcanic Upland. In B. Veale & J. G. Innes (Ed.), Ecological research in the central North Island volcanic plateau region: Proceedings of a NZ forest workshop (p. 2–17). Rotorua: Forest Research Institute.
37. Nugent, G., Parkes, J. P., & Tustin, K. G. (1987). Changes in the density and distribution of red deer and wapiti in Northern Fiordland. New Zealand Journal of Ecology, 10, 11–21.
38. Nugent, G. (1992). The conservation role of commercial deer hunting. Christchurch: Forest Research Institute.
39. Nugent, G., Fraser, K. W., & Sweetapple, P. J. (1997). Comparison of red deer and possum diets and impacts in podocarp-hardwood forest, Waihaha catchment, Pureora Conservation Park. Science for Conservation, 50, 1–61.
40. Nugent, G., Fraser, K. W., & Sweetapple, P. (2001). Top down or bottom up? Comparing the impacts of

arboreal possums and 'terrestrial' ruminants on native forests in New Zealand. Biological Conservation, 99, 65–79.
41. Nugent, G., & Whitford, J. (2006). Animal health board project no. R-10577: Part 3: The short-term effect of Possum control on TB prevalence in pigs. Lincoln: Landcare Research Contract Report LC0607/002.
42. Nugent, G. (2014). Personal communication.
43. Nugent, G. unpublished.
44. Parkes, J. P., Nugent, G., & Warburton, B. (1996). Commercial exploitation as a pest control tool for introduced mammals in New Zealand. Wildlife Biology, 2, 135–141.
45. Parkes, J. P. (2005). Feral goat. In King, C. M. (Ed.), The handbook of New Zealand mammals, 2nd edition (p. 374–392). Melbourne: Oxford University Press.
46. Parliamentary Commissioner for the Environment (2011). Evaluating the use of 1080: Predators, poisons and silent forests. Wellington: Parliamentary Commission for the Environment.
47. Pfeiffer, D. U., Hickling, G. J., Morris, R. S., Patterson, K. P., Ryan, T. J., & Crews, K. B. (1995). The epidemiology of *Mycobacterium bovis* infection in brushtail possums (*Trichosurus vulpecula* Kerr) in the Hauhungaroa ranges, New Zealand. New Zealand Veterinary Journal, 43, 272–280.
48. Powlesland, R. G., Knegtmans, J. W., & Marshall, I. S. J. (1999). Costs and benefits of aerial 1080 possum control operations using carrot baits to North Island robins (*Petroica australis longipes*), Pureora Forest Park. New Zealand Journal of Ecology, 23, 149–159.
49. Powlesland, R. G., Knegtmans, J. W., & Marshall, I. S. J. (2000). Breeding biology of North Island robins (*Petroica australis longipes*) in Pureora Forest Park. Notornis, 47, 97–105.
50. Powlesland, R. G., Knegtmans, J. W., & Styche, A. (2000). Mortality of North Island tomtits (*Petroica macrocephala toitoi*) caused by aerial 1080 possum control operations, 1997–98, Pureora Forest Park. New Zealand Journal of Ecology, 24, 161–168
51. Pracy, L. T. (1974). Introduction and liberation of the opossum (*Trichosurus vulpecula*) into New Zealand. Wellington: NZ Forest Service.
52. Smale, M. C., & Kimberley, M. O. (1993). Regeneration patterns in montane conifer/broadleaved forest on Mt Pureora, New Zealand. New Zealand Journal of Forestry Science, 23, 123–141.
53. Smale, M. C., Beveridge, A. E., & Herbert, J. W. (1998). Selection silviculture trials in North Island native forests: impacts on the residual forest and their implications for sustainable forest management. New Zealand Journal of Forestry, 43, 19–30.
54. Sweetapple, P., Nugent, G., Whitford, J., & Knightbridge, P. I. (2002). Mistletoe (*Tupeia antarctica*) recovery and decline following possum control in a New Zealand forest. New Zealand Journal of Ecology, 26, 61–71.
55. Sweetapple, P., & Nugent, G. (2009). Possum demographics and distribution after reduction to near-zero density. New Zealand Journal of Zoology, 36, 461–471
56. Sweetapple, P., & Nugent, G. (2011). Chew-track-cards: A multiple-species small mammal detection device. New Zealand Journal of Ecology, 35, 153–162.
57. Sweetapple, P. J., & Fraser, K. W. (1997). Assessment of red deer and possum kills during an aerial 1080 control operation in the Rangitoto Range. Lincoln: Landcare Research Contract Report LC9697/139.
58. Veltman, C. J., & Westbrooke, I. M. (2011). Forest bird mortality and baiting practices in New Zealand 1080 operations from 1986 to 2009. New Zealand Journal of Ecology, 35, 21–29.
59. Veltman, C. J. (2014). Personal communication.
60. Walker, A., & Cooke, R. (2003). Waimiha: People of character. Waimiha: Waimiha Reunion 2001 Committee.
61. Warren, A. (1984). The effects of 1080 poisoning on bird populations in Tihoi, Pureora State Forest Park, winter 1983. Auckland: New Zealand Forest Service.
62. Westbrooke, I. M., & Powlesland, R. G. (2005). Comparison of impact between carrot and cereal 1080 baits on tomtits (*Petroica macrocephala*). New Zealand Journal of Ecology, 29, 143–147.
63. Wilkes, O., Gaukrodger, J., & Ritchie, N. (2011). Chronology of Pureora. Hamilton: Department of Conservation.
64. Williams, D. S., & Leathwick, J. R. (1990). Remeasurement of ecological transects in some central North Island forests. Rotorua: Forest Research Institute

Protecting the Wildlife from Introduced Predators

13

C.M. King, J.G. Innes and J.R. Hay

Abstract

This chapter describes the arrival in Pureora Forest Park (PFP) of introduced predators (especially rats and stoats), the research by the Forest Bird Research Group during the logging moratorium; the effects of introduced predators on native fauna; the intense research work that finally identified the key causes of the decline in kokako; and the healing effects of systematic predator control work in kokako habitat.

Keywords

Rats · House mice · Hedgehogs · Brushtail possums · Mustelids · Stoats · North Island kokako · Forest Bird Research Group · Research by management · Mapara SF · Brodifacoum · Mesopredator release · Waipapa/Pikiariki Restoration Project

Understanding Predators

There have always been predators in New Zealand, so sudden deaths of adults and losses of eggs and young are not new experiences for the native fauna. The difference is that, in prehuman times in the North Island, the predators that were most threatening to diurnal birds were harrier hawks and falcons, which operated by sight during the day (the massive Haast's eagle lived only in the South Island, as far as we know). The only nocturnal predators (the laughing owl, morepork and tuatara Fig. 2.11) hunted mainly for other, generally small prey (lizards, frogs, bats, and large invertebrates) [76, 78].

Hence, the long-established endemic birds evolved effective defences against daytime aerial

C.M. King (✉)
School of Science, University of Waikato, Private Bag 3105, Hamilton 3240, New Zealand
e-mail: cmking@waikato.ac.nz

J.G. Innes
Landcare Research NZ Ltd, Private Bag 3127, Hamilton 3240, New Zealand
e-mail: innesj@landcareresearch.co.nz

J.R. Hay
Department of Conservation, Te Papa Atawhai, Private Bag 4715, Christchurch 8140, New Zealand
e-mail: rhay@doc.govt.nz

C.M. King et al. (eds.), *The Drama of Conservation*,
DOI 10.1007/978-3-319-18410-4_13

attack. Successful defences included the large body size allowed for by loss of flight in the moa, adzebill, takahe and many others; the nocturnal ground-feeding habits of the kiwi and kakapo; and the brilliant camouflage of kakapo, the burrows of kiwi and the carefully placed nests of kokako that made them all hard to spot by eye from above. Useful strategies that enhanced contacts with mates, such as body scent, or loud nocturnal calling, or that saved energy, such as feeding and nesting on the ground, carried no serious penalties in that ancient world. The long-resident short-tailed bats could safely behave quite unlike other bats, by tucking up their wings and scuttling across the forest floor a little bit like mice.

The game changed when mammalian predators arrived, because these can hunt at night and by scent, and not only on the ground but also in the trees. The kiore (Polynesian rat) was the first to arrive, and as the first mouth to the table the kiore undoubtedly did much damage. The consequences could not be documented at the time (since then, some of them have been reconstructed by comparing the surviving native fauna of offshore islands that were, or were not, colonised by kiore [71, 73]). Today, kiore have given way to ship rats, possums and stoats as ubiquitous and unwelcome residents of the forests.

Introduced predators, including human hunters, have always been assumed to have a huge and ongoing effect on the productivity and survival of birds, but identifying which predators have most impact, and eliminating other concurrent explanations such as the loss of normal food supplies and fragmentation of habitat, is never easy. The native species are not all equally vulnerable to the same predators, or in all seasons, all places, or in every year, and cannot all be observed equally easily, or at all times of day.

PFP is, unfortunately, well stocked with mammalian predators. It has or has had populations of all four rodents (rats and mice: Fig. 13.1), and all three mustelids (stoats, weasels and ferrets:

Fig. 13.1 The four species of commensal rodents known from Pureora Forest Park. The kiore (*top left*) arrived in New Zealand with Polynesian colonists in the late 1200s AD; it was replaced throughout the country by the Norway rat (*top right*), which arrived with the first Europeans after the mid 1700s; it in turn was replaced by the ship rat (*lower left*, showing two of three colour variations) after the mid 1800s, except in towns, sewers, rubbish dumps etc. Wild house mice (*lower right*, with two colour variations) reached the North Island after the 1830s. Not to scale. *Paintings by P. Barrett, from King (ed) (2005)*

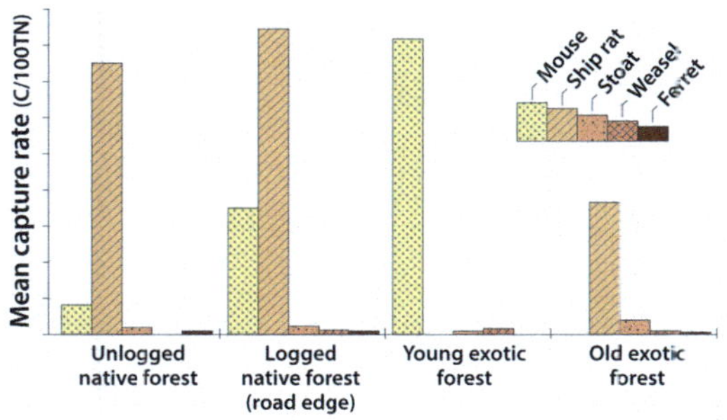

Fig. 13.2 The relative abundance of five species of small introduced predators (rodents and mustelids) by habitat in Pureora Forest Park. Indices of abundance (captures per 100 trapnights) are averaged over the 5 years 1983–87 and do not reflect the substantial variation between years within habitats—e.g., the population irruption of mice in the young exotic forest of the Pikiariki block (planted in 1978) in 1984, followed by a detectable increase in number of weasels caught there. Hedgehogs were more common than expected but not routinely sampled by the methods used. *From King et al. (1996: fig. 11)*

Fig. 13.4) that have become established in New Zealand, plus hedgehogs and cats. By the time PFP was created, all of these species had been present, some widespread, in the North Island for decades. Fortunately, and contrary to common assumptions, few of them are now of any but occasional and local concern in PFP (Box 13.1).

The kiore has disappeared from PFP totally. Only one colony of Norway rats has been found in PFP, a mere remnant of its former huge distribution [43]. Weasels, cats and ferrets are all relatively rare, and none of them hunt birds often enough to justify targeted management. Ship rats and stoats remain the most numerous predators of vertebrates (Fig. 13.2), plus possums (Chap. 12), while mice and hedgehogs add to the havoc among invertebrates and lizards.

Norway Rats

Ever since Cook's first visit in 1769–70, the masters of visiting European ships have often sought out sheltered waters where they could bring their vessels close inshore for maintenance, hull-cleaning and re-provisioning. All ships of the period were infested with Norway rats, which were strongly encouraged to leave. (One American whaler ended his voyage in pre-colonial New Zealand, when he was suffocated while his ship was being fumigated for rats: [77: 185]). Hence, Norway rats were almost certainly the first of the European species to arrive [5], spreading inland from coastal settlements in huge numbers from the early 1800s or earlier. From then on, life for the native fauna became more perilous than before, and for humans, more uncomfortable (Chap. 4).

Box 13.1 Mammalian predators of PFP

A systematic survey of small mammals in PFP over the five years 1983–87, using a strictly standardised regime of traps operated four times a year, collected a massive total of 1793 ship rats, plus smaller numbers of seven other introduced predators. Estimates of their abundance varied with habitat and trap type, as shown in the following table.

The two most significant predatory species sampled were ship rats and stoats.

Ship rats lived "wall-to-wall" throughout all native forest types, logged or not, but they were scarce or absent in exotic forest. Of 179 sites used throughout this programme at which traps were set for rats and stoats, only 7 failed to catch at least one ship rat in five years: at 14 sites, 20 or more rats were caught. Analyses of this abundant material have provided us with baseline information on relative abundance, population structure and breeding biology of both species [36].

Long-term changes in the abundance of ship rats are usually estimated in terms of a trapping index (captures per 100 trap-nights, corrected for occupied traps). This index varies between years, seasons and habitats, and is strongly affected by trap type, food supplies and the success of the previous breeding season for rats. More labour-

Local variation in mean abundance over five years (1983–87 inclusive) by habitat (as captures per 100 trapnights, C/100TN) of all species of small ground carnivores in Pureora Forest Park by habitat type. Values in any row followed by the same letter do not differ significantly at $P = 0.05$; rows without letters were not tested because of small sample size. All means are controlled for all other variables using the Generalised Linear Model (GLM) [42]

	Unlogged *C/100TN*	Logged *C/100TN*		Exotic *C/100TN*	
		Interior	Road edge	1978	Older
Mouse	0.79 a	1.71 b	3.51 c	8.18 d	–
Ship rat					
Rat traps	4.91 a	8.02 b	9.18 ab	0.09 c	–
Fenn traps	7.53 a	–	8.49 a	0 c	3.66 b
Norway rat					
Rat traps	0	0	0	0	–
Fenn traps	0.56 a	–	0.10 b	0 c	0 c
Stoat	0.16 a	–	0.19 ab	0.05 a	0.37 b
Weasel	0	–	0.09	0.14	0.07
Ferret	0.08	–	0.05	0	0.02
Feral cat	0.06	–	0.01	0.10	0.01
Hedgehog	1.51 ab	–	0.10 c	1.06 b	2.02 a

For further information on all these species, see King [43]

intensive methods operated over shorter terms show that the general average density of ship rats in North Island forest in January is about 4/ha (400 per km^2), compared with 0.03/ha (3/km^2) for stoats [32].

In 24,272 trap nights using Fenn traps set over five years and equally good at catching all three, the survey collected 13 ferrets, 18 weasels and 63 stoats. Only the stoats provided enough data for population analysis. They were more at home along roadsides and in older exotic forests, where they were more likely to find mice and rabbits. Their average age at capture was 15 months, and their average annual mortality rate from independence to one year old was about 76 %. The proportion of stoats in the PFP sample aged over a year old declined over the five years of the collections, from 52 % of 21 collected in summer and autumn of 1983 to only 27 % of 22 collected in the same two seasons of 1984–87. These results suggest that the natural density and productivity of stoats in PFP are both relatively low, and a determined trapping campaign has some chance of exceeding their natural mortality rate and temporarily reducing their local density.

Norway rats are good burrowers and swimmers, and aggressive predators of small ground fauna. At first they invaded all habitats up to the tree line, and for decades they ravaged the ground fauna of invertebrates, lizards, tuatara, kiore and ground-nesting birds, the ground-feeding short-tailed bats and the fallen fruits of forest trees. After the easiest foods available at ground level were exhausted, Norway rats were less able to reach the resources that remained up in the canopy, including the tree-fruits and the arboreal fauna that had survived the attentions of the kiore. That is because, although Norway rats can climb, they are not agile in trees [20], so the economics of arboreal hunting are against them [45]. They became scarcer in forests after the mid-late nineteenth century [74]. PFP is one of few forests in which an apparently permanent small population of Norway rats is known to have survived at least to the mid-1980s [36].

Fig. 13.3 A ship rat attacking a fantail on its nest, which had been built on the end of a small branch so was relatively inaccessible to less agile predators. *David Mudge, Nga Manu Images*

Ship Rats

From the 1860s, the common rat on board European ships was the ship or black rat [4], and this new species was just as quick as its predecessor to escape to shore whenever possible. One reason suggested by Ian Atkinson for the otherwise unexplained later arrival of ship rats in New Zealand could be that Norway rats could thrive in the cold, wet conditions typical of the many sailing ships that followed James Cook to New Zealand after 1769; but the tropical origin and smaller size of ship rats made them better adapted to life in the narrower, warmer spaces within and between the holds of steam ships arriving from the late 1850s onwards.

Since then, ship rats have largely replaced the Norway rat in most mainland forests. They have become hugely abundant, in part because they are very efficient at foraging above the ground [45], so could get immediate and easy access to the arboreal resources that had been relatively safe during the long period of dominance by Norway rats.

Ship rats climb rapidly and well, especially up the smooth staircases provided by tangles of supplejack vines, and they find dry, secure nests in tree holes and epiphyte clumps throughout the canopy. It probably helps that they do not have to compete for these desirable habitats with any native specialist tree-dwelling rodents such as squirrels, and any resident birds they encounter can easily be made to provide board and/or bed.

Ship rats are omnivorous generalists, and are by far the largest single cause of losses of eggs, chicks and sitting birds in non-beech forests on the main islands, including in PFP, for two reasons. First, ship rats are ubiquitous predators in their own right, capable of reaching almost any birds' nest, even those at the far ends of small branches (Fig. 13.3), and they are abundant in North Island forest (averaging 3–6 per hectare in summer) every year. Only in southern beech forests after a heavy seedfall can mice reach higher numbers (for a short while in some years) than ship rats in North Island podocarp forests do virtually all the time. Second, they are the principal prey of stoats and cats in podocarp forests, so they add indirectly to the toll by supporting the numbers of other predators.

Ship rats also destroy large numbers of invertebrates, forest fruits and seeds [9], accumulating hundreds of empty berries of hinau and miro in caches, each with the kernel neatly chewed out from one end. The caches are usually found in the hollow bases of old podocarps (commonly matai) or under large fallen logs.

The relentless predatory toll by ship rats on the birds and their food supplies, accumulated over decades, has disrupted the ancient interactions between the trees and the birds that pollinated their flowers and dispersed their seeds, with serious consequences for the long term regeneration of the forest (Chap. 2) [9, 10, 15, 39]. The scale of the destruction wrought by ship rats in forests throughout New Zealand may be compared with the dramatic and well-publicised consequences of the oil spill from the *Rena*. That was a shocking, one-off event that killed up to 20,000 seabirds in the Bay of Plenty over the summer of 2010–11. During the breeding season in New Zealand nationally, predators, especially rats and possums, kill more than that number of eggs and chicks of native forest birds *every hour*—a total of around *26 million a year,* according to a back-of-an-envelope conservative estimate made by Graeme Hill [26] from research data analysed by John Innes.

Mice

In their native lands in the northern hemisphere, house mice are mostly commensals, living only in and among human habitations in Europe and Asia. They did not come to New Zealand with the Polynesians, to whom they were unknown, but with European traders and settlers. Mice were certainly present in settlements in the Bay of Islands by about 1830, and in Whanganui and Manawatu by the 1840s.

> About 1844, one of my children came running to say that an extraordinary little animal had made its appearance in the verandah, and enquired whether it might not be a frog: on going to see, I found it was a mouse, the first we had seen in the place [Whanganui]; after a short time it disappeared, but in about three or four days afterwards, a large colony of, perhaps, a hundred came and took possession of the house, which they have retained ever since [69].
>
> [In Moutoa, Manawatu, in 1851] The house was literally swarming with mice. The prepared flax was stacked against the sides of the house, covered with a light roof; this formed a cover for the vermin, and I watched them running all over my blankets till I fell asleep [63: 36].

There is no way to tell when mice arrived in the Pureora area, but they certainly came with human help, stowed away in supplies carried in from the coast. The Maori had their own view on who was responsible for bringing them in: "… they assign to the missionaries the credit of introducing mice" [37: 335].

The earliest European settlement in the Taupo area was the mission station of Thomas Grace at Pukawa, established in 1855 and abandoned in 1863. The neighbouring village of Te Rapa was the seat of Tuwharetoa paramount chief Te Heuheu, and the scene of several visits by European explorers (Chap. 4). Grace complained of the difficulty of carting supplies inland over unformed roads and unbridged rivers, but did not mention mice among the many other troubles that dogged his family there [12].

House mice in New Zealand, free of competition from the larger native wood mice and voles of Europe, occupy a wide range of habitats including farmland, pastures, gardens, roadside verges, and both native and exotic forest. They can reach considerable numbers wherever they can find food and escape from larger predators; they especially favour thick vegetation which provides cover from predators and a productive, moist habitat.

Mice are not always thought of as predators, and are not generally abundant in podocarp forest (average density 4.5 mice/ha: [32]). But where litter-dwelling invertebrates are abundant, such as under the thick ground cover on the 1978 young plantation at PFP (Fig. 13.2, Box 13.1), and in any forest after a heavy tree seedfall [1], all that extra protein fuels occasional mouse population irruptions. Mice in large numbers eat many invertebrates, and also support higher than usual numbers of bigger predators that hunt birds. Sudden increases in numbers of mice are a predictable and mostly unavoidable side effect of successful removal of rats, in PFP and elsewhere [30]. An ingenious field experiment conducted by Lucy Bridgman in PFP showed why: ship rats are active predators of mice [11], so where rats are common, mice are not.

Hedgehogs

Hedgehogs first arrived in the North Island in 1905, and were seen in Te Kuiti in 1930 [49], but were still absent from the west Taupo forests in the early 1950s [14]. They are nocturnal insectivores, best known for snuffling noisily through the undergrowth in search of slugs, snails and other juicy invertebrates.

When the mean earth temperature falls below 10–11 °C, hedgehogs construct large untidy nests of leaves and grass, and hibernate in them. Hibernation is stressful and hazardous, so hedgehogs survive best in times and places where it is not necessary. Hence, the length of the hibernation period and the proportion of the population attempting it depend on the severity of the winter.

In much of the lowland North Island, mild winters and generous food supplies are kind to hedgehogs, and some do not hibernate at all. But at Pureora, the soil temperature falls below 11 °C on 187 days a year on average, so hibernation is necessary. Under the shelter of the forest canopy, or in a nest under a thick layer of litter, cold days might be less challenging, but still sufficient to discourage most hedgehogs from venturing out in mid winter [42]. Hedgehogs live throughout PFP, especially in the older exotic plantations.

Although hedgehogs are a lot more agile than they look, they cannot climb trees, so arboreal fauna are safe from them. They are, on the other hand, very efficient predators of almost any animal food available on the forest floor, especially invertebrates. John Innes et al. [32] estimated that an average of some 740 g of invertebrates per hectare per night is eaten by introduced mammalian predators (Fig. 13.6), of which a massive 660 g/ha/night is taken by hedgehogs (assuming that the only available figure for the density of hedgehogs in North Island podocarp-broadleaved forest, 5.5/ha at one site, is typical).

Many native animals also depend on invertebrates, especially the kiwi, which immediately focuses attention on the hedgehog as a species of conservation concern. As Innes et al. [32] point out, competition for food with hedgehogs is a less serious matter for adult kiwi than is predation on their chicks by stoats, but if kiwi chicks can grow faster in areas rich in invertebrates, they may be able to reach a safe weight that much sooner.

Mustelids

Three species of mustelids (Fig. 13.4) have been brought to New Zealand, the last of the specialist carnivores and the only ones legally and deliberately introduced with Government assistance. They were released on rabbit-infested pastures across both main islands, and spread out from there with or without human help. They arrived much later than most other introduced mammals. None of the mustelids could have been present in PFP before the middle 1870s.

For many decades before the first mustelids set foot on our shores, kiore, Norway and ship rats, cats, dogs, pigs and human hunters had devastated the native fauna. No mustelids ever met a takahe, kakapo or tuatara in the central North Island (Chap. 2), nor any of the long list of extinct birds on islands reached by rats but not by mustelids: all of these lost populations fell to other predators.

The decline of forest birds in the Hauhungaroa Ranges after the 1856 Hinana feast (Chap. 3) followed the spread of ship rats across the North Island, but preceded the arrival of stoats by at least 2–3 decades, probably longer. A very early survey of the west Taupo forests referred to the rich birdlife of that area, except that:

Fig. 13.4 The three species of mustelids introduced into New Zealand are all present in Pureora Forest Park. Weasels (*below*) prefer anywhere mice are abundant; stoats (*top*) are widespread in all habitats but especially in any kind of forest; and ferrets (*centre*) live mainly on the fringes of the forest and in adjacent farmland. In cold climates, stoats occasionally turn partially white in winter, but fully white "ermine" are extremely rare in Pureora. Not to scale. *Paintings by P. Barrett, from King (ed) (2005)*

> ……kiwi were now very rare or absent, and that once numerous wekas were ….said to have all disappeared following the arrival of stoats in the region about 1912 [14].

Ferrets are semi-domesticated rabbit specialists gone wild, and probably followed the rabbits along the line of the Main Trunk Railway as it extended its associated bow-wave of forest clearance through the King Country (Chap. 5). Ferrets are common on farmland around Pureora: in 2006, 31 ferrets lived on a 25 km^2 area of pasture adjacent to the eastern boundary of PFP, overlooking Lake Taupo [44], and ferrets have found their way into anywhere within PFP that supports rabbits.

Stoats are agile climbers, totally at home in the forest, and known to travel long distances in a short time, so they could have dispersed across country to Pureora from other release sites in any direction. They prefer forest to open country, where they might be at risk of a bruising encounter with a cat, ferret or a harrier hawk. The favourite prey of their British ancestors were rabbits and voles, and the thick grass in occasional clearings and along roadsides through the exotic plantations at Pureora harbours both rabbits and mice, which is probably why a 1983–87 survey caught most stoats there (Box 13.1).

Some native birds with large and vulnerable flightless chicks, including all species of kiwi and the blue duck and kaka, are at special risk of stoat predation, and conservation management for them must aim to protect them from mustelids. It was reassuring to have some evidence specifically from PFP (from the declining numbers of older stoats collected over 5 years) that it is possible to exceed the natural mortality of adult stoats in that habitat, and also to remove most of the relatively small numbers of young stoats produced each year [41].

Weasels are the smallest of the three, and are specialists on mouse-sized rodents. Mice are found in PFP only in patchy and unpredictable numbers under thick vegetation, such as road edges and in young plantations established close to the protest areas of 1978 (Chaps. 9–10). So the long history of human management at PFP has provided a mosaic of different vegetation types meeting all the different habitat preferences of the three mustelid species.

Cats

Feral cats have probably been present in PFP since very early European times, as we can tell from travellers' comments on their wide distribution and predatory habits.

> The cat (Ngeru, Tori) probably introduced itself from some early ship. Like every other alien animal introduced into this teeming country, it has taken full possession of the soil; and, like the dog, has multiplied beyond due limits. Ages of domesticity have failed to eradicate the native wildness of the cat: opportunity serving, puss soon relapses into feline barbarism, and forsakes the hearth to prowl the forest [28: 127].
>
> The common cat is at present found in a wild state in the forests of New Zealand, and is another

> cause of the extermination of indigenous birds. It is remarkable that these wild cats soon resume the streaky grey colour of their original stock—the wild cat of northern Europe [3: 55].
>
> On 13 November [1843], north of Taumarunui—about 90 km from the sea and even further from pakeha settlement, the guides [accompanying Richard Taylor and George Selwyn] killed a wild cat. Its stomach contained the intact bodies of 30 lizards [79: 58].

During the nineteenth century, farmers released many cats onto rabbit-infested pastures adjacent to PFP in the (vain) hope that they might keep down the rabbits. Feral cats remain widespread, though not common, throughout the protected area. They easily adapt to independent life in the forest, and become very wary of people and hard to catch, although some get caught accidentally in possum traps. They do not despise carrion, so may be killed while scavenging carcases off the road or by secondary poisoning along possum lines, or after aerial 1080 drops.

Feral cats are opportunist predators, preferring possums, rabbits and rodents where these are abundant but adding lizards and birds when opportunity offers. One survey collected 15 cats from PFP during the mid 1980s, and found that the principal items in their guts were rabbits, possums, mice and invertebrates; only three of these cats had eaten birds. One adult female was lactating, and six were juveniles, so there was clear evidence that these cats belonged to a breeding population [41].

Feral cats can maintain a self-perpetuating population in the forest (regrettably, augmented occasionally by irresponsible humans), and they do occasionally kill birds and ship rats, but the extent to which they affect bird populations is unknown.

Possums

Possums had long been known to eat the seeds and fruits of many forest trees which are also important foods of many forest birds, most critically the kokako. The implication was considered serious: what remained of kokako habitat after logging was being further impoverished by possum browsing [48] But, as if competition for food from abundant possums was not already bad enough for the kokako, worse effects were discovered later.

In 1993 a landmark paper published in *Notornis* gave conservationists their biggest and most unpleasant surprise in years. Long before, in 1980, monitoring of a kokako nest that had been destroyed by an unknown predator detected signs normally associated with possums, but the raiders were never caught in the act. Then Kerry Brown et al. [13] began pioneering the new technique of monitoring birds' nests by time-lapse video cameras. Their tapes confirmed the previously unproven suspicions that possums did not merely compete with kokako for food, but actively raided their nests and destroyed both eggs and chicks (Fig. 13.5). This was unarguable evidence that possums are the only invasive pests that cause significant direct damage *both* to the forest vegetation *and* to the wildlife of PFP.

Why Are New Zealand Birds so Vulnerable?

Contemporary visitors to protected forests expect to see abundant birds, and if they do not, they often blame introduced predators—especially, and often justifiably, the two most efficient killers, ship rats and stoats. To give only one example:

> I pace a silent bush and curse the memory [of those who introduced mustelids to Fiordland], for deer, rabbits and opossums have not robbed us of one fraction of our heritage that the stoat has [27: 122].

Time to remember then, that many early European explorers, deeply impressed by the great forests that still covered vast areas of the lowlands, and by the strange and beautiful birdlife of New Zealand, also often commented on how quiet the forests were.

> Amid the gloom of these Antipoceal forests, there reigns a solemn and almost unbroken stillness [2: 23].
>
> There is a silence peculiar to the New Zealand forest which must be felt to be understood...... It is the absence of all living things which renders the silence and solitude of the woods so oppressive.

Fig. 13.5 Possums were once thought of as strictly herbivorous, until video monitoring detected them eating birds' eggs and chicks, including those of kokako. *David Mudge, Nga Manu Images*

> Occasionally a pair of Kaka parrots may be seen wheeling high above the hill tops with harsh discordant cries, or the melancholy note of the great New Zealand pigeon comes booming through the woods; but except at early morning the traveller may often wander for hours, I had almost said for days together, through the gloom of these woods where the sun's rays can scarcely penetrate, and the breeze passing over the tree-tops through the uppermost whispering boughs may be seen and heard, but cannot be felt. Not a sparrow–not a mouse to be seen; it seems the silence of death, or more properly the stillness of the yet unborn; the gigantic Moa and one or two other extinct species of birds which, even in historic times had their home in New Zealand, used to shun the gloomy shades of the forest and cleave to the flat marshy lands [51].

One possible reason is that the bush often *is* in fact very quiet during the times and seasons when birds are least active and detectable, even when a full complement of species is actually present. For example, most forest bird species become quiet when they go into post-breeding moult, typically in late summer at the time when many people are visiting the forests. Birds were indeed relatively more abundant in the mid-nineteenth century than now, but that was not always the impression the early explorers got.

More significantly, modern survey methods making allowance for low detectability prevent simple conclusions like "I didn't see any, so there are none there". So a silent bush does not always and everywhere necessarily count as evidence of the ravages of introduced predators.

The whole issue of estimating the extent of the effects of predation is eye-wateringly complex even for researchers who have studied it all their lives. Many other factors (such as exotic diseases, habitat loss, and disruption of food supplies) can affect bird numbers in some places. The usual assumption that predation is to blame is often correct, but to be valid, that conclusion has to take all those other factors into account too.

The exemplary study that did exactly that was presented by Innes et al. [32] to a 2007 landmark symposium on historic biodiversity changes in New Zealand. They reviewed a huge list of studies, taking into consideration all possible explanations for the drastic historic decline in diversity and numbers of native birds. Their assembled data confirm that, while other dangers are not to be ignored, predation by introduced mammals usually is the primary cause of declines of birds in remaining large tracts of native forest (Fig. 13.6). Without predator management, the numbers of native forest birds on the New Zealand mainland cannot be prevented from further decline.

Mustelids are commonly and incorrectly blamed for much more of the historic destruction of native wildlife than they are really responsible

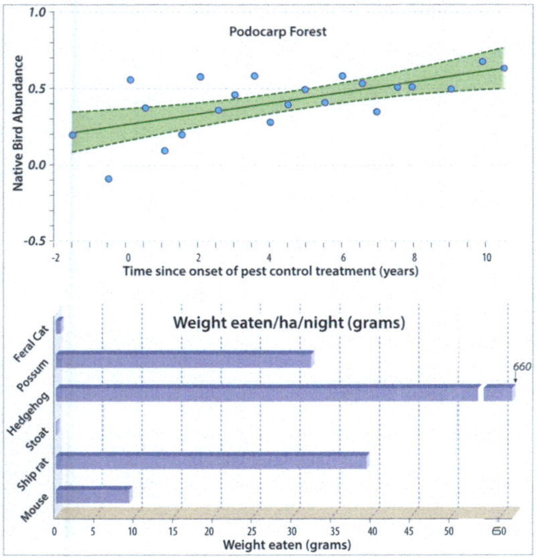

Fig. 13.6 A comprehensive review published in the NZ Ecological Society's symposium, *Feathers to Fur*, considered the effects of predation on both vertebrate and invertebrate faunal transformations in New Zealand. *Above* Positive numbers in the left axis show the increase in relative abundance of native forest birds in podocarp forests with time since the onset of a repeated pest mammal control programme (each point gives the difference in bird abundance between a pair of survey groups, one inside and one outside a pest control area). The same comparison for exotic birds was not significant. *Below* The likely extent of predation by six species of pest mammals on litter invertebrates in podocarp forests The toll exerted by hedgehogs is more than 8 times as heavy as that of all five other species together. *Redrawn by Max Oulton from Innes et al. (2010: Fig. 2a and Table 6)*

for—not because they are not fully capable of it (they certainly are), but because (a) historically, they arrived long after hordes of rats had already done more than a hundred years of terrible damage, and (b) now, because mustelids are much less abundant than are rats. This idea seemed counter-intuitive when first suggested [40], and at first sight still is, but the evidence accumulated over the 30 years since then is gradually changing many old assumptions.

Of course, mustelids cannot be ignored. Native birds have certainly been at risk from stoats since at least the 1890s if not sooner, and the large, vulnerable flightless chicks of kiwi, blue duck, and kaka are especially at risk from stoats. But rats have done more and worse damage for much longer, and are still many times more abundant, so better conservation outcomes can usually be achieved by targeting ship rats first.

Protecting the Kokako

The people of a nation tend to identify themselves with the iconic native animals and plants unique to their country. Anything that threatens the existence of endemic species is often taken as a threat to the people. Hence the ancient Maori regarded with foreboding the prospect that they themselves might be doomed, as were the species with which their ancestors shared the forests—the moa and the kiore (Chap. 3).

In our own time, the kokako and kiwi bring out such reactions more than do most other birds. With their ghostly grey plumage and haunting, flute-like song, kokako seem like magical spirits mourning the departed forests and their formerly teeming companions. The high national value of this beautiful bird is illustrated by New Zealand's $50 banknote, which bears on its reverse an image of a kokako, with Rod Hay's picture of Pureora Forest (the northern margin of Waipapa EA) in the background.

The kokako is endemic at family level, which implies that its ancestors arrived here millions of years ago—long enough to have produced a group of unique and irreplaceable birds found nowhere else. It belongs to the Callaeatidae, the New Zealand wattlebirds, which is one of the most ancient elements of the avifauna and has no close affinities with any other group of birds in the world [65].

All the New Zealand wattlebirds were once abundant and widespread, but the huia is certainly extinct. Sporadic but unconfirmed records of the South Island kokako suggest that it has been virtually if not actually extinct for years, and the two subspecies of saddlebacks were confined to offshore islands until recent (2002–09) translocations to pest-fenced mainland sanctuaries. The North Island kokako (Fig. 13.7) is the only member of this ancient group still surviving in the remains of its original distribution

Fig. 13.7 An adult kokako at its nest, with two chicks. The pink wattles of chicks at this age, about 20 days, gain the blue of adults after fledging at about 3 months old. *Crown Copyright, Department of Conservation Te Papa Atawhai. Photographer Dick Veitch (1969)*

on the mainland. The kokako symbolises the ancient forests and the losses of the ancient trees and birds of Pureora, which evoke in many people a love and affection for old New Zealand; a world that has almost disappeared.

Although the historic decline of kokako was very clearly linked to habitat destruction (Fig. 13.12), that alone did not explain why kokako continued to disappear from forests that still stood. Once the primary need for habitat protection was widely accepted and the logging of its habitat stopped (Chap. 10), attention turned to the next problem, the woefully low production of chicks by the surviving adult birds.

One obvious problem is due to the birds themselves, not to human mismanagement: the natural behaviour of kokako makes them vulnerable to nest disturbance for longer over each year than are many other forest birds. Kokako lay their eggs late in the season, between November and February, so their chicks are still in their nests (Fig. 13.7) just at the time the season's crop of young predators is emerging in increasing numbers.

Kokako clutches are small (2–3) and their incubation period (16–20 days) is long. After hatching, the chicks remain in the nest for a month, and after the fledglings leave the nest they remain in their parents' territory for another 9–12 weeks, clumsy at first but learning to be independent. Young kokako are exposed to the risk of encountering a rat or possum for proportionately longer than birds with earlier nesting habits, larger clutches, shorter incubation and pre-fledging periods, and faster post-fledging growth.

On the other hand, kokako are long-lived, competent and extraordinarily forgiving parents, if given relief from predation. Only 33 nests had been documented between 1880 and 1989, but most of the few protected from predators had succeeded [34]. Research to pin down exactly why, and what to do about it, was done in three successive phases over the twenty years following the end of the logging in PFP.

The first phase concentrated on diet and habitat use. The key ground-breaking work was done partly in PFP by a team led by Rod Hay during the logging moratorium in 1978–81, backed up by two independent studies of habitat use at Puketi [8, 58]. The second phase was a small but significant predator control trial at Pikiariki in 1982–83 (Fig. 13.16), plus the development of standard survey techniques for monitoring mortality rates of rats through aerial 1080 pest control operations [30]. The third phase in 1989–97 comprised an extraordinary eight-year experiment on whether and how it might be possible to manage the nesting success and population dynamics of

Fig. 13.8 A meeting of members of the Forest Bird Research Group, in their office in Rotorua in 1980. *Left* to *right* Rod Hay, Neil Howie, John Leathwick, Bob Patterson, John Stevens and Alan Saunders. *Photo courtesy Rod Hay*

kokako. Its conclusions are central to the successful recovery of kokako populations.

Phase One: The Forest Bird Research Group, 1978–81

Soon after the Taupo seminar (Chap. 9), a Kokako Research Advisory Committee was formed, and during the three years of the logging moratorium it supervised an intensive study of the ecology of the kokako (along with a study of the effects of logging on general bird populations), undertaken by a specially convened group of scientists entitled the Forest Bird Research Group (FBRG) (Fig. 13.8).

The Royal Forest & Bird Protection Society (RFBPS), New Zealand's leading conservation society, launched an appeal for NZ$30,000 to support the group and finance the research, and specifically the full-time salary of the only non-government field scientist of the group, Rod Hay. Alan Saunders and John Leathwick were seconded fulltime to the project from NZWS and NZFS. The other members of the group, and the many other people who contributed their time and expertise to its work in between other duties, are listed in Box 13.2.

Box 13.2 Contributors to the FBRG research project on the kokako, 1978–81 [25]

Kokako Research Advisory Committee Christoph Imboden, Malcolm Crawley, Keith Prior, David Black, David Field, Tony Beveridge, Rob Guest, David Collingwood, Dave Stack.

Forest Bird Research Group scientists: Rod Hay, Alan Saunders, Malcolm Harrison, John Leathwick.

Field assistance: Neil Howie, Robert Patterson, Alison Davis, Les Renney, John Stevens, Elaine Marshall.

Trapping kokako: Don Merton, Gideon Anderson, Leon Cooke.

Government officers: Alice Fitzgerald, Tony Robinson, Kevin Hackwell, Mick Clout, David Dawson (all of DSIR); Ralph Powlesland, Phil Moors, Noel Hellyer (NZWS); John Innes, Jack Walker, John Gaukrodger, Alan Champion (NZFS)

The terms of reference given to the group were based on NZFS's request for information on the habitat requirements of kokako, and specifically on the effects of selection logging on their long-term survival. Work started on 9 January 1979 [75], not only in Pureora (Pikiariki, Waipapa and a cutover block in northern Pureora) but also in other forests at Rotoehu and Mapara. Companion studies looked at the other forest birds, the vegetation, forest structure and species composition in logged and unlogged forest at PFP and at Whirinaki, and the phenology of many plant species, and incorporated other work on the diet of possums for comparison with the diet of kokako [48].

In 1972, there were known to be surviving groups of kokako scattered between the last few large patches of forest in the North Island (e.g., Mamaku, Rotoehu and Urewera National Park), but the total number of breeding adults remaining at that time was quite unknown.

The kokako team concentrated on the habitat use of kokako, their food habits, competition with introduced mammals, basic population parameters, improving survey techniques to facilitate future studies, and recommendations to ensure the long-time persistence of the kokako [25]. This was a huge list of tasks for a short and minimally funded study, and Hay's 168-page report is testimony to the urgency and intensity with which it was carried out. Hay's pioneering research at Pureora, Mapara (King Country) and Rotoehu (Bay of Plenty) unravelled key details of kokako behaviour and ecology for the first time.

Kokako were observed to use a very wide range of foods, many of them seasonal. Fruits, especially the berries of fivefinger, raukawa, broadleaf, putaputaweta and kaikomako, and matai were very important in the autumns of the years in which they fruited (not every year). Invertebrates such as the sixpenny scale (*Ctenochiton viridis*) were important in summer, and leaves in spring, when insects and fruit were less available (Box 13.3).

The results showed that kokako were not so much generalists as sequential specialists, and the removal of any of these seasonal supplies was a serious matter. Worse, many of these were the same items that were favoured at the same seasons by possums, which had arrived less than 20 years previously (Chap. 12) but were now much more numerous than kokako.

Box 13.3 Overlap in the diets of possums and kokako at Pureora

These data were documented by the Forest Bird Research group working at Pureora, Mapara and Rotoehu during the moratorium [48]. The Pureora study area was a 40 ha block of slightly-modified podocarp forest where possums have steadily increased in numbers since they arrived in the 1960s.

There is considerable overlap between the diets of kokako and possums; the leaves and/or fruit of some species are eaten by both. Browsing has reduced the abundance of preferred kokako food plants in much of the remaining kokako habitat. At Pureora, major food items for kokako included fruit of fivefinger, raukawa, kaikomako, putaputaweta, matai and supplejack, and leaves of mahoe, while possums included leaves of fivefinger, raukawa, mahoe and supplejack as important items in their diet.

The present distribution of kokako in the North Island suggests that their decline has been caused not only by forest clearance and introduced predators, but also by impoverishment of habitat resulting from the introduction of browsing mammals (Fig. 13.9).

Habitat use by kokako was largely correlated with the distribution of these resources. Kokako tended to sing from tree tops, but fed mainly in sub-canopy epiphytes and fruiting hardwoods, avoiding the lower understory plants. They were poor fliers but still highly mobile among the forest trees. They would climb rapidly up tree trunks and lianes on their long legs; run through the canopy with an agile movement worthy of any squirrel; glide from high trees across gullies with a looping, rather clumsy flapping flight; and descend by diving (Fig. 13.10).

Kokako lived in and defended their large (7–11 ha) territories, either as single birds or

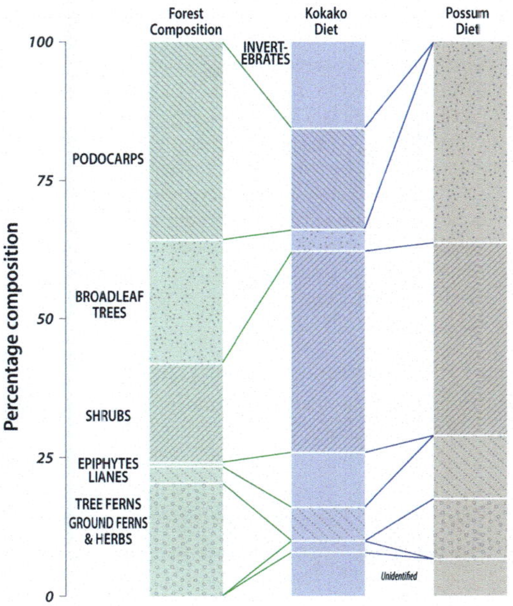

Fig. 13.9 The diets of kokako and possums in relation to forest composition at Pureora. The kokako diet was recorded from December 1978 to May 1981 and includes leaves, buds, flowers, fruit and insects; possum diet was recorded during May, July, August and December, 1980 and refers to leaf diet alone. See the original diagram for a breakdown of individual species if they comprised greater than 1 % of the total quantity. *Simplified by Max Oulton from Leathwick et al. (1983: fig. 1)*

pairs, all year round, so they needed to find all their food within it, and could not respond to logging by moving to quieter places elsewhere. Hence kokako need large areas of continuous forest to permit dispersal between breeding groups and to maintain their populations.

Kokako numbers were highest in dense podocarp forest (averaging 0.10 birds/ha), but they were much less common in unlogged medium-density podocarp/hardwood and tawa/hardwood forests (0.06 to 0.01 birds/ha); none were recorded in recently logged medium-density podocarp forest. Kokako had colonised some older cutover forests, which had been logged before possums arrived and which had recovered to some extent, but recent selection logging had removed the largest trees carrying epiphytes and important food resources, and favoured the growth of wineberry and ferns that provided little short-term substitute food [25].

Kokako reproductive output can never be large, because they start breeding later in the season than many other forest birds and take a long time to incubate and fledge their chicks—and even after fledging, the young remain dependent on their parents for weeks. In years of abundant fruit they might nest twice in succession, but their

13.10 Kokako move through the forest subcanopy with an athletic action known as "squirrelling", spend a minimal time on the ground, then use their long legs to climb rapidly back to the canopy by "laddering", followed by flapping or gliding between emergent crowns, and returning to the subcanopy in a "controlled plummet". *Redrawn by Max Oulton from Hay (1981: 81)*

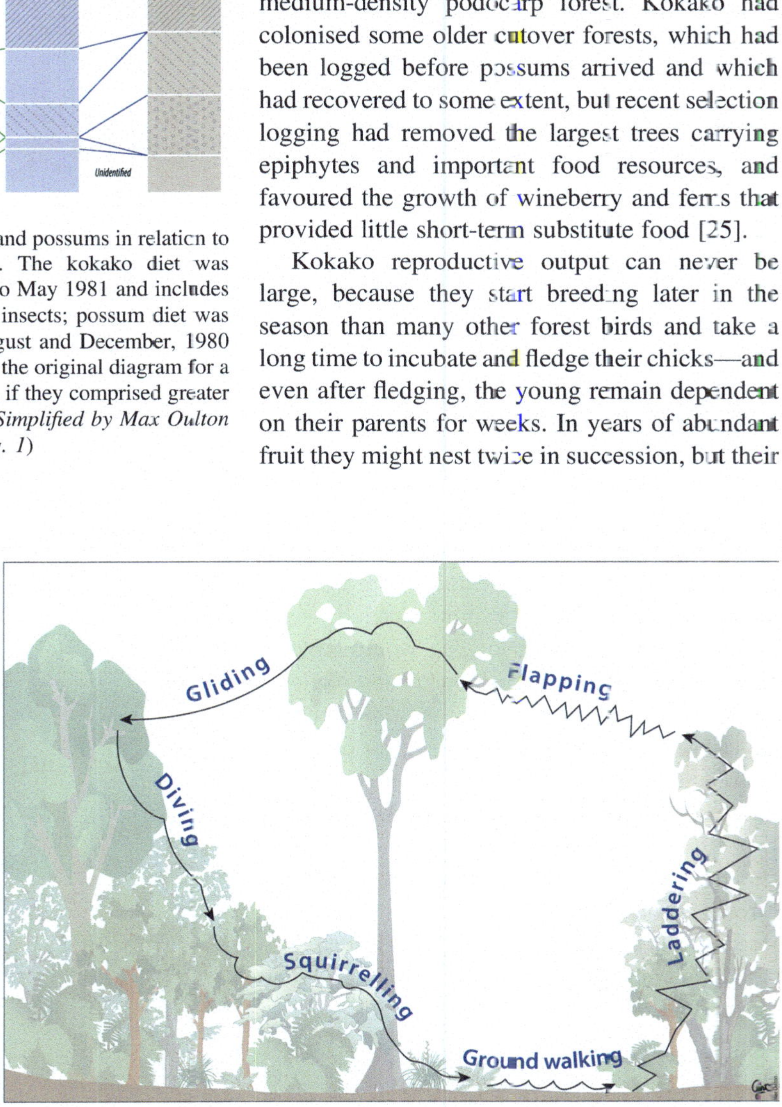

potential output even in ideal conditions is only about *four young per year per pair*.

In the breeding seasons of 1979–80 and 1980–81 combined, the remaining adult kokako in Pureora produced only *three young between 15 pairs* [25: 76]; predators destroyed the rest. Of the five nests found in the three FBRG study areas, three were destroyed at the egg stage, and two at the fledgling stage. Most causes of nest losses could not then be confirmed, and few juveniles were seen.

Naturally, the FBRG study included a first attempt to monitor numbers of predators, both in the known kokako nesting areas and in unlogged compared with logged forest. In winter, spring and summer, up to 20 % of tracking tunnels were marked by rats or mice, but in autumn the indices for both rodents more than doubled, dropping back to previous levels by the following spring. That reflects a known and expected annual cycle of productivity of rodent pups—the surprise came when the results for logged and unlogged forests were plotted separately.

In logged forests, the autumn peak in tracking rate for mice reached over 50 %, twice the rate for unlogged forests, whereas the autumn peak for rats was smaller in both. This was the first clue, not understood at the time but later amply confirmed (e.g., by a 1983–87 survey: Fig. 13.2), that thick cover on the ground (e.g., logging slash) favours mice more than rats. Mustelid tracks appeared, but only at a fraction of the rates for rodents.

FBRG concluded that logging would remove the large trees that supported the greatest loads of epiphytes and lianes that were key food sources for kokako, and that (in the presence of deer and possums) the vegetation that colonised logged areas was not favoured by kokako. Accordingly, their primary recommendation was that 'logging should not proceed in areas of the North Block of Pureora State Forest Park containing kokako populations'.

FBRG finished their report with the sober statement that the kokako

> ….faces extinction locally and perhaps nationally…..[it] is not right at the brink…..[as it] has withstood a certain degree of habitat diminution and impoverishment, and predation, and….still survives in moderate numbers in some areas. Knowledge of those factors operating against kokako welfare enables management techniques to be implemented in anticipation of a critical situation rather than as a last resort [25: 90].
>
> Survival in the face of…adverse factors may be possible until the addition of further stresses renders the population non-viable. Some of these events (e.g., possums) are of recent origin in some kokako habitats. Survival under conditions of heavy modification in the past does not mean that survival under light modification today is necessarily possible [25: 98]

FBRG's recommendations [25: 98–101] were presented to the Government in October 1981 (Box 13.4).

Box 13.4. Recommendations of the Forest Bird Research Group [25]

1. Logging should not proceed in areas of the North Block of Pureora State Forest containing kokako populations.
2. Waipapa Ecological Area should be extended northward to include areas of virgin forest.
3. A high priority should be given to control of browsing mammals in areas of kokako abundance.
4. An experimental control programme for predators is recommended, and has already been initiated.
5. Periodic monitoring of kokako populations should be undertaken to check trends and allow for management if necessary. They should be carried out five-yearly in the areas used in the 1978–81 study (Pikiariki and Ngaroma blocks of Pureora, plus Rotoehu and Mapara).
6. Surveys of kokako breeding success should be carried out in a number of areas. Any known kokako population can be surveyed using the techniques from this study by interested groups or individuals.
7. Positive management procedures should be instituted in certain key kokako areas [such as Mapara and

Puketi, under a plan for control of predators and competitors and planting of food trees based on the "game-keeper" concept]. One person should be employed at Mapara to carry out such work as the plan dictates.

Phase Two: Pikiariki, 1982/83

In the summer of 1982/83, FRI scientists attempted to remove predators from a block of the Pikiariki EA using a range of toxins, and monitored the number of nests destroyed. They took advantage of the newly developed toxin brodifacoum, made up in attractive 'Talon' baits (Box 13.5). No-one likes the idea of poisoning sentient animals, even if they are common pests, so it is important to find the most humane methods available. In New Zealand it is even more important to use whatever means we can to protect the eggs and young of the irreplaceable endemic fauna.

Toxic baits were laid in a 40 ha block in part of the Pikiariki Ecological Area [30], and the non-treatment area was 1.1 km away in the same forest (mapped in Fig. 13.17). From 25 January to 22 March 1983, 339 bait tunnels containing Talon baits were laid out on the ground in 22 parallel lines averaging 31 m apart, covering 35 ha.

Box 13.5 Use and properties of brodifacoum ("Talon") toxic baits

In many other countries of the world, the very idea of large-scale use of toxic baits to protect native fauna is rejected with horror. New Zealand is one of the very few countries in which it is a feasible management option, because the ecological costs of using toxic baits are provisionally regarded as much smaller than the damage suffered by vulnerable wildlife if toxins are not used [31]. Nevertheless, it is not done lightly: the advantages and disadvantages of using toxic baits for pest control have been debated for many years. Research has concentrated on minimising the levels of their use and maximising their efficiency.

Brodifacoum is a second generation anticoagulant, with a delayed action that ensures that target animals consume more than enough to kill them, so are unlikely to survive a sub-lethal dose after taking only one bait. Its effectiveness, however, comes with a cost—longterm persistence in the environment and very high risk of by-kill [57].

Brodifacoum is licensed for killing possums and rats. Like 1080, it will kill stoats and cats that feed on poisoned animals. It has been successfully used in aerial operations to completely eradicate possums and rats and stoats on many offshore islands and fenced 'mainland islands' that are now sanctuaries for endangered animals [19].

On the islands where it has been distributed from the air, brodifacoum has clearly increased populations of native species because it has eradicated the pests that prey on them. An example is Ulva Island off Rakiura/Stewart Island. DOC cleared Ulva Island of rats in 1997, and since that time populations of rare birds like tieke (South Island saddlebacks), toutouwai (Stewart Island robin) and mohua (yellowhead) have been successfully established on the island.

The Department of Conservation does not spread brodifacoum from the air on the mainland, except to clear pests from within newly-fenced sanctuaries. Rats and mice appear to prefer 'takeaway' foods, which they can pick up and carry away to a safe place to eat, so DOC has to reduce the risks of by-kill from spilt toxic bait by confining it to ground bait stations. Brodifacoum should be avoided where more humane alternatives would be as efficient [19, 50]. One of its few benefits is that accidental poisoning can be treated with Vitamin K1.

Brodifacoum is more cost-effective than first generation anticoagulants when used in ground operations, because bait stations do not need to be replenished nearly as often. But it takes a very long time to break down in soil and water, and accumulates in the tissue of exposed animals for years. At least 21 species of native birds including kiwi, kaka, kakariki and tui are known to have been killed by brodifacoum. Department of Conservation standard operating procedures also require that an area where brodifacoum has been used must be closed to hunting for three years after the operation. In comparison, an area must be closed for at least five, up to eight months following an aerial or ground 1080 operation.

Contrary to common perception, sodium monofluoroacetate (1080) is a much safer and kinder alternative [57]. Also, aerial distribution of 1080 will often be far cheaper than an operation employing ground crew to use brodifacoum or other toxins. At the moment, both 1080 and anticoagulants other than brodifacoum are used in PFP, but public debate surrounding all and any use of 1080 on principle continues (Chap. 12).

Rat populations targeted repeatedly by control operations using brodifacoum or any other single toxin soon develop an inbuilt behavioural protection against poison baits. Studies on wild Norway rats (which may apply in part to ship rats) show they will eat almost anything edible, so long as they have learned as youngsters what foods are safe. Pups learn from their mothers, and from watching other rats feed, whether to accept an unfamiliar food or not, and as adults they follow each other to good feeding sites. They will destroy any accessible fruits, seeds, invertebrates, small vertebrates and their eggs, but those that learn to avoid poison baits quickly enough will be the only ones to survive, and become the parents of young that do the same.

So the first baiting programme using a proven toxin in a new area is usually very successful, but repeated operations need to switch toxins frequently in order to combat learned bait-shy behaviour. This will be one of the most serious future problems for park management faced with the issue of long-term control of ship rats, especially if public objections remove one of the best current weapons, aerial distribution of 1080—despite all reasoned argument in favour of it [57].

The results were monitored by recording footprints in tracking tunnels, both on the ground and in trees (Figs. 13.11 and 13.12) and catches in steel Fenn traps (designed for stoats, but equally effective in catching rats) set in both treated and untreated blocks, on the ground and up trees. Two weeks after the baits went out, the number of rats caught in Fenn traps dropped to zero, and the number of tracking tunnels recording rat footprints declined by 73 %. The methods used and results obtained in the Pikiariki trial were later applied to other areas and developed into a standard operating procedure, using robust bait stations attached to trees (Fig. 13.11). It was so effective that, several years later, John Innes and his team were able to conclude that 'most ground-based and aerial poisoning operations (also killing possums) reduced indices of ship rat abundance by at least 90 %' [30].

In 1983, the team had the thrill of watching two chicks grow from hatching on 8 February to partial independence by 15 May [34]. The nest, a large structure of twigs labelled no. 29, was built after failure of an earlier nest and was placed in a typical position, 7 m up a mahoe and under dense overhead cover. The behaviour of the chicks and the adults was observed for on average 7 hours a day for 25 days between 31 January and 12 March. It was the first time kokako chicks had hatched from a nest in a whole protected territory, although they had previously hatched and learned to fly from nests in heavily protected individual trees [17].

Fig. 13.11 Routine pest management techniques. **a** A member of DOC field staff (Ces Koia) refilling a bait station in Waipapa EA containing toxic pellets for possums and rats. *Crown Copyright, Department of Conservation Te Papa Atawhai. Photographer: John Mason*. **b** Footprint tracking plates are a standard method of monitoring the effectiveness of any control operation. A removable tray protected from rain by a simple tunnel carries a central ink pad, from which animals transfer footprints to papers on either side. *Crown Copyright, Department of Conservation Te Papa Atawhai (1996). Photographer: Ralph Powlesland*

This careful study led to the conclusion that the kokako are good at hiding their nests from the eyes of the diurnal avian predators searching from overhead, such as the harriers and falcons with which they had evolved. The kokakos' problem is that their natural defences are no longer appropriate.

Dense vegetation and lianes of bush lawyer and supplejack leading up from the ground, which kokako use to hide and assist access to their nests, can be climbed equally well by rats and possums, so the kokako's normal nest-choice behaviour gives them no protection against nocturnal introduced mammal predators [34].

Fig. 13.12 FRI scientist John Innes climbing a tree to check tracking tunnels set in the branches. *Crown Copyright, Department of Conservation Te Papa Atawhai. Photographer: John Mason*

Phase Three: Research by Management, 1989–97

By June 1988, when a workshop at FRI in Rotorua reviewed the progress of ecological research on the kokako, the major recommendations of FBRG had been largely met. Most significantly, the logging moratorium had been made permanent, which had halted one of the obvious reasons for kokako decline. The next thing to do was to summarise current progress and set future objectives for research and management of kokako [29].

There was no easy answer, because the reasons for continued disappearance of kokako from the remaining forests were still debatable. One was the problem of habitat fragmentation, which had chopped the formerly-continuous forests into patches supporting small isolated groups of surviving birds. The science of population genetics had recently come up with a rule of thumb predicting that a minimum of between 500 and 1000 breeding adults is needed to maintain a healthy population without inbreeding over the long term [21].

The problem with small remnant populations is that it takes a certain number of unrelated partners to maintain the genetic health of a breeding unit. Below that threshold, mating with relatives becomes inevitable as the number of unrelated mates available declines, so inbreeding increases, recessive genes are expressed, the genetic diversity of the stock declines rapidly, and survival probabilities fall, reducing the breeding stock still further. This dismal downward spiral is well known to population geneticists, and is the common fate of many small populations.

The remnants of the west Taupo forests were among the last nine known locations in the King Country where kokako were known to survive, and they all represented the largest patches of forest available in the region [56]. Kokako had been present but had disappeared within the previous 20 years from a further 12 patches. Other forest remnants were far too small to support a self-sustaining population (Fig. 13.13).

If so, as pointed out by Colin Ogle, kokako living at the maximum local density found by FBRG, 0.23 birds/ha, need more than 2000 ha of suitable habitat to survive long term. At worst, if 1000 birds are needed for an effective population of 500 pairs, and if these are living at the minimum density found by FBRG (0.16 birds/ha), then possibly there might be no patch of forest left anywhere in the King Country large enough

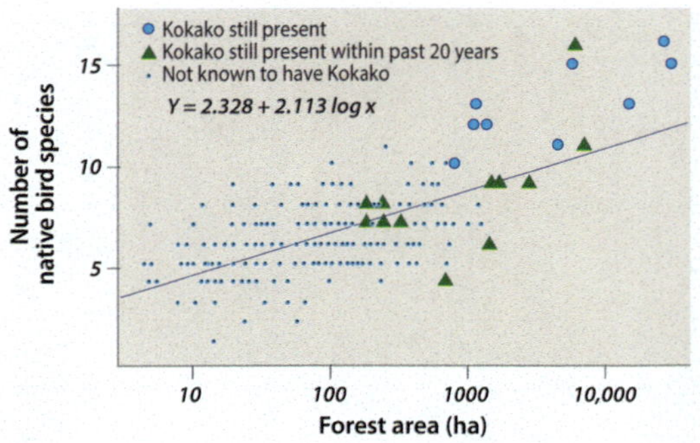

Fig. 13.13 The distribution of forest birds in forest remnants of the King Country, as surveyed in 1982–83. Only the largest nine fragments still supported kokako populations, and at that time they were not secure even there. At another 12 sites, kokako had disappeared from forest fragments where they had been known to live within the previous 20 years, including two sites at which they were gone as recently as 1979 and 1980. *Redrawn by Max Oulton from Ogle (1987: fig. 2), courtesy Surry Beatty Ltd*

to ensure the long term survival of the kokako [29: 17].

On the other hand, small populations had been rescued before, and the history of conservation in New Zealand provides plenty of inspiring examples. The black robin was brought back from a single breeding pair, and no-one believed the kokako had got to the near-terminal stage that the black robin had been in.

So, while acknowledging the real problem of habitat fragmentation, the urgent question for the workshop was how to distinguish between the two current hypotheses proposed to explain why kokako were disappearing from the forests that still stood: (1) competition with browsing mammals for food, and (2) predation.

All that was known at the time of the workshop was that the current total number of kokako nests monitored to a known outcome was 23, of which 7 were successful, 10 destroyed by predators, and 6 failed for other reasons [29]. These figures were bad enough, but they were still not really reliable estimates of the true predation rate, since predators may have been attracted to nests by the visits of observers, and they did not count deaths away from nests, such as adults killed by falcons (like the one observed by Bill Drower: Chap. 8).

The overwhelming need right now, John Innes told the meeting, is to work out whether removing browsing mammals or predators was the more urgent priority to increase kokako productivity. He proposed a series of controlled experiments over five years, which were debated at length. In due course he saw their conclusions inform the next decade of kokako research and management, which eventually brought the kokako back from the brink.

Research funding had always been tight, but under the new rules accompanying the creation of DOC in the previous year, it was now to be decided by competitive bidding. Of 500 bids submitted to DOC for the 1988/89 year, eight concerned kokako, of which two were ranked highly enough to be funded [16].

Workshop participants were invited to help DOC decide on another eight management projects. There was general agreement that, now that the issue of habitat protection was universally accepted, discussion needed to focus on *which* competitors and predators to control, and *to what level*.

Management operations against different pests need to use the right method at the right time of year to be effective. If the kokako are *failing to try* to breed, their nutritional condition, perhaps correlated with the damage by browsing animals to the quality of their habitat, could be the main cause. If they are attempting to breed but *failing to produce* surviving offspring, then predators targeting eggs, chicks or fledglings could be responsible.

Answers to both these questions need to be compared with the potential productivity of kokako living relatively free of both browsers and predators. To pin down the details, it was generally agreed that a "research-by-management" (RbM) approach concentrating on working out the relative threats posed by browsers and by predators could potentially disentangle the various possible causes of decline of the kokako. Adaptive management made use of routine large-scale pest control in a co-ordinated experiment to directly test the pest-limitation hypothesis, enabling researchers and managers to investigate the cause of decline and to increase populations simultaneously.

It sounds simple to say that the effects of rats, possums, mustelids and cats can be distinguished by removing them separately and testing which action permits the best increase in kokako numbers. The reality is much more complicated. Possums are not only browsers but also predators (Fig. 13.5), and ship rats are predators that also eat fruits; in addition, possums and ship rats interact with each other as well as kokako, so removing possums (and getting a consequent improvement in vegetation) tends to increase the numbers of ship rats, which may in turn favour cats and stoats…and so on. In nature, managers cannot change only one thing.

So the RbM approach was both visionary and pragmatic, accepting that it relies on significant variables that are not isolated or independent. Critics pointed that out during robust discussions at the workshop. Yet, with all its limitations, the RbM experiment was agreed to be the simplest way to find a cost-effective management regime that could maintain kokako populations over the long term.

The next problem was to decide *where* and *when* to do the work. With the help of the feedback from the workshop, Gretchen Rasch of DOC led a group compiling a *Recovery Plan for North Island Kokako*, the first of a series of such plans to be produced by DOC's Threatened Species Unit [62]. A team led by John Innes was commissioned to run an eight-year RbM experiment to be done in three North Island forests [35, 66]. The nearest of the three experimental areas to PFP, Mapara, is <50 km to the west (Fig. 11.6), and the results from there are shown in Fig. 13.14.

Pest control was switched on and off as experimental treatments in the three mainland forests in different years, and Little Barrier Island was used as a further reference area, to document the breeding success of kokako living free from all of the introduced predatory species.

The work was demanding and time-consuming, because at that time the only method of getting precise data on kokako numbers and productivity was by territory mapping, a skilled

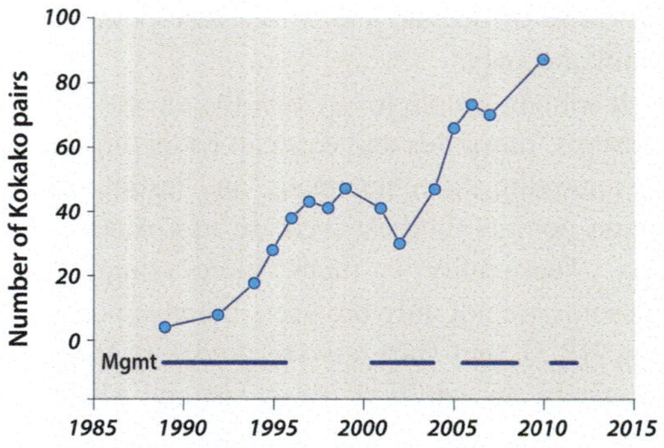

Fig. 13.14 Reversible effects of predator control on kokako breeding success documented at Mapara. Over nearly three decades, regular monitoring of the effects of intermittent pest management has shown that the increase in number of resident kokako pairs during periods of freedom from predation predictably slips back into a decline when management is switched off, but that effect can be reversed when management starts again. The managed 'years' in this sense were the summer breeding seasons starting in the calendar years of 1989–96, 2001–03, 2006–07, 2010–11. The apparent delayed response to management arises because kokako juveniles recruit into the breeding population after 1–4 years. *Updated from Innes et al. (1999), courtesy John Innes*

and exhausting job. Worse, the RbM approach required rigorous comparisons between treatment and non-treatment areas, and before- and after-treatment surveys which demanded that management be deliberately withheld or switched *off* at certain scheduled times and places, meaning that researchers, managers and the public sometimes had to stand by and watch some kokako suffer without spoiling the experiment by rescuing them.

Inevitably, public criticism of this strategy made the researchers feel worse. In the long term, the heartbreak of documenting the carnage in the non-treatment areas was fully justified, because it provided conclusive proof that, despite the overlap in diets (Box 13.3), competition for food from browsing herbivores is a relatively minor current cause of the decline of kokako compared with predation.

How was this vital conclusion reached? In spring each year, predators were targeted with trapping or, more frequently, poisoning. Sometimes aerial 1080 was used. Alternatively, 50 m grids of poison bait stations for ship rats (50 cm sections of yellow plastic "Novacoil" drainage pipes, which excluded non-target species such as birds) were laid out in known kokako territories. Each contained three Talon (WB50) poison baits, checked weekly and renewed as necessary. DOC files record the estimated cost of setting up the first (1989/90) season as NZ$4813, equivalent to >NZ $8500 in today's money, not including labour.

Introduced browsing and predatory mammal pests were effectively removed from two forest areas, and the team monitored pest abundance, kokako chick output and adult density in the managed forests and in an unmanaged non-treatment block. The declining numbers of rats were monitored by counting the numbers of baits taken and the number of tracking tunnels recording footprints of rats. Possums were sometimes targeted by fur trappers using leg-hold traps and cyanide. Best results were obtained when possums and ship rats were driven down to very low threshold levels (<1 % residual trap catch for possums; <1 % tracking rate for ship rats, using particular indexing techniques: Chap. 12) at the onset of the kokako nesting season. Rats were always back to previous numbers well before the next season, and possums over 3–4 years, but mustelids and feral cats were not specifically targeted at all after the first few years, and kokako food supplies were not supplemented.

These results showed conclusively that the number of kokako breeding pairs at Mapara (and not only at Mapara, but generally, for example within PFP around the Mangatutu Track: Figs. 13.15 and 14.9), increased steadily when and only when the numbers of nest predators were sufficiently reduced. The output of kokako chicks and the density of adults increased significantly in all protected study populations, due mainly to the greater success of nesting attempts. That then increased the number of pairs attempting to breed, initially as newly recruited young females formed pairs with residual single males.

These experiments, accompanied by video-monitoring of nests, clearly identified ship rats and possums as the primary predators of nesting kokako. Although these were not the only causes of loss, the kokako could cope with the other perils of their lives for as long as they were

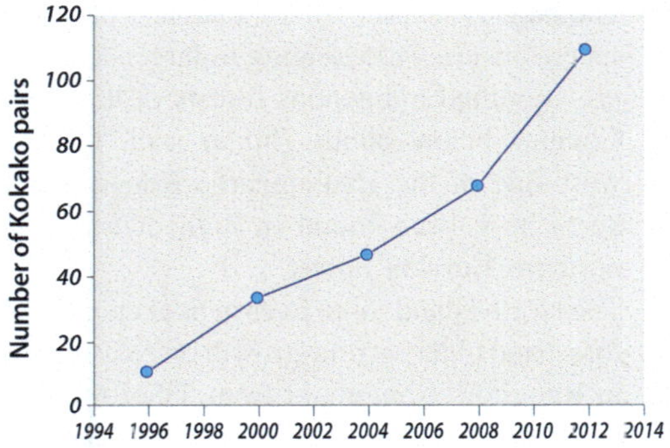

Fig. 13.15 Numbers of kokako pairs under intensive protection in the Mangatutu catchment of Pureora Forest Park, 1996–2012. *Data courtesy JG Innes, Landcare Research, and Tertia Thurley, DOC*

protected from rats and possums. Freedom from rats and possums is therefore one of the main conditions needed by kokako to achieve increased productivity and density. Where that condition is met, kokako can breed well and build up their numbers despite occasional encounters with the normally much less abundant mustelids or cats.

The clinching evidence was the chilling observation that this benefit could be measured, predicted and simply switched on and off at human will. At Kaharoa, increased productivity stopped abruptly, and at Rotoehu, four years of dismal results suddenly became three years of successive increases, as management operations swapped from one to the other [35]. The team concluded that predation is a more immediate cause of current kokako declines than is competition for food, although that is a known complication.

The actual calibration of threshold pest tracking rates against kokako breeding success has turned a research result into a predictive tool. Future field managers can tell at once if the numbers of rats and possums have been reduced enough to boost the productivity of the kokako for the current season, and if not, they can run the operation again while there is still time.

Implementation of these methods in high-priority protected areas is now routine, and kokako populations are recovering in many areas. At Pureora in 2014, there were 115 pairs of kokako in the unlogged dense podocarp and podocarp/hardwood forest in the Waipapa Ecological area; 108 pairs in the Mangatutu catchment that is managed by volunteers from the Howick Tramping Club (Fig. 13.15), and 13 pairs in the Tunawaea catchment that lies between these two. The Kokako Recovery Group now regards these as one population, the largest in New Zealand.

Other Native Wildlife

FBRG was interested in the whole bird community, not only the rare ones. Hay's report [25] was accompanied by a second, written by Malcolm Harrison and Alan Saunders from NZWS [24], presenting the results of an intensive statistical analysis of >6000 five-minute bird counts made during the moratorium (1978–81) in three areas of podocarp/tawa forest.

One of the areas analysed was in Pureora North Block, where the impact of recent 30 % selective logging on bird populations could be estimated by comparison with Waipapa, an adjacent area of unlogged forest previously recognised (Chap. 9) as an area of outstanding wildlife value (total 670 ha).

Despite the acknowledged limitations of five-minute bird counts recorded over a short period (and the variation in conspicuousness of 15 indicator bird species was tested with mathematical models), no significant differences could be detected except that tuis appeared to favour the unlogged block. RFBPS, a co-funder of FBRG and normally its staunchest supporter, criticised this conclusion [72], and demanded an independent audit in case NZFS used it to justify continued logging. Harrison & Saunders simply re-emphasised the need to assess the impact of predators on bird populations over a longer term.

The third report of the three produced by FBRG concentrated on the vegetation of the seven study areas studied for kokako and other bird species in PFP and other central North Island indigenous forests [47]. It reviewed the impact on forest structure of selection logging trials, the phenology of the main species of shrubs and trees, and the damage done to them by browsing mammals, particularly possums.

PFP had also figured largely in the results of the extensive survey of the indigenous wildlife of the King Country in 1977, after which NZWS had reported on the distribution and habitat requirements of eleven species of wildlife considered to be 'more sensitive' or in 'more critical' condition (Chap. 2), and hence most vulnerable to future changes in land use [7].

Most of the locations these species occupied were in the Rangitoto and Hauhungaroa Ranges, within or near the present boundaries of PFP. Nine critical species or groups were discussed and their 1977 distributions mapped. Over the years since then, new species have been added and further information collected on the original nine.

1. Hochstetter's frog is one of four very rare and primitive endemic frog species that have no tadpole stage and no equivalents elsewhere outside New Zealand. Hochstetter's frog was still present on the Rangitoto Range in 1983/84 [46].
2. North Island brown kiwi were still widespread in the King Country south of Pureora in the 1950s and 60s. E&B mill workers living along the Waione River used to talk of kiwi keeping them awake at night. The Otorohanga Zoological Society were transporting brown kiwi from Northland into the Pureora North Block in 1970. There may still be kiwi in the Waihaha, Tihoi and perhaps the Maramataha areas; a search for kiwi in the Waihaha area found none in January 1994 [6], but the blue duck survey of 2011 picked up kiwi sign (feathers and footprints) in Tihoi, and a live chick in the Waihaha.
3. Blue duck were present in streams on both sides of the Hauhungaroa Range from Pureora south, and up into the North Block (Chap. 12). A reporting system started in the 1980s, in collaboration with NZWS, now shows that the blue duck population in PFP has grown over the last few decades.
4. The New Zealand falcon was almost common (the King Country includes the best falcon habitats in the North Island). Reporting systems aiming to help with territory mapping started in the 1980s.
5. The New Zealand parrots (kaka and parakeet species) were wide ranging in larger tracts of less modified indigenous forests of the King Country below about 750 m a.s.l. On the west side of the Hauhungaroa Range, parakeets have been found in high densities in northern Hurakia Forest.
6. The North Island robin lives in pockets of high forest and older second-growth on easy terrain in Rangitoto Range and from Tihoi to Waihaha on the eastern side of the Hauhungaroa Range, plus in three areas of Hurakia Forest.
7. Fernbirds live in widely scattered areas of swamp, scrub and heathland throughout PFP.
8. Long-tailed bats are present but with unknown distribution. Bats were reported in Tihoi SF in September 1975, and on the south side of the Kokokataia Stream in the Maraeroa C block and in Pureora [75]. They were present at Ngaherenga campground at Pureora in the 1980s and 1990s, and at Kakaho campground in the mid 2000s, and throughout the Pikiariki Ecological Area [16].
9. Short-tailed bats were found in 1996 at three sites at Pureora during an intensive study on the native root parasite *Dactylanthus taylorii* [18]. This was a very significant first re-discovery of live short-tailed bats, then believed to be extinct in the area. The flowers of this strange parasitic plant are adapted for pollination by short-tailed bats, which often feed on the ground (Fig. 2.14).
10. The common skink, tree skink, copper skink, and green gecko are all present; the copper skink is regarded as 'fairly common' in Pureora Village [46].

Complications: Mesopredator Release

A systematic five-year survey of small ground predators in PFP (Box 13.1) showed that stoats are much more abundant than are weasels, ferrets, or feral cats, so forest birds are much more at risk from stoats than from any other true carnivores. But for most North Island forest birds (except kaka, parakeets, blue duck and kiwi), even stoats are a small risk compared with ship rats, which are not in fact primarily carnivorous. This is nothing to do with predatory prowess, it is simply a matter of *relative numbers*. Vulnerable native fauna are hundreds of times more likely to encounter a ship rat than a stoat. In typical North Island mixed forests, an area of 40–65 ha might supply home ranges for on average 300–500 rats, and one stoat [64].

But since stoats also eat rats, an urgent question arises, very relevant to management policy to protect the kokako from predation: would effective stoat control risk increasing the numbers of rats? Such an unwelcome consequence, called mesopredator release, describes a situation in which a top predator such as the stoat normally keeps down the numbers of a mid-ranked or mesopredator such as the ship rat. Removal of stoats alone would then be a real risk if it permitted rats to increase to numbers far higher than stoats ever reach, with proportionately drastic consequences. The risk depends critically on how far predation by stoats does in fact control the numbers of rats.

It is easy to assume that specialist predators such as stoats must always have a devastating effect on their prey, and indeed, when the prey are confined to a small space or have no anti-predator defences, that is true. Very many naïve endemic birds on small islands have met tragic ends that way, but on a large mainland area, things are different.

If a stoat in PFP comes face to face with a rat, the odds of it getting a meal are not certain, but probably fairly good. But just on the grounds of probability alone, if a stoat has to search every day up and down throughout the large three-dimensional space occupied by a hectare of podocarp forest with a canopy averaging 25 m tall (say, 260,000 m^3) in order to find one of less than half a dozen rats living there, the odds are on the rat.

According to recent estimates, stoats living in podocarp forest ate on average only one rat a week [38]. Given that stoats always live at very low density—around 1–4 stoats or fewer per km^2 except after a massive seedfall in a beech forest —this very small harvest of rats by a small number of stoats is nowhere near enough to cause any decrease in the large numbers of rats

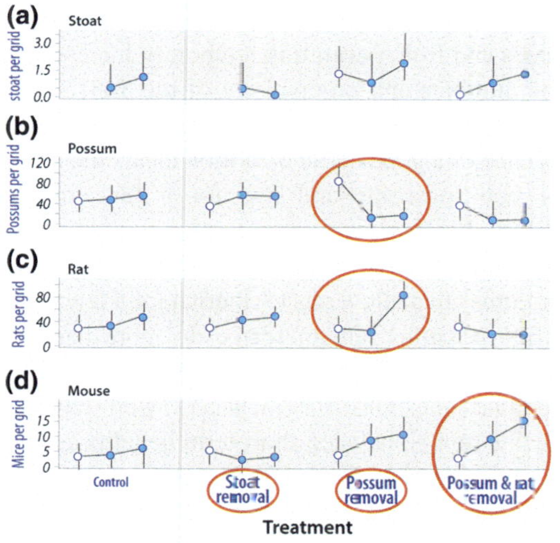

Fig. 13.16 Contrary to common belief, pest control targeting stoats alone has no significant effect on the numbers of ship rats, i.e., removing stoats does not "release" ship rat numbers—stoats simply cannot catch and eat enough rats to make any difference to their numbers (col 2). But when competition and interference from possums is minimised, a significant increase in numbers of ship rats predictably follows (col 3). Mice benefit when protected from both possums and rats (col 4). *Redrawn by Max Oulton from Ruscoe et al. (2011: fig 1)*

present and capable of producing dozens of descendants per season per female.

The important conclusion for the protection of kokako comes from controlled comparisons of the consequences of removing stoats alone, or possums alone, or possums and rats but not stoats (Fig. 13.16): stoats were simply too few, and ate too few rats, to make any difference to rat numbers. But, as an unwelcome twist to this puzzle, rats did respond with increased numbers when they were protected from competition from possums [64].

One study monitored ship rat abundance in a podocarp-hardwood forest over 14 years (1990–2004), including through two aerial possum-poisoning operations (1994, 2000). The mean abundance indices for ship rats increased nearly fivefold after possum control, and remained high for up to 6 years after the 1994 poisoning. Rat fecundity was high (50–100 % of adult females breeding), even during winter, and young animals dominated the population (73 % in age classes 1–3) in 2001–02 when rat numbers were increasing. The conclusion was that the removal of possums reduced competition for seeds and fruit, permitting temporary increases in the numbers and productivity of rats [68].

Although the immediate short-term consequence of a good aerial operation clears the forest of both possums and rats for a few months (Chap. 12), rats recover much more quickly [35]. So can widely spaced possum control programmes that allow rats to flourish for a few years while possums remain scarce do more harm to birds than good? This is one of many uncomfortable questions for managers who have to decide how to balance short-term benefits against longer-term risks for the birds, not to mention the ever-present problem of operational costs.

We must also ask the opposite question: what effect might rats have on the numbers of stoats? Field experiments and mathematical models have offered different answers, still not consistent or definite but suggesting that rats may have more effect on stoats than stoats have on rats. In the field, rat poisoning operations in Pureora were followed by lower catches of stoats, since stoats are easily killed by eating poisoned rats [55]. In a computer model, the predicted fledging rates of kokako improved best when *both* ship rats and possums were removed, partly because the model (developed before the publication of estimates that stoats eat on average one rat a week) incorporated an additional indirect effect of ship rat control in reducing stoat abundance [59].

Once more we return to the conclusion that management of ship rats, or preferably rats, possums and stoats together, is best for general ecosystem recovery in most places, including PFP. But it is not simple.

The Waipapa/Pikiariki Restoration Project

The area designated as Pikiariki EA in 1984 had been badly damaged by Cyclone Bernie at Easter 1982. High winds and heavy rain caused 138 windfalls among the trees disturbed by selection logging, and further damage was predicted round the margins of the reserve where formerly continuous forest was exposed by clear felling up to 1978 [54].

To mitigate this problem, Timberlands (a private forestry company) allowed DOC to manage part of the exotic forest plantation on both sides of the Pikiariki EA as a buffer zone. Later DOC management permitted logging of these areas, for a small economic return and to the detriment of the reserve.

Although it is still true that the extent, shape and design of the Pikiariki EA are deficient, that the forest (dominated by podocarps) and birdlife are in an unstable condition, and that future management must include replanting of native species and intensive predator control, there are compensations. The easy access and history of research effort make Pikiariki a nationally significant site of scientific interest, which fully justifies the current restoration project.

Because the reserve of podocarp forest at Pikiariki EA is irregular in shape, and parts have been selectively logged [54], the quality of kokako habitat there is much diminished. Nevertheless, the outcome of the 1982/83 predator programme in Pikiariki (Fig. 13.17) had been enormously encouraging. The problem was that there was no funding for a routine programme of predator control to be implemented anywhere in PFP at that time.

Ten years after the original survey by FBRG, during the summer of 1990–91, the southern third of unlogged forest of the much larger Waipapa EA was surveyed for kokako [52]. It covered an area of 4375 ha of easy terrain at 500–600 m a.s.l., not counting an enclave of private land in its centre. It is dominated by rimu/tawa forest, where two 'walk through surveys' followed 20 transect lines marked at 300 m intervals in January and March, and tapes of kokako calls were played at 250 m intervals.

The team found 34 kokako territories supporting 17 pairs of birds in the Waipapa EA, of which four pairs (23 %) fledged seven juveniles. The conclusion was that there had been a decline of up to 60 % in kokako density since 1980–81 [52]. Active pest management had been prescribed for Waipapa by the FBRG, but had been done only sporadically, in 1990 and 1993.

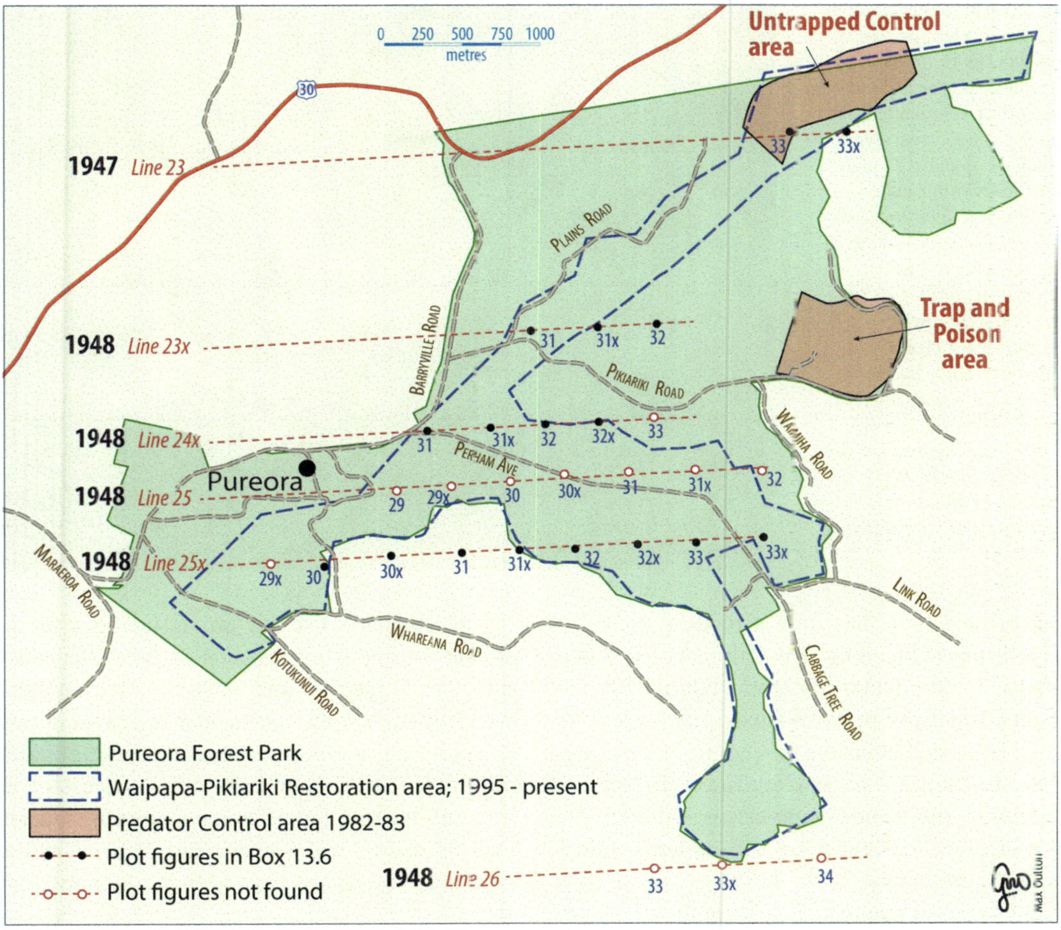

Fig. 13.17 The Pikiariki Restoration Area. The National Forest Survey lines (Chap. 6) passed through this area in 1947–48, and the plot data are summarised in Box 13.6. The treatment and non-treatment areas used in the first experiment in predator control for the protection of kokako by Innes et al. in 1982–83 are marked. *Redrawn by Max Oulton from Murphy (1984), by permission Barry Murphy*

The next kokako survey in the Waipapa was scheduled for the summer of 1993–94 [60, 61]. Poison baits put out in October and November reduced the number of rats by 83 %, rather less than the target of 95 %—perhaps a result of bait shyness. Fourteen kokako were located, but uncertainties inherent in the survey technique meant that there may have been only eight individual birds, including two pairs.

A proposal for funding a regular management operation to rescue and protect the Waipapa/Pikiariki areas for kokako was put together over the 1994/95 season by a team led by Tony Roxburgh, and presented by John Gaukrodger to a meeting of the Deputy Directors of DOC in 1995. It was turned down, but later rescued when John Holloway found the NZ$220,000 required from another fund. The Waipapa Restoration Project began in 1995 [67], and still continues. For example, in 1997, a juvenile male plus an adult and a juvenile female kokako were captured at Mapara (lured into a mistnet by playback of kokako calls) and translocated to Pikiariki for release (Fig. 13.18).

The two juvenile kokako travelled to and from the adjacent Waipapa EA for the next 5–9 months, as newly released birds usually do before settling into territories [33]. But a group of three

Fig. 13.18 Translocations of kokako from managed populations to protected sites has become a standard method of supplementing small groups of survivors, which are usually all males. Here, DOC staff have brought a female kokako from Mapara for release in the Pikiariki EA in September 1997. *Left* to *right* Phil Bradfield (Mapara), Ra Hepi (Pureora), Billy Steiner (Mangakino) and Hazel Speed (Pureora). *Crown Copyright, Department of Conservation Te Papa Atawhai (1997). Photographer: John Mason*

translocated birds has only a modest chance of establishing a breeding population. There is now only a single kokako living in Pikiariki EA, but about 250 nearby in the Waipapa EA [70].

DOC started intensive pest control in Waipapa with the aim of improving forest health and increasing the numbers of native wildlife. The project 'was designed as a long-term animal control programme with the objective of providing maximum practicable benefit to species, biological processes, and ecosystems under threat from possums and rats, with minimum risk to these values' [67]. Bait stations covering 2500 ha in the Waipapa EA are filled with poison bait between September and April each year, targeting possums and rats. The Waimanoa Ecological Area is used as an untreated control (Fig. 11.7).

Today the possums have all but gone, and rat numbers are controlled in a two-years on- and-off pulsed regime. Routine work includes monitoring of possum and rat populations, monitoring of fruitfall and seed fall in traps placed beneath fuchsia, fivefinger and raukawa, and a phenology study of these species and kamahi, mahoe and pate. Breeding success of robins, tits, New Zealand pigeon, moreporks and kokako is regularly monitored.

The Waipapa EA has proved to be a fertile ground for important research on New Zealand's native bird species, and intensive management has resulted in some spectacular increases in rata flowering and in bird populations there. Possum control by DOC continues to be successful in reducing possum impacts on canopy tree species such as mahoe and kamahi and also on native mistletoe. The results have been vitally important not only for kokako, as reported above, but for other species as well.

Robins and Tomtits

Conservation authorities are required to make good management decisions about iconic species, so they need reliable information about population density and trends. The Waipapa EA is an ideal location to do this important work. The high nesting success (around 80 %) and abundance of North Island robins in the Waipapa area has enabled several translocations of robins to other sites within the Waikato region and further afield. These robins (and their relatives the tomtits) have also been tolerant and forgiving subjects of necessary research on the effects of 1080 operations on bird populations (Chap. 12).

Kaka

Kaka are large and noisy birds well known to Pureora residents, who have for decades enjoyed observing their flight (Chap. 9). They nest in tree holes with only one entrance, and only the female kaka incubates the brood. Stoats, possums and ship rats are all good climbers that can easily reach the nests. In unprotected areas, where sitting females are vulnerable to being caught on the nest by one or other of these predators, nesting females may suffer such high mortality that the sex ratio of the kaka population becomes strongly skewed towards males. In the Waihaha EA, in PFP on the eastern side of the Hauhungaroas, this imbalance reached three males to one female in the summer of 1994/95. The slow breeding rate of kaka combined with the shortage of female mates is a serious threat to the long term survival of the kaka population within the Waihaha EA [23].

A large-scale study led by Terry Greene and Ron Moorhouse compared the nesting success and mortality rate of adult females in protected versus unprotected nests. Three of six study areas had ongoing pest control, each paired with the same arrangement of three areas without [53]. The Waipapa EA was one of the managed areas where regular toxic baiting of possums by various methods including 1080 began in 1995. Of 31 kaka nesting attempts, 27 produced a total of 70 chicks fledged (2.3 per nest: Fig. 13.19).

Fig. 13.19 Terry Greene using abseiling gear to climb to a kaka nest with his sampling bag during a monitoring progamme in Pureora. *Crown Copyright, Department of Conservation Te Papa Atawhai. Photographer: Shaun Barnett. Inset:* Well-grown kaka chicks at Nest 64, with an adult, 14 January 2004. *Greg Martin*

At Whirinaki, a comparable but unmanaged forest 100 km away, 5 of 13 nesting attempts succeeded, producing a total of 14 fledglings (1.1 per nest). Over all six areas, the success of protected nests was at least 80 %, more than double that of the unprotected ones (38 %), and the number of adult females killed was 5 % compared with 65 %. Stoats were a significant threat to kaka nesting in beech forests, because the study spanned three seedfall years when their numbers temporarily soar, but that did not apply to PFP where there is virtually no beech (Chap. 2).

How sure can anyone be of what is the real density of these mobile, elusive birds? One team made 13 systematic point counts using distance sampling as part of a national research programme covering Pureora, Whirinaki and Nelson Lakes (South Island). They confirmed the method as reliable, and found that the estimated October density of kaka in the Waipapa EA was about one per 2 ha, and roughly constant over the eight years 2000–07 [22]. The Waipapa EA now supports over 800 kaka, the largest protected population of wild kaka on the main islands of New Zealand).

Box 13.6 Pre-logging podocarp forest of the Pikiariki area, 1946–47

Data on number of podocarp stems per acre (0.4 ha) in the Pikiariki EA from the National Forest Survey, showing pre-logging forest

Species	Line/Plot number																
Line	23		23X			24X				25X							
Plot	33	33x	31	31x	32	31	31x	32	32x	30	30x	31	31x	32	32x	33	33x
Matai	4	8	45	14	14	34	69	31	41	55	43	32	13	8	25	17	6
Totara	–	1	14	1	1	10	2	19	10	7	4	6	–	–	3	–	7
Rimu	4	2	8	13	1	–	3	3	2	17	38	6	10	16	21	27	22
K'tea	–	5	5	3	1	4	–	8	6	1	1	6	–	–	–	1	9
Miro	2	–	4	3	–	2	–	4	–	4	–	2	–	–	6	–	11
Total	10	16	76	34	17	50	75	65	59	84	85	52	23	24	49	44	65

structure (except for Lines 25 and 26). Compiled by Murphy [54]. Lines mapped in Fig. 13.17.

References

1. Alley, J. C., Berben, P. H., Dugdale, J. S., Fitzgerald, B. M., Knightbridge, P. I., Meads, M. J., & Webster, R. A. (2001). Responses of litter-dwelling arthropods and house mice to beech seeding in the Orongorongo Valley, New Zealand. Journal of the Royal Society of New Zealand, 31, 425–452.
2. Angas, G. F. (1847). The New Zealanders illustrated. London: Thomas M'Lean.
3. Angas, G. F. (1866). Polynesia. London: Society for Promoting Christian Knowledge.
4. Atkinson, I. A. E. (1973). Spread of the ship rat (*Rattus r. rattus L.*) in New Zealand. Journal of the Royal Society of New Zealand, 3, 457–472.
5. Atkinson, I. A. E. (1985). The spread of commensal species of *Rattus* to oceanic islands and their effects on island avifaunas. In P. J. Moors (Ed.), Conservation of island birds (p. 35–81). Cambridge, England: International Council for Bird Preservation (Technical Publication 3).
6. Barnett, S. C., & Speed, H. J. (1994). Kokako and kiwi survey, Waihaha Ecological Area, Pureora Forest Park. Hamilton: Department of Conservation.
7. Best, H. A. (1977). The King Country land-use study: Indigenous wildlife. Wellington: New Zealand Wildlife Service, Fauna Service Unit report 8B.
8. Best, H. A., & Bellingham, P. J. (1990). A habitat survey of North Island kokako in Puketi Forest, Northland. Department of Conservation Science and Research Series, 17, 1–25.
9. Beveridge, A. E. (1964). Dispersal and destruction of seed in Central North Island podocarp forests. Proceedings of the New Zealand Ecological Society, 11, 48–55.
10. Beveridge, A. E. (1973). Regeneration of podocarps in a central North Island forest. New Zealand Journal of Forestry, 18, 23–35.
11. Bridgman, L. J., Innes, J., Gillies, C., Fitzgerald, N. B., Miller, S., & King, C. M. (2013). Do ship rats display predatory behaviour towards house mice? Animal Behaviour, 86, 257–268.
12. Brittan, S. J., Grace, G. F., Grace, C. W., & Grace, A. V. (Eds.) (1928). A Pioneer Missionary among the Maoris 1850–1879: Being letters and journals of Thomas Samuel Grace. Palmerston North: G H Bennett & Co.
13. Brown, K., Innes, J. G., & Shorten, R. (1993). Evidence that possums prey on and scavenge birds' eggs, birds and mammals. Notornis, 40, 169–177.

14. Bull, P. C. (1952). Birds and mammals of Western Taupo. New Zealand Science Review, 10, 91.
15. Clout, M. N., & Hay, J. R. (1989). The importance of birds as browsers, pollinators and seed dispersers in New Zealand forests. New Zealand Journal of Ecology, 12, 27–33.
16. Department of Conservation. Archives. Hamilton: Department of Conservation.
17. Dominion, 21 May 1983.
18. Ecroyd, C. E. (1996). The ecology of *Dactylanthus taylorii* and threats to its survival. New Zealand Journal of Ecology, 20, 81–100.
19. Fisher, P., & Fairweather, A. (2010). Brodifaccum. A review of current knowledge. Version 2.6. Hamilton: Department of Conservation.
20. Foster, S. P., King, C. M., Patty, B., & Miller, S. D. (2011). Tree-climbing capabilities of Norway and ship rats. New Zealand Journal of Zoology, 38, 285–296.
21. Franklin, I. R. (1980). Evolutionary change in small populations. In M. E. Soulé & B. A. Wilcox (Ed.), Conservation biology: An evolutionary-ecological perspective (p. 135–149). Sunderland, MA: Sinauer Associates.
22. Greene, T., Jones, A., Dennis, G., & Sachtleben, T. (2010). Distance sampling to determine kaka (*Nestor meridionalis septentrionalis*) density within Waipapa Ecological Area, Pureora. New Zealand Journal of Ecology, 34, 297–305.
23. Greene, T. C., & Fraser, J. R. (1998). Sex ratio of North Island kaka (*Nestor meridionalis septentrionalis*), Waihaha Ecological Area, Pureora Forest Park. New Zealand Journal of Ecology, 22, 11–16.
24. Harrison, M., & Saunders, A. J. (1981). A comparison of bird populations in logged and unlogged indigenous forest areas within Pureora and Whirinaki Forests, North Island, New Zealand. Wellington: New Zealand Wildlife Service, Forest Bird Research Group.
25. Hay, J. R. (1981). The kokako. Wellington: New Zealand Wildlife Service, Forest Bird Research Group.
26. Hill, G. (2012). Why I enjoyed the Rena disaster. Forest & Bird, 346, 49.
27. Houghton, P. (1974). Hidden water. Auckland: Hodder & Stoughton.
28. Hursthouse, C. (1857). New Zealand, or Zealandia, the Britain of the South, Vol 1. London: Edward Stanford.
29. Innes, J., Owen, K., & Smale, S. (1988). Proceedings of the FRI kokako workshop. Rotorua: Forest Research Institute.
30. Innes, J., Warburton, B., Williams, D., Speed, H., & Bradfield, P. (1995). Large-scale poisoning of ship rats (*Rattus rattus*) in indigenous forests of the North Island, New Zealand. New Zealand Journal of Ecology, 19, 5–17.
31. Innes, J., & Barker, G. (1999). Ecological consequences of toxin use for mammalian pest control in New Zealand. New Zealand Journal of Ecology, 23, 111–127.
32. Innes, J., Kelly, D., Overton, J., & Gillies, C. (2010). Predation and other factors currently limiting New Zealand forest birds. New Zealand Journal of Ecology, 34, 86–114.
33. Innes, J., Molles, L., & Speed, H. (2013). Translocations of North Island kokako, 1981–2011. Notornis, 60, 107–114.
34. Innes, J. G., & Hay, J. R. (1995). The nesting of the North Island kokako (*Callaeas cinerea wilsoni*)—Review of accounts from 1880–1989. Notornis, 42, 79–93.
35. Innes, J. G., Hay, R., Flux, I., Bradfield, P., Speed, H., & Jansen, P. (1999). Successful recovery of North Island kokako *Callaeas cinerea wilsoni* populations, by adaptive management. Biological Conservation, 87, 201–214.
36. Innes, J. G., King, C. M., Flux, M., & Kimberley, M. O. (2001). Population biology of the ship rat and Norway rat in Pureora Forest Park 1982–1987. New Zealand Journal of Zoology, 28, 57–78.
37. Jameson, R. G. (1842). New Zealand, South Australia and New South Wales: A record of recent travel in these colonies, with especial reference to emigration and the advantageous employment of labour and capital. London: Smith, Elder & Co.
38. Jones, C., Pech, R., Forrester, G., King, C. M., & Murphy, E. C. (2011). Functional responses of an invasive top predator *Mustela erminea* to invasive meso-predators *Rattus rattus* and *Mus musculus*, in New Zealand forests. Wildlife Research, 38, 131–140.
39. Kelly, D., Ladley, J., Robertson, A. W., Anderson, S. H., Wotton, D. M., & Wiser, S. K. (2010). Mutualisms with the wreckage of an avifauna: The status of bird pollination and fruit-dispersal in New Zealand. New Zealand Journal of Ecology, 34, 66–85.
40. King, C. M. (1984). Immigrant killers: Introduced predators and the conservation of birds in New Zealand. Auckland: Oxford University Press.
41. King, C. M., Flux, M., Innes, J. G., & Fitzgerald, B. M. (1996). Population biology of small mammals in Pureora Forest Park: 1. Carnivores (*Mustela erminea, M. furo, M. nivalis* and *Felis catus*). New Zealand Journal of Ecology, 20, 241–251.
42. King, C. M., Innes, J. G., Flux, M., Kimberley, M. O., Leathwick, J. R., & Williams, D. S. (1996). Distribution and abundance of mammals in relation to habitat in Pureora Forest Park. New Zealand Journal of Ecology, 20, 215–240.
43. King, C. M. (Ed.) (2005). The handbook of New Zealand mammals, 2nd edition. Melbourne: Oxford University Press.
44. King, C. M., McDonald, R. M., Martin, R. D., & Dennis, T. (2009). Why is eradication of invasive mustelids so difficult? Biological Conservation, 142, 806–816.
45. King, C. M., Foster, S. P., & Miller, S. (2011). Invasive European rats in Britain and New Zealand: Same species, different outcomes. Journal of Zoology, 285, 172–179.
46. Krzystyniak, S. (1984). Pureora North Block animal survey, summer 1983. Auckland: New Zealand Forest Service.

47. Leathwick, J. R. (1981). The vegetation of kokako and general bird study areas in some central North Island indigenous forests. Wellington: New Zealand Wildlife Service, Forest Bird Research Group.
48. Leathwick, J. R., Hay, J. R., & Fitzgerald, A. E. (1983). The influence of browsing by introduced mammals on the decline of the North Island kokako. New Zealand Journal of Ecology, 6, 55–70.
49. Leigh, A. H., & Clegg, S. E. (1989). Maniapoto District wild animal management plan, 1989 to 1999. Hamilton: Department of Conservation.
50. Littin, K. E., O'Connor, C. E., Gregory, N. G., Mellor, D. J., & Eason, C. T. (2002). Behaviour, coagulopathy and pathology of brushtail possums (*Trichosurus vulpecula*) poisoned with brodifacoum. Wildlife Research, 29, 259–267.
51. Meade, H. (1870). A ride through the disturbed districts of New Zealand. London: John Murray.
52. Meenken, D., Fechney, T., & Innes, J. (1994). Population size and breeding success of North Island kokako in the Waipapa Ecological Area, Pureora Forest Park. Notornis, 41, 109–115.
53. Moorhouse, R., Greene, T., Dilks, P., Powlesland, R., Moran, L., Taylor, G., Jones, A., Knegtmans, J., Wills, D., Pryde, M., Fraser, I., August, A., & August, C. (2003). Control of introduced mammalian predators improves kaka *Nestor meridionalis* breeding success: Reversing the decline of a threatened New Zealand parrot. Biological Conservation, 110, 33–44.
54. Murphy, B. (1984). Pikiariki Ecological Area, its pattern, processes and management. BForSc thesis, University of Canterbury.
55. Murphy, E. C., Clapperton, B. K., Bradfield, P. M. F., & Speed, H. J. (1998). Effects of rat-poisoning operations on abundance and diet of mustelids in New Zealand podocarp forests. New Zealand Journal of Zoology, 25, 315–328.
56. Ogle, C. (1987). The incidence and conservation of animal and plant species in remnants of native vegetation within New Zealand. In D. A. Saunders, G. W. Arnold, A. A. Burbidge & A. J. M. Hopkins (Ed.), Nature conservation: The role of remnants of native vegetation (p. 79–87). Chipping Norton, NSW: Surrey Beatty & Sons.
57. Parliamentary Commissioner for the Environment (2011). Evaluating the use of 1080: Predators, poisons and silent forests. Wellington: Parliamentary Commission for the Environment.
58. Powlesland, R. (1987). The foods, foraging behaviour and habitat use of North Island kokako in Puketi State Forest, Northland. New Zealand Journal of Ecology, 10, 117–128.
59. Ramsey, D. S. L., & Veltman, C. J. (2005). Predicting the effects of perturbations on ecological communities: What can qualitative models offer? Journal of Animal Ecology, 74, 905–916.
60. Ranum, A. (1994). Effect of a 1080 poison operation on the abundance of ship rats (*Rattus rattus*) in the Pikiariki ecological area, Pureora Forest Park. Hamilton: Department of Conservation.
61. Ranum, A. (1994). Survey for North Island kokako (*Callaeas cinerea wilsoni*) in the Pikiariki ecological area, Pureora Forest Park. Hamilton: Department of Conservation, Waikato Conservancy.
62. Rasch, G. (1991). Recovery plan for North Island kokako. Wellington: Department of Conservation, Threatened Species Unit.
63. Rochfort, J. (1853). The adventures of a surveyor in New Zealand. London: David Bogue.
64. Ruscoe, W. A., Ramsey, D. S. L., Pech, R. P., Sweetapple, P. J., Yockney, I., Barron, M. C., Perry, M., Nugent, G., Carran, R., Warne, R., Brausch, C., & Duncan, R. P. (2011). Unexpected consequences of control: Competitive vs. predator release in a four-species assemblage of invasive mammals. Ecology Letters, 14, 1035–1042.
65. Shepherd, L. D., & Lambert, D. M. (2007). The relationships and origins of the New Zealand wattlebirds (Passeriformes, Callaeatidae) from DNA sequence analyses. Molecular Phylogenetics and Evolution, 43, 480–492.
66. Sinclair, A. R. E., Innes, J., & Bradfield, P. (2006). Making endangered species safe: The case of the kokako of North Island, New Zealand. New Zealand Journal of Ecology, 30, 121–130.
67. Speed, H. J., & Bancroft, F. J. (1996). Waipapa restoration project annual report no. 1, October 1995–May 1996. Hamilton: Department of Conservation.
68. Sweetapple, P. J., & Nugent, G. (2007). Ship rat demography and diet following possum control in a mixed podocarp–hardwood forest. New Zealand Journal of Ecology, 31, 186–201.
69. Taylor, R. (1855). Te Ika a Maui, or, New Zealand and its inhabitants. London: Wertheim and MacIntosh.
70. Thurley, T. (2014). Personal communication.
71. Towns, D. R. (2009). Eradications as reverse invasions: Lessons from Pacific rat (*Rattus exulans*) removals on New Zealand islands. Biological Invasions, 11, 1719–1733.
72. Waikato Times, 14 February 1982.
73. Whitaker, A. H. (1973). Lizard populations on islands with and without Polynesian rats, *Rattus exulans* (Peale). Proceedings of the New Zealand Ecological Society, 20, 121–130.
74. White, T. (1890). On rats and mice. Transactions and Proceedings of the New Zealand Institute, 23, 194–201.
75. Wilkes, O., Gaukrodger, J., & Ritchie, N. (2011). Chronology of Pureora. Hamilton: Department of Conservation.
76. Wilson, K.-J. (2004). Flight of the huia. Christchurch: Canterbury University Press.
77. Wolfe, R. (2005). Kororareka: Hell-hole of the Pacific. Auckland: Penguin.
78. Worthy, T. H., & Holdaway, R. N. (2002). The lost world of the moa. Bloomington: Indiana University Press.
79. Young, D. (2004). Our islands, our selves. Dunedin: University of Otago Press.

14 Recreation in Pureora Forest Park

D.J. Gaukrodger, N.A. Ritchie and C.M. King

Abstract

This chapter discusses the recreational facilities of Pureora Forest Park (PFP), including the provision of hunting permits; the management of the Recreational Hunting Area (RHA); the exploitation of deer, goats, possums and pigs; the development of tracks and huts, campgrounds and toilets; the observation tower, and educational facilities for schools; the tracks for off-road vehicles and mountain bikes; and the 83 km Timber Trail, a new trail for cyclists and walkers partly incorporating historic sawmill tramlines, with extensive interpretation signage providing a living illustration of PFP as described in this book.

Keywords

Recreational hunting · Red deer · Feral pigs · Feral cattle · Brushtail possums · Pureora Hunting Competition · Outdoor education · Observation tower · Walking tracks · Huts and campsites · Mountain biking · Cycle tourism · The Timber Trail · Sawmill history

Recreation can be defined as "The pleasurable and constructive use of leisure time" [4]. The very purpose of Forest and Conservation Parks is to provide for "public recreation and enjoyment" so long as "natural and historic resources are protected". So developing recreational facilities has been an important part of PFP management from the beginning.

The geographic central point of the North Island lies squarely within Pureora Forest Park. The stone marking the spot (at a point 5 km along Link Track northeast of Pureora (Fig. 14.1) is a point of interest for visitors, and a reminder that recreation in the park does not include anything to do with the sea.

A report presented to the Taupo seminar [4] (Chap. 9) discussed the current and potential uses of PFP for recreational purposes. It recognized three roughly distinct groups of recreational users

D.J. Gaukrodger
3a Te Angi Street, Turua, RD4, Thames 3574, New Zealand
e-mail: beejaygauk@xtra.co.nz

N.A. Ritchie
Department of Conservation, Te Papa Atawhai, Hamilton Office, Private Bag 3072, Hamilton 3240, New Zealand
e-mail: nritchie@doc.govt.nz

C.M. King (✉)
School of Science, University of Waikato, Private Bag 3105, Hamilton 3240, New Zealand
e-mail: cmking@waikato.ac.nz

C.M. King et al. (eds.), *The Drama of Conservation*,
DOI 10.1007/978-3-319-18410-4_14

Fig. 14.1 This obelisk marking the geographic centre of the North Island (Fig. 0.1 in Preface) is not far from the track between Pureora and Titiraupenga (Fig. 14.9). *Crown Copyright, Department of Conservation Te Papa Atawhai. Photographer unknown*

of forest lands in general. First, there is a small, active group interested in the more traditional uses of tramping and hunting, and only this minority is likely to venture into the interior of the forest far from any facilities. The needs and interests of hunters are well known, because hunting is a long-established sport requiring permits that could be listed and analysed, and some hunters are good at providing feedback to forest managers.

Second, there is a far larger group of people from all walks of life who are attracted to the periphery of the forest, for more passive forms of recreation including picnicking, short walks, photography, nature study, ornithology and rock hounding. Because Pureora is ecologically richer than any other forest in the North Island, and easily accessible from all the main urban centres of the Waikato, Bay of Plenty, Auckland and Wellington (Fig. 0.1 in Preface), this group is likely to grow with any increase in provision of facilities. These and other activities are important, but need no permits so cannot be traced.

Third, there is a wider group of distant admirers, who may never actually visit the park but gain pleasure (=recreation) simply from giving it "existence value"—the satisfaction of knowing that it exists. To cater for such a wide range of user interests, Russell Dale and Fleming [4] suggested that PFP should be divided into recreation zones, ranging from the inaccessible and unmodified areas reserved for those interested in remote experience, through to recreational development and education zones. Facilities such as car parking, picnic tables, toilets, information boards, sign-posted nature walks, information centres, camp sites, and lodges for educational groups, would need to be developed. As a first response, NZFS developed an Information Centre offering historical displays and a supply of maps and brochures (Figs. 14.2 and 14.3).

Over the next few years, much of the work of John Gaukrodger, the last NZFS OC (1981 to 1987), and John Mason, the Environmental Ranger, was focused on the future development of PFP, especially recreational facilities. They spent many hours "scrambling around the area planning roads and tracks". The team focused on access, accommodation and information including brochures and signage, in order to establish Pureora as a prime destination for outdoor recreation, starting with hunting.

Hunting

New Zealand Pigeons

Hunting of pigeons and other native birds had been part of Maori culture for centuries (Chap. 3). But

Fig. 14.2 After 1978, the NZ Forest Service began to convert Pureora State Forest 96 from a production forest into a publicly accessible Forest Park. The immediate need for a visitor centre was met by converting a former NZFS cookhouse, and providing an information display on the history of the area and on the new recreational facilities available. *Crown Copyright, Department of Conservation Te Papa Atawhai (1980). Photographer unknown*

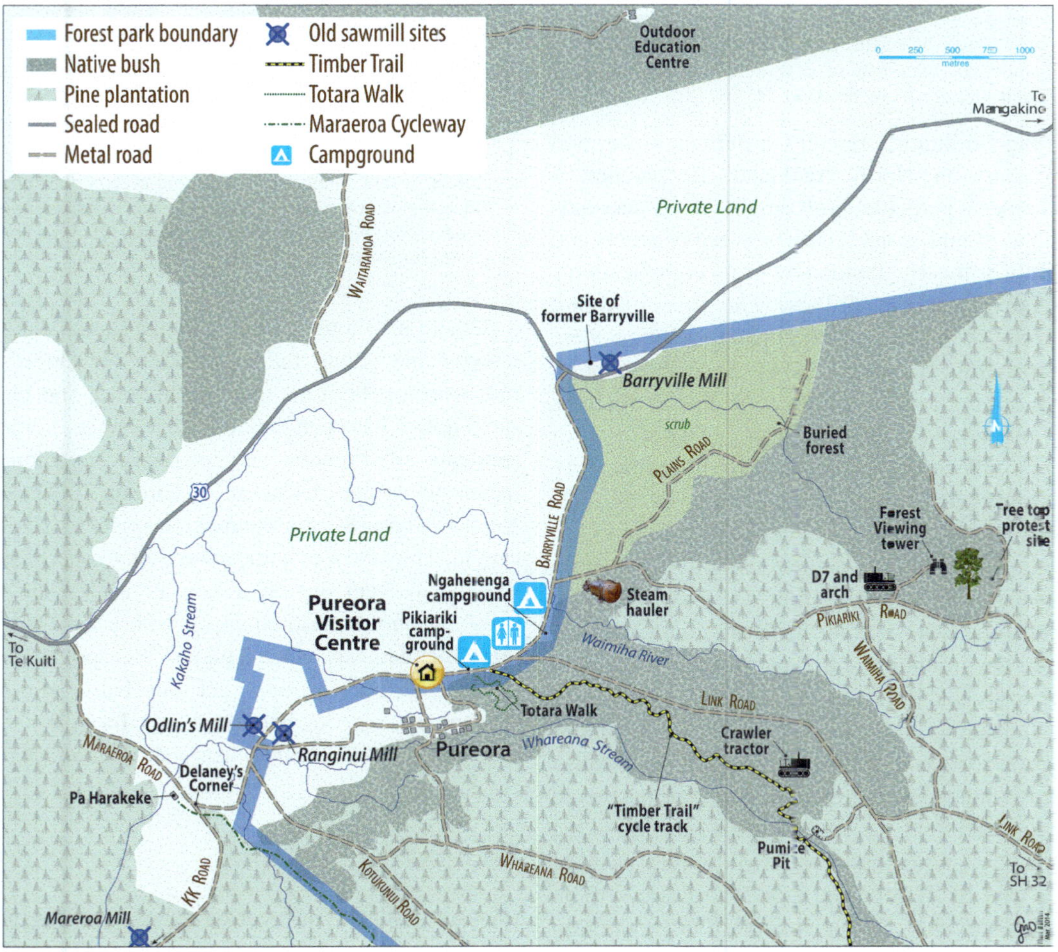

Fig. 14.3 Present location of the Pureora Forest Park HQ, and of local historic sites and visitor facilities. *Redrawn by Max Oulton from DOC visitor information sheet, with additions*

under various Acts of Parliament dating back to 1895, unauthorised killing of pigeons was restricted, and eventually became a criminal offence [7]. Pigeons are now absolutely protected under an amendment to the 1922 Animals Protection Act, although some Maori continue to insist that the taking of pigeons is not illegal poaching, but their right under the Treaty of Waitangi.

In North Auckland it was not uncommon to come across evidence of native pigeon shooting. In the autumn when the miro trees were laden with berries, poachers would make a clearing around a tree to provide a clear view and then shoot the pigeons as they came into eat.

> .... we came across a miro tree adjacent to which the poachers had set up a tent camp complete with fireplace and all the comforts of home. We duly chopped up the tent and all the utensils before leaving the area. Justice was done and nothing more was heard of the incident. The poachers could not complain of course without revealing their identity [Ivan Frost, in 14: 167–8].

Pigeon-hunting remains a difficult issue, and not only in North Auckland. In practice it becomes a perpetual problem that often descends into a cat-and-mouse game between poachers and wildlife officers, attended by many complications.

For a start, not all those who cheerfully shot dozens of pigeons in a single expedition were Maori (Fig. 14.4) [31: 151]. Second, shooting was not the only cause of the decline in numbers of pigeons. Hence, reprimanding members of their staff caught poaching raised contradictory questions and disagreements among NZFS officers. For example, after Ivan Frost transferred to Pureora as a junior tutor in cruising, he caught a poacher and reported him to his boss, Eric Johnstone, who was all for letting the culprit go—on the grounds that the logging was destroying the birds' habitat anyway.

Johnstone and many others saw the contradiction between passing legislation to protect birds whilst allowing their habitat to be destroyed by clearfelling—as was the usual practice at the time. Not surprisingly, some Maori responded to requests by Europeans, that Maori stop killing pigeons, with a similar plea that Europeans stop cutting down the bush [7: 28].

Fig. 14.4 A group of hunters posing in about 1900 with their bag, a large pile of native pigeons. Pigeon hunting, legal then, is now totally prohibited, but pigeon poaching is still a major headache for land managers. *Fairey and Plum shooting party, The Nelson Provincial Museum, Tyree Studio Collection, 176893*

Other OCs thought differently: the station diary records one young fellow working for Odlin's mill who lost both his gun and his bird, and the same policy has been continued ever since. Some residents of Pureora village found ingenious means to escape detection. According to Dave Yanko (Chap. 8), the cunning ones seldom got caught. They devised some good tricks, especially those who went out on a Sunday walk with a baby, pushing a pram equipped with a false bottom. The plan was, when a pigeon was spotted, to go through the actions (get out baby, shoot pigeon, put pigeon, baby and gun back in the pram) as fast as possible, then walk home [32].

Feral Pigs

Pigs were by far the earliest of the introduced domestic species, and the first large mammals to run wild throughout the North Island bush. From at least 1800 the outward spread and subsequent enormous increases in density of feral pigs made them an integral part of the Maori trading economy well before any Pakeha visited the future PFP, and a very important source of meat for Maori and for European explorers alike (Chaps. 3 and 4). Pigs have remained widespread in the back country for two hundred years (Fig. 14.5), although not now in the numbers typical of the early nineteenth century.

The clearing of forest and advance of farming in the 1920s and 30s helped to remove the cover needed by pigs, and so curtailed their distribution. Pig hunters have always tried to correct this problem by capturing pigs alive and releasing them into suitable areas to build up a hunting resource. This practice is of course illegal, but that does not stop it [10].

Feral pigs come in all shapes, colours and sizes, some with traces of their domestic ancestry, while others have reverted to the long-legged, shaggy black-coated, bad-tempered type wrongly labelled "Captain Cookers" (Chap. 3).

From the earliest days of Pureora Forest village, many residents were keen hunters, unless their weekends were tied up with rugby, and most hunters were interested primarily in bringing home meat. In the late 1940s and early 1950s, when pigs and feral cattle were abundant in the forest and permits unknown, fresh meat was easy to get, and a very acceptable form of barter for goods and services.

Fig. 14.5 Feral pigs have provided a vital source of meat and sport in the King Country for 200 years. This hunter (Cyril Rutland) is bringing out a carcase from the Mokau valley in the 1930s. *National Library PA Collection 0001-2-1-022*

At that time, Rusty Russell (Chap. 8) not only drove trucks to bring in logs from the bush, but he also had to make frequent trips to collect supplies from the Mangapehi rail station. "The guys would have a bit of furniture or something arriving by train, and they'd come to me and say, 'Aw Rusty, can you pick up such-and-such for me'. And every second Sunday morning there'd be a leg of pork or a lump of beef or 2–3 birds left on the copper, so I never needed to go hunting .... I thought, this place isn't too bad!".

One man remembered by Kitch Pedder in the 1950s went out every Sunday, and always came back with a pig. Most of the regular local hunters were responsible, and appreciated the advantages of a permit system. The cowboys and the Auckland mavericks who enjoyed putting bullet holes in bulldozers came later, especially after the Mangakino road went through (Fig. 14.3).

Wild pigs were a pest to farmers during the 1940s and 50s, because they rooted up pastures and killed many lambs, so hunting for meat overlapped with hunting for pest control. A bounty system on snouts and tails was run by the Department of Agriculture from 1930 to 1957, and then by NZFS until 1970 [13]. Counting buckets-full of rotting snouts submitted for bounty payment was nobody's favourite job [13]. Since the 1980s, pig populations have been variable but relatively low, and official control is not now considered necessary.

Pigs eat almost anything—fruits and seeds; the roots and rhizomes of bracken, supplejack, thistles and nettles; fungi and grasses; worms, snails, frogs, lizards, birds, eggs, young mammals, and carrion. Their rooting and wallowing can churn up the forest floor and retard regeneration on their favourite habitats—the broad ridge tops, and the warm sunny slopes and bush flats with deep soils and litter that are often full of worms.

Juicy and vulnerable, slow-moving endemic species that were once quite safe on the ground, such as giant weta, birds, short-tailed bats, lizards, giant land snails and frogs, were once rich and easy prey, but most have long since gone from the mainland. Hochstetter's frog was no doubt once common in the Rangitoto Range, where a limited search in 1983 located a single surviving population [17].

Since the late 1950s, recreational pig hunting has become a definitive back-country sport, locally important, fiercely defended and now facilitated by improved hunting equipment and access. Wild pig carcasses have been saleable as game meat since the 1960s, but not at high prices (in 1988, carcases over 45 kg were fetching NZ $2/kg). The number of pigs taken (Box 14.1) is less important than their size, quality and fighting ability. Pig-hunting is regarded as a fiercely competitive game in which hunters, dogs and pigs are pitted against each other. Boars were highly valued, while sows and piglets avoided.

The noisy chase enjoyed by parties of pig-hunters with dogs is a very different strategy from that used by deer hunters who often work alone, silently stalking their quarry. Pig hunters tend to be dedicated specialists, perhaps because they need to invest in one or more pig dogs (the good ones are very valuable) and obtain permission to bring them onto conservation land. They may put in as many as 80 hunting days a year (averaging 40 days) in companionable groups usually of two or three hunters plus up to four dogs per party. Pig hunters also tend to be among the most vocal opponents of the use of 1080 (Chap. 12), largely because of the danger to their dogs.

Wild pigs can be large animals, and the boars have razor-sharp tusks, which make pig-hunting a dangerous sport for men and their dogs. The Te Kuiti Pig Hunting Club weigh-in for 1987 recorded an average weight for 119 boars of 54 kg, and the largest 120 kg. Sows are smaller and less often hunted [18].

The unfussy diet of feral pigs makes them vulnerable to picking up and spreading bovine TB from scavenging dead possums. From 1975 that made them, like ferrets, additional targets for the attention of monitoring by the Animal Health Board (AHB), because while populations of pigs and ferrets might not be able to sustain TB as possums can, both are useful sentinels that will quickly detect the presence of TB in an area. But national-scale TB control programmes sometimes conflict with the hunting of pigs and other game for sport, requiring local negotiations.

Feral Cattle

Feral cattle were very large and unpredictable animals, and some could be positively dangerous. The risk of meeting an aggressive wild bull meant that field staff always had to carry a rifle during the 1946–55 National Forest Survey [22] and for years afterwards. Harry Bunn was out in the bush one day in about 1956 with Colin Sutherland, when a big white bull appeared, pawing the ground only 40 m away. It was the same one which, a week earlier, had taken on a bulldozer along Carter's (Morningside's) tram line (Fig. 7.4), and had given Ken Seymour a bit of help to scramble up a bank with one horn in his rump.

Now, Harry looked around, there was nowhere to go, what could he do? At that moment, Colin, out of sight, shouted to Harry, and when he didn't answer yelled again. The bull could not work out where the voice was coming from. Colin yelled again, the bull tossed a bit more dirt in the air and then took off. "They got him not long after that", remembered Harry with satisfaction. Dave Yanko asserts that it took 15 bullets to fell him (they had only a .22 rifle)—one guy would fire, and then throw the rifle to the next one and run.

Feral cattle and pigs were easy sources of free meat for years, but in due course the cattle were replaced by rising numbers of deer.

Deer

By the time red deer reached the west Taupo forests in the early 1950s, they were already classified as noxious pests and were unprotected, so they were legitimate targets for recreational hunters right from the start.

After the early 1960s, wild-killed venison could be sold at sixpence a pound. Then all other game were forgotten and everyone was into the deer. “I went into town”, said Dave Yanko,” and spent my savings on a rifle. In summer you could go hunting before and after work, and see 10–20 deer at a time—I used to go with Johnny Bull in his little Austin. He never missed”. Dave and other NZFS staff used their venison money to buy a car, or go on overseas trips.

Deer meat in 1960 (before decimalisation of the NZ currency) fetched about 20 pence a kilogram, or £6 for a 30-kg carcass (equivalent to NZ$260 in today’s money). The price rose sharply in the late 1960s and early 1970s, sparking even greater interest from hunters. This of course stimulated increasing numbers of applications to NZFS for hunting permits—68 in 1966/67, 110 in 1967/68 alone [33].

Game-recovery depots and processing plants were set up, inspected by the Ministry of Agriculture and Forestry, to which hunters took the carcasses. Venison hunters were regarded by NZFS staff as a mixed blessing. Some of them caused park staff quite a lot of trouble, as several frank comments in the Pureora station log book show.

> 1971: Many [venison hunters] can only be classed as those who do not hesitate to wreck gates, locks or anything that gets in their way of gaining a dollar [33].
>
> 1972: With the high prices being paid for venison, hunting is now a deadly serious business and it is little wonder that spotlighting [supposedly prohibited in RHAs] and poaching is commonplace [33].

Eventually, strict inspection requirements were imposed on hunters presenting venison meat for sale (which meant that the lungs, liver and kidneys had to be carried out of the bush). In the difficult economic conditions of 1974/75, that did not reduce the numbers of hunters applying for permits [33], and applications still ran up to 5000/year in the mid 1980s (Box 14.1).

The Pureora herds had rapidly developed a reputation for large body size and good trophy heads, especially over the 20 years after the mid 1950s. By then, deer populations elsewhere were already declining, but at Pureora the colonising process was still accelerating or was only recently complete.

Inevitably, the sizes of the trophy heads declined as the deer population passed its peak, from a 17-pointer recorded in 1953, through 10 heads of 12 to 15 points from 1955 to 1982, down to the two smallest of 10 points taken in 1988 [18]. Occasional 12-pointers still turned up into the 1990s. Since 2000 the average score of the top ten heads presented to the judges at the Hunting Competition has improved steadily.

NZFS had announced its intention to create Pureora Forest Park as far back as 1975 (Chap. 6). At that stage, the responsibility for managing it fell on NZFS staff based at Te Kuiti. Their early annual reports seldom included much information about recreational use of the forest, because virtually the only demand for access was for recreational hunting. NZFS was responsible for managing the register of hunting permits—issued not for money, since access was free, but simply because it was an important safety precaution for everyone to know how many people with firearms were in the forest, and where.

The Recreational Hunting Area

Although the toll on deer exerted by the commercial helicopter crews in the late 1970s (Chap. 12) was minimal in the Pureora area, it was extensive enough to raise protests from local hunters on the grounds that their opportunities for recreational hunting were being restricted. They supported other recreational hunters around the country in demanding that their sport be recognised. The official response was to develop the concept of Recreational Hunting Areas (RHAs).

Altogether, ten national RHAs were created in New Zealand between 1980 and 1986, although two have since lapsed. The total area of the remaining eight RHAs is about 178,000 ha, which is approximately 2 % of the conservation land administered by DOC [9].

The Waikato Branch of the New Zealand Deerstalkers Association (NZDA) proposed in

1980 that one of the RHAs should be established in part of PFP north of SH 30. The National Recreational Hunting Advisory Committee recommended that the area was suitable, and in July 1981 some 18,750 ha was gazetted for this purpose [23].

Control of red deer and wild pigs within the RHA became primarily the responsibility of recreational hunters alone, while all commercial operations (aerial and ground) were confined to the southern part of the Park, as described in Chap. 12 (Fig. 12.5) [8].

The establishment and management of RHAs was provided for under Part 3 of the Wild Animal Control Act (1977), which defines them as areas "where hunting as a means of recreation is to be used to control… the numbers of wild animals". The Act defined recreation as a pastime, a means of relaxation without gaining any financial reward. Under "Conditions of Use", therefore, the Act required NZFS to "discourage and endeavour to prevent" the removal of any carcase in saleable condition [23: 30].

On the face of it, the management of an RHA requires maintenance of sufficient numbers of red deer to keep hunters satisfied, whereas management of an EA to protect native vegetation requires removal of browsing ungulates to the lowest level possible. Hence the designation in 1980 of an RHA in PFP, including two areas north of SH 30 already proposed as EAs at the Taupo seminar in 1978 (Fig. 11.7), produced a difficult conflict of interests for NZFS managers.

In practice, NZFS had no choice but to accept these completely contradictory aims, but they still remain and are managed by a pragmatic combination of recreational hunting supplemented by official efforts within the EAs.

> In brief the RHA cannot be managed for the prime objective stated in the Wild Animal Control Plan ("To manage for recreational hunting the red deer and wild pigs present within the RHA"), but despite these limitations, hunting in the RHA can and will be fostered as part of the general policy for recreational hunting management in the Maniapoto District [18: 73].

PFP is the only significant hunting ground for red deer in the northwestern half of the North Island [2]. Its central position places it within half a day's travel for nearly half of New Zealand's population [23]. Not surprisingly, then, hundreds of hunters make for PFP on their annual hunting trips, especially during the March–April rutting season (Fig. 14.6).

PFP was divided into 15 administrative blocks based on the major catchments. Originally permits for specified blocks (not only in the RHA) were issued from the Te Kuiti and Pureora offices of NZFS [23], but now they are available from DOC online, and cover the whole Park (Box 14.1, Fig. 12.5). The blocks were used for analysis of hunter return data until 1986/87, when hunters killed on average one deer every seven days of hunting effort, although in some years fewer than half of them returned their kill data. In the early 1990s, permits were issued for 4-month periods, so annual totals were estimated as the sum of the returns for the Roar, averaging 2100–2500 a year, plus those for the two separate 4-month periods of winter and summer, of about 1500–2000 a year each, adding up to a mean total of 5823 permits a year [8].

Box 14.1. Recreational hunting statistics for Pureora

Hunting effort across the Park varies between years according to local changes in numbers of game species, and when and where hunters need to avoid pesticide operations, official goat control work, commercial hunters, or adverse conditions of public access. Similar factors affecting other hunting destinations (e.g., Kaimanawa and Kaweka Forest Parks) also affect hunting intensity in PFP [28].

For locations of hunting blocks, see map Fig. 12.5.

Year	Permits issued (% kill returns[a])	Number of hunters	Deer	Pigs	Goats	Total
RHA native forest blocks (Okahukura, Ngaroma, Ranginui, NW Outlier)						
1982/83	1145 (54)	2334	219	91	126	436
1983/84	1296 (54)	2705	196	39	187	422
1984/85	2747 (24)	2747	209	54	106	369
1985/86	2384 (27)	2384	180	49	82	311
1986/87	2255 (50)	2255	202	50	77	329
1987/88	1719 (44)	1719	148	55	31	234
North Block pines (removed from RHA in 1987)						
1982/83	176 (37)	381	18	4	0	22
1983/84	173 (46)	384	23	8	2	33
1984/85	277 (15)	277	20	0	1	21
1985/86	236 (21)	236	12	6	0	18
1986/87	194 (30)	194	8	0	0	8
Pureora excluding the RHA[b]						
1982/83	2350 (43)	2971	741	321	134	1196
1983/84	2156 (45)	4594	737	197	169	1103
1984/85	5207 (21)	5207	835	175	166	1176
1985/86	4521 (25)	4521	756	206	151	1113
1986/87	3933 (51)	3933	750	168	87	1005
1987/88	3269 (46)	3269	580	155	69	804

[a]Because so many hunters did not report their kills, these figures are probably underestimated by 50 % or more. Permits were not issued individually (one per hunter) until after the 1983/84 season.
[b]Recreational hunters were also allowed to hunt in the commercial hunting zone, and usually took many more deer per year than did the helicopter operators ([18: 18], and see Box 12.4).

A. **All hunted species, 1982–88**
   Data from Leigh and Clegg [18: figs. 5.2 and 8.1].
B. **Deer only, annual reported and estimated kills in Pureora FP, 1989–93**

Annual periods run from winter (June–Sept) through summer (Oct–Jan) to (Feb–May, the "Roar"). Data from Fraser [8: tables 1 and 3].

Period	Number of permits issued	Hunter diary return rate (%)	Reported deer kills	Estimated total deer kills
Winter 1989–Roar 1990	5734	42.0	1223	2911
Winter 1990–Roar 1991	5504	50.4	1275	2530
Winter 1991–Roar 1992	6086	66.2	1340	2024
Winter 1992–Roar 1993	5863	69.8	1220	1748

Fig. 14.6 Red deer stag during the autumn rut, roaring out his challenge to all comers. During "the roar", the best stags with the largest antlers are easy to locate even in thick forest, which in turn makes autumn the most popular season for deer hunters (Box 14.1B). *Photo Copyright* www.rodmorris.co.nz. *All Rights Reserved*

NZFS had for years attempted to encourage recreational hunting by providing more tracks, publicity and hunter support, as well as ongoing monitoring of kill returns and vegetation re-measurements. Their statutory responsibility to do this, at least during the short critical period until NZFS handed over management of the park to DOC in 1987, ensured the resources (funding, equipment and experienced operators) to do the necessary work. That was when the roading system was extended and upgraded (Chap. 11), and repeated surveys assessed the relative densities of deer, possums and goats in the North Block [6].

Recreational hunters appreciated and used these facilities, but could not reduce the numbers of deer enough to get a real reduction in browsing damage, especially in the more remote areas far from easy access. "In Pureora", commented John Mason, "hunters do not have to move far from the roads to shoot a deer" [32].

In 1987 the Conservation Act made DOC responsible for the management of deer (and other introduced mammals) in PFP. Some 1327 ha of pine plantations that were part of the original Pureora RHA in 1986 were not included in the transfer of land management authority from NZFS to DOC the following year (Fig. 11.4), but hunting is still permitted there.

The dilemma facing DOC then and now is that Section 6e of the Conservation Act 1987 allows DOC lands to be managed for recreation only "to the extent that the use of any natural or historic resource for recreation or tourism is not inconsistent with its conservation".

In fact, recreational hunting is unable to meet DOC's mandate for its estate, including the RHAs, "to manage for conservation purposes all lands… and protection of resources for the purpose of maintaining their intrinsic values" [18]. For example, they could not reverse the effects of browsing on the wildlife of the Waipapa EA [17].

This is certainly a concern. But given the wide range of other urgent conservation problems faced by DOC, comments Fraser [9], recreational hunting may be the only long-term low-cost animal control mechanism available to DOC over large chunks of its estate. Within PFP, recreational hunters have at least been able to maintain animal numbers at stable levels.

More difficult problems arose with the increasing use of 1080 to control possums in the Park, in order to reduce the transmission of bovine TB from possums to cattle (Chap. 12). The conflict of interests between animal health managers, farmers and hunters rapidly escalated through the 1980s, and Park managers were caught in the middle. In the firing line were NZFS/DOC staff John Gaukrodger, John Mason, and Andy Leigh, Senior Environmental Ranger from Auckland Conservancy.

Serious trouble was prevented by a carefully designed communication plan, aiming to develop working partnerships with hunters and greater co-operation between all interested parties. Regular meetings between managers and hunters, and improvements to roads, tracks, huts and campsites (often by the hunters themselves, especially members of the NZ Deerstalkers Association, NZDA, in Te Kuiti, Te Awamutu, Waikato and Taumarunui) all contributed to significantly improved relationships. Research on the effects of 1080 on bird populations (Chaps. 12 and 13) helped to address the criticisms of the environmental lobby.

Don Verity and his team of willing helpers put in hours of voluntary labour on road and track maintenance, weed spraying, and the building and furnishing of the shelters at the Piropiro campsite (Fig. 14.12). Their efforts made a great contribution to developing the continuing positive relationships between hunters and land managers, and have been recognised by a memorial plaque at Piropiro.

Preparation of the Maniapoto Wild Animal Management Plan [18] was another strategy to increase communication, as the work of preparing it required a lot of fruitful consultation with stakeholders. The President of the NZDA, in his address to their Annual Conference in Auckland in 1988, remarked on the way that hunters and NZFS land managers had been able to work together successfully at Pureora, but not in other prime hunting areas around the country. Why was this so? he enquired. John Gaukrodger, in the audience, took this as significant recognition for the team at Pureora.

The current DOC website specifically states that hunters may access the entire Park. In the late 1980s they took out on average 960 deer a year, usually more than the commercial hunters took from the helicopter zone (compare Boxes 12.4 and 14.1). Commercial hunting may be allowed in the RHA if and when the efforts of recreational hunters fail to reduce browsing damage, except during the autumn "roar".

Hunting effort is generally greater in the northern blocks, even though selling the carcases taken from the RHA is prohibited, and both recreational and commercial hunters operate in the southern blocks. But the recreational harvest is biased towards older deer (averaging almost 4 years for both sexes), especially trophy stags during the roar when their antlers are fully developed, whereas commercial hunters remove more hinds and young (averaging 2.3 years for stags and 2.9 years for hinds), so are more effective in reducing the population [8].

The Pureora Hunting Competition

The Pureora hunting competition, the first of its kind in New Zealand was planned by John Gaukrodger, John Mason and Ron Bevidge, a member of the Pureora Forest Park advisory committee and president of the Te Kuiti branch of the NZDA at the time. The competition officially started in 1988, and the first prize-giving day was held in late April. The free beer on offer during the first year probably contributed to its success. It provided an opportunity for hunters and DOC staff to get together and discuss matters of mutual interest in an informal, convivial setting, and now attracts participants from far and wide.

The competition still runs every year (without the free beer), and the main reward for the participants is the mana (prestige) of bringing in the largest set of antlers (Fig. 14.7). The organisers of the competition have kept records of the numbers and quality of deer shot within the Park. Ron Bevidge acted as official measurer on behalf of NZDA until he retired from this duty in 2013.

The Game Animal Council

The Game Animal Council is a statutory body established on 28 November 2013 under the Game Animal Council Act. It represents a significant change in official attitudes towards introduced game animals in New Zealand. Since at least the 1930s, forest conservationists have seen them only as pests whose browsing damage has to be controlled (Chap. 12); hunters have always seen them as a valuable game resource to be managed. The new Council will take both concerns into account, represent the interests of

Fig. 14.7 The annual Pureora Hunting Competition. (**a**) Kira Hughes Jr, twice winner (in 1991 and 1992), with a 12-pointer stag. *Crown Copyright, Department of Conservation Te Papa Atawhai. Photographer: John Mason.* (**b**) Ron Bevidge (NZ Deerstalkers' Association) scoring deer antlers submitted by competitors. *Crown Copyright, Department of Conservation Te Papa Atawhai. Photographer: John Mason*

recreational hunters, and improve the management of hunting resources while contributing to positive conservation outcomes.

The Council's job is to advise the Minister of Conservation on hunting issues, provide information and education to hunters, promote safety initiatives, conduct game animal research, and manage certain designated herds of special interest. It will be interesting to see how the Council manages the strongly differing opinions surrounding the new legal status of game animals in the park.

Dogs in PFP

Dogs are prohibited from all public conservation land except where authorisation has been granted by DOC. Visitors to PFP must have a permit to take a dog for a walk there, unless it is always on a leash.

A standard open hunting permit for the Pureora Forest/Western King Country allows a hunter to take up to two dogs, or a party of hunters a maximum of three dogs, into public conservation land. Hunters with dogs but without a firearm still need a hunting permit. Dogs must be kept under control and not be allowed to stray onto surrounding farm land or cause distress to other recreational users of the park. Avian aversion courses are available, which train hunting dogs to avoid kiwi.

Well-trained dogs make excellent hunting partners and can greatly improve hunting success, but poorly controlled dogs can pose a major threat to adult kiwi and blue duck [27].

The Pureora Forest Park Lodge and Outdoor Education Centre

A key recreational development in the park is the Pureora Forest Park Lodge and Outdoor Education Centre, an accommodation and conference facility built on the edge of the Waipapa EA (Figs. 14.3 and 14.8). A set of disused Ministry of Works buildings was moved from the Poro-o-tarao Tunnel camp

Fig. 14.8 The Pureora Forest Park Lodge and Outdoor Education Centre. *Crown Copyright, Department of Conservation Te Papa Atawhai. Photographer: Neville Ritchie (2014)*

(once housing the workers building the new, enlarged tunnel on the North Island Main Trunk Railway, opened in 1980) and converted to a lodge by members of the Lions Clubs of Benneydale, Piopio and Te Kuiti [30]. It turned out to be a major project costing NZ$200,000, and the bill would have been much more without the large team of willing helpers who did most of the work, mostly local farmers led by Pat Neal and Ron Bevidge and supported by NZFS staff.

The site north of the village across SH 30 was offered by PFP Advisory Committee, and NZFS widened and metalled the road leading to it. The Lodge and its surrounding tracks have developed rapidly into significant educational resource.

It is an ideal location for organized groups of all ages: professional staff from the Pureora office are available to give talks; there are tramping tracks, farms, power stations, a pa, a former coal mine and Pureora Mountain all within easy reach for day visits; and the native forest and its bird life all around.

Small, ecologically-oriented conference groups have taken full advantage ever since. The Lodge and its associated track systems developed by NZFS have become part of DOC's infrastructure serving all users. Sited as it is on the doorstep of a significant protected area, the Lodge has become critical to the success of the Waipapa/Pikiariki Restoration Project (Chap. 13).

The Observation Tower

The idea of building an observation tower in the forest was first advanced in mid 1981 by Rob Guest, then the District Forester at Te Kuiti [11]. He pointed out that the PFP Recreational Development Plan aimed to foster extensive use of the ranges for people to discover the great abundance and variety of wildlife in the Pureora district.

The level of interest in such a structure was clear when Television New Zealand's Natural History Unit constructed a temporary tower in the autumn of 1981, mainly to film kokako and the work of the FBRG (Chap. 15). They erected a steel scaffold tower with platforms at 40 and 60 ft, which provided a fascinating view of the canopy and sub-storey structure of the bush. A number of people used it for photography and observation after the filming was finished and before the tower was dismantled.

The idea of building a similar structure in the Pikiariki educational area started as a tea-time challenge. Arthur Cowan and the PFP Committee picked it up, and with the support of the "three Johns" (Gaukrodger, Innes and Mason), treated timber was obtained locally. With the help of a fourth John, DOC works officer John Stock, a new engineered design was commissioned.

Construction by DOC carpenter Tony Williams was completed in sections, and now a 12 m tower stands a mere 10 minutes' walk from the Bismark Road car park (Fig. 14.3). It is listed on the DOC web site as a good spot for observing native birds such as native pigeon, kakariki, and kaka.

Walking Tracks

DOC produces a visitor information brochure which lists, maps and describes 15 walking tracks in the park, ranging from short walks suitable for

Fig. 14.9 Tracks, huts and campsites in Pureora Forest Park, managed by the Department of Conservation for public recreational use. Ngaroma road (marked by a 4 × 4 symbol) is no longer a through road to the old Ngaroma mill and village beyond the northern park boundary. The Timber Trail, a spectacular 83 km track for cyclists and walkers, opened in 2013. *Redrawn by Max Oulton from Department of Conservation information*

wheelchairs and families to serious trails requiring hours or days to traverse (Fig. 14.9).

For families, Pureora has two of the best short family walks in New Zealand. Totara Walk is a well-graded footpath close to the village, and the Select Loop Road (accessible by cars and wheelchairs) offers a drive through a former selection logging area in the Waipapa EA. Both pass through tall podocarp forest—awe inspiring for children and parents alike. The trees and shrubs, the range of ferns, and the diversity of birdlife make these walks a wonderful experience.

The Totara Walk (Fig. 14.3, wheelchair accessible) starts near the DOC Field Base, and the Select Road Loop starts close to the Pureora Forest Park Lodge. The first can be completed in half an hour, but with children it is likely to take significantly longer; the second takes about an hour. Both are even more interesting at night. In the Pureora bush, there is always the chance of seeing glow-worms alongside damp flushes, or on upturned tree roots. Glow worms are probably more common than most people realise, as few people walk around the bush in the dark, and especially without using a torch. Occasionally the rotting wood on the floor of the bush may glow with phosphorescence.

Trampers were first allowed access to the top of Pureora in the 1958/59 year, to the distress of some remaining tangata whenua who remembered how it was once protected by tapu (Chap. 3). Much worse, in the early 1960s, there was an infamous two-bulldozer race to the summit straight up from the lower slopes, an illegal wet-day stunt which did extensive damage in crushing montane vegetation, and almost led to a sacking. It eventually caused significant erosion of the fragile pumice topsoil and alpine vegetation in the upper montane zone. Extensive remedial measures were necessary in the following years (Figs. 1.10 and 14.10).

Fig. 14.10 Helicopter pilot Allan Murtagh flies timber up to the summit of Pureora Mountain for repair work on the infamous bulldozer track. Erosion of the fragile pumice soil is a constant problem (See Fig. 1.10). The damage to the lower slopes still shows in the background. *Crown Copyright, Department of Conservation Te Papa Atawhai (1984). Photographer: John Mason*

An alternative route was built to enable foot access to the summit from the western side, and named the Toitoi Track, but it was never the best option. Now, an easily graded new track starts from the saddle on Link Road, by which trampers can take an hour and a half to reach the summit from the northern side. The effort is rewarded with spectacular views east to Lake Taupo, southeast to the volcanoes of Tongariro National Park, and southwest to the cone of Taranaki. From there the Pureora Mountain track heads south to link up with the Hauhungaroa Track and the Timber Trail.

Visitors to the YMCA camp just to the east of Titiraupenga often climb the Arataki Track to just below the summit, which offers sweeping views to the east, south and west. Since the B9B block to the north of it, including the peak itself, was returned to Maori ownership in 2000 (for reasons explained in Chap. 11), there has been not only a radical change in the view, but also a clearer statement that people climbing the track could be faced with prosecution if they do not stop short of the summit [24].

Today, the DOC web site lists the following information about the Mt Titiraupenga Track.

> Time: 8 h return to Link Rd car park. This track crosses many small streams before climbing a southern ridge of Mt Titiraupenga and joins the Arataki Track just below the summit rock. The summit of Mt Titiraupenga is Maori land and a taonga (treasure) to local iwi. Please respect these values and the landowners' wishes by not climbing the summit rock (http //www.doc.govt.nz/).

Huts and Campsites

DOC encourages outdoor activities and short-term stays in PFP. It is quite possible to visit Pureora for a day from a number of locations, but to fully appreciate the area, an overnight stay is best.

At the time of writing, there were four public camping sites (Figs. 14.3 and 14.9). The Pikiariki site is adjacent to bush, near Pureora Village and opposite the entrances to the Totara Walk and Timber Trail; the Ngaherenga site is nearby; the

Piropiro site is on the western side of the park, near the midpoint of the Timber Trail; and the Kakaho site is on the eastern side of the park, off SH 32 and Kakaho Road. In addition, the YMCA and the Tihoi Venture School operate camps for organised groups.

There are also seven self-contained rental cabins in Pureora village, plus four bush huts for trampers (Figs. 14.9 and 14.11) and an increasing variety of comfortable, privately-operated accommodation businesses that have been established nearby since the opening of the Timber Trail.

Family Recreation

Pureora is a great place for family holidays. When Rob Guest was the NZFS District Forester based in Te Kuiti, he often took his family there.

> For more than thirty years, the Guest family has enjoyed the recreation opportunities at Pureora—indeed four generations have visited since we first started visiting in the early 1980s with our children. We still travel back with the family, now extended to include our grandchildren.
>
> We started camping at the Ngaherenga camping site shortly after it was designated and the family tradition continues—although the basic facilities are now much enhanced. As the site abuts the Pikiariki podocarp forest, we have found much of interest within easy walking distance of the tent. In fact it hasn't always been necessary to leave the campsite. One attraction of camping there, particularly for children, is the open fire—constant brews [for 'gumboot' tea], camp oven bread, toasted marshmallows etc. Even when we didn't camp there, we would often take visitors out there for day trips. My mother (now 97) still talks about her visit when we boiled the billy at lunch time to make a brew.
>
> Young children are fascinated by insects, and Pureora is a great place to find interesting specimens. The shrubs which we planted in the early 1980s to demarcate the camping pitches have grown so much, they partially overhang the sites, and are inhabited by stick insects and giraffe weevils. On the way to the top of Pureora Mountain (a longer walk for more energetic families) we found some vegetable caterpillars. These are the mummified remains of the underground larvae of native moths which have been taken over by a parasitic fungus. The fungus feeds on the caterpillar, and eventually produces its fruiting bodies on long stalks above the ground.
>
> The campsite is also a great place to see one of New Zealand's least common mammals. Long tailed bats regularly fly out from the bush edge at dusk, right over the campsite. The rarer short tailed bat is also present in the area, but is less easily

Fig. 14.11 Bog Inn hut was originally built in 1960 close to the Ongarue mire, to support FRI scientists doing research on the forests nearby. NZFS ranger Doug Widdowson was one of the six men who took about ten days to build the hut from split mountain totara planks. Heavy rolls of roofing iron, nails and fittings were air-dropped to the site, but had to be dug out of the adjacent swamp, hence the name "Bog Inn". The hut has been upgraded, and is still a popular destination for trampers. *Crown Copyright, Department of Conservation Te Papa Atawhai (1990). Photographer: John Mason*

seen. We used to tell the children that we went to bed after we had seen the bats flying and we got up in time to hear the kokako as the sun hit the top of the trees in the morning.

By day, small freshwater koura (crayfish) can be caught in the creeks, but children delight in visiting the water supply creek after dark, shining a torch into the water and seeing the bright red eyes of the small koura as they move around the bed of the creek.

Other magnets for children easily accessible from the campsite are the old steam hauler and caterpillar tractor (Figs. 7.1 and 14.3), relics of the timber industry now preserved and well signed. A short walk from the Bismark Road car park (just off Pikiariki Road) is the 12 m high Observation Tower, giving children an exciting climb, a different perspective of the bush and the chance of seeing bush birds at close quarters.

The Buried Forest at the end of Plains Road is an interesting site which can form the goal of a morning's walk from Pureora (Chap. 2). It is still possible to see, in the bottom of one of the drainage ditches, the podocarp logs still in situ buried by volcanic pumice, and even leaves and litter still remaining from that time [12].

Since then, DOC have moved two more historic relics of the logging era—an old D7 bulldozer and a logging arch—into position as permanent displays on Pikiariki Road.

Tracks for Off-Road Vehicles and Mountain Bikes

DOC is required by its governing legislation to foster appropriate recreational activities on the public lands it manages, subject to requirements for environmental protection. Hence it has encouraged PFP's obvious potential as a unique and attractive venue for off-road vehicles and mountain bikes.

As logging work in the native forest tapered off, the well equipped NZFS roading unit was redirected to improving and maintaining the roads formerly used by timber trucks. By the mid 1980s there was a network of 284 km of roads suitable for most vehicles. Hunters appreciated this access to the back country, especially as the road network linked to kilometres of old logging tracks suitable for stalkers, 4 × 4 vehicles, quad bikes, mountain bikes and horses. The Link and Ngaroma Roads were used as part of the International Rally of NZ during its early years. More recently, members of the 4 × 4 fraternity have established a purpose-built off-road track with basic amenities at what is now the end of Ngaroma Road (Fig. 14.9).

Mountain bikers have explored the Pureora Forest Park since the late 1980s, because the numerous forest roads and tracks offer many challenging routes for adventurous riders. By the early 1990s club groups and individuals from Taupo, Rotorua, Hamilton, and New Plymouth and beyond began to visit the area regularly.

Recreational and competitive mountain biking has evolved significantly over the last two decades. Many new cycle trails are being built, including the privately-owned Maraeroa Cycleway on the border of PFP (Fig. 14.3). The best trails have been widely publicized in a popular guidebook produced in successive editions since 1991, *Classic New Zealand Mountain Bike Rides* [16]. So DOC commissioned a report outlining the present and future opportunities for mountain biking in the park [1].

An increase in visitor numbers brings economic benefits to the local communities, and also provides opportunities for wider education about the natural and historical features and the environmental values of the PFP. So DOC has developed a collaborative approach for the development of mountain biking facilities in PFP, and has defined several goals for the future. Hubs for biking activities will be developed in each major area of the Park, including Pureora Village, Okahukura Block, Kakaho Camp, Piropiro Camp, Mangakahu Valley and South Pureora. The demand for multi day trails will be met as funds permit, starting with the opening of the Timber Trail.

The Timber Trail

Development of the Timber Trail

In the mid 1990s, DOC archaeologist Neville Ritchie proposed the development of a 'short walk' from the Mangakahu Valley Road up the former Ongarue tramway to the spiral and tunnel, which were significant historic features interesting to visitors. But since the end of native logging in that area, and the closure of the tramway in 1958, the route had become overgrown, the drainage channels had blocked up through lack of maintenance, and some sections had become impassable. Only parts of the route had been kept open informally by hunters seeking access to the deer and pig populations in the Hauhungaroa Ranges. So the idea was put on hold, but not forgotten.

Over the five years after 2000, DOC field staff surveyed some 30 km of the boggy and overgrown tramway up to its terminus on the plateau above the Maramataha River gorge. They recognised that it had the potential to be developed into a great tramping and mountain biking experience. The main problem was that so many streams (some deeply entrenched) would have to be bridged, so the project would be very expensive to develop.

Undeterred, the field survey morphed from a proposal to establish a 45 km cycleway between the Mangakahu valley and DOC's established campsite at Piropiro (Fig. 14.12), into an even more ambitious proposal to extend the trail right through to Pureora village, using a mix of old logging roads, bulldozed logging tracks and a completely new track round the southern flanks of Pureora Mountain.

The proposal was actively promoted to DOC's senior management and to tourism promoters, District and Regional Councils, iwi, interested locals and the wider community. It was dubbed the Central North Island Rail Trail (CNIRT), a direct counterpoint to the hugely successful Otago Central Rail Trail (OCRT) established nearly a decade earlier. But there was no immediate prospect of getting the level of funding required to build it, despite a compelling business case [15] predicting benefits of NZ $7 million a year, mostly of direct value to the local community and wider Central North Island region. An independent report rated the project as the No. 1 Priority tourism infrastructure project in the King Country [29].

In 2007 the Minister of Conservation requested DOC to review future rail trail options on Conservation estate. The review concluded that the Ongarue tramway afforded the best medium term possibility for establishing a new premier rail trail on lands substantially managed by DOC. The route is largely unmodified; the equivalent of millions of dollars-worth of construction had already been done by Ellis and Burnand; it had the potential to be developed into an expansive network affording multiple recreation opportunities; and the tramway in its heyday played a unique social-historical role which would give the trail a special character.

The new National Government supported the idea of building a national cycleway (first suggested at an economic forum in February 2009),

Fig. 14.12 Piropiro Camp in 2013, the halfway point along the Timber Trail. *Marama Shearer*

to be known as 'Nga Haerenga (The Journeys)—The New Zealand Cycle Trail'. The project was run by a dedicated project team from official and partnership organisations. In addition to the Government's investment of NZ$50 million, local communities generated a further NZ $30 million of co-funding.

Almost overnight, the CNIRT became the largest trail construction project ever undertaken by DOC, completed over two years (2011–2012). The project team was led by Project Director John Gaukrodger, Project Construction Manager John Stock, and archaeologist/historian, Dr Neville Ritchie. Hoz Barclay was contracted to survey for the optimum route through the challenging country around the flank of Pureora Mountain and from Piropiro across the Maramataha River gorge to the terminus of the Ongarue tramway (Fig. 14.9). Once the route was settled, Project Construction Foreman Mac Waretini had day-to-day oversight of the work teams. After prolonged debate about the name of the trail, it was officially renamed the Timber Trail in 2012.

Major challenges to be met included negotiating easements over neighbouring private and iwi lands; large-scale restoration work at the spiral, including diverting the Mangatoi Stream out of the tunnel and back to its original course; forming, designing and building more than 30 bridges and hundreds of culverts; all in relatively remote bushclad mountainous terrain. The Mangatukutuku viaduct, dismantled in 1958 [21: 205], has been replaced by a spectacular 88 m suspension bridge, the fourth largest of eight big suspension bridges on the Trail (Fig. 14.13).

From the outset the Timber Trail was planned to create a major new recreation experience in the Pureora Forest Park in the form of a family-friendly 2–3 day cycling adventure trail with easy gradients, good riding surfaces, and high level river crossings that saved riders the effort of having to climb down into each stream bed and then back up out again.

To maintain the easy grade, the Trail gradually grew in length from an initial 60 km to a final 83 km of high standard track through rugged back country. Associated objectives were to provide recreational, ecological, cultural and historic information by high quality trailside interpretation; and to provide economic development opportunities for local people and businesses.

The Ongarue Tramway

Rail trails are simply abandoned rail corridors redeveloped for recreational uses, mainly cycling and walking. There are now over 2500 rail trails world-wide, including six in New Zealand.

Former rail and tramway routes are ideally suited for recreational trails because they were always built with gentle grades; they connect towns, settlements and logging sites (the reason for their existence); and they provide safe and direct access to scenery and views which are not visible from public roads. The trails themselves have an inherent historical character derived from their rail heritage along the route. These include embankments, cuttings, benching round steep faces, bridges, tunnels, watering stops, stations, camps, and other railway infrastructure.

Once there were around 1000 bush tramways in New Zealand, with a total length of about 5000 km, much longer than the railway system. Until the late 1940s tramways were the principal method of transporting logs to sawmills (Chap. 5). Unless the tram routes were later turned into roads, they were largely obliterated when the land was converted to farming or plantation forestry.

The southern half of the Timber Trail incorporates the former Ongarue Tramway (Fig. 7.6), a narrow-gauge logging railway built by Ellis and Burnand. Construction started in 1922, and the tramway was in continuous use until 1958. Tramway historian Paul Mahoney has emphasised the high national heritage value of the Ongarue Tramway [19, 20].

Ellis and Burnand was one of the largest companies working in the era of native timber logging in New Zealand. Their four major logging operations in the King Country, at Ongarue, Mangapehi and Manunui on land they owned adjacent to the North Island Main Trunk line, plus at Maraeroa on Maori land near Pureora, played a large part in a key pioneering industry in

Fig. 14.13 Cyclists on one of the high suspension bridges on the Timber Trail. *Marama Shearer*

New Zealand, and subsequently in the story of PFP (Chaps. 5 and 7).

Because Ellis and Burnand were not reliant on the State Forestry allocation system, the Company was able to invest capital in large scale operations with quality equipment that could repay investment over a lifecycle of 40 years or more. The Ongarue tramway was the paramount example of this; well engineered and with one of the best locomotives (a Climax: Figs. 5.11 and 5.12) and best log haulers (two Washingtons) in New Zealand. At its greatest extent the Ongarue tramway was 45 km long. If the many branches are counted, the aggregate length of the system would be around 130 km [34].

During its operating life, Ellis and Burnand's Ongarue tramway was easily the most publicised and well-known bush tramway in New Zealand. The company ran an annual picnic train for many years, carrying many locals and visitors in a special carriage for a day on the line. A comprehensive national history of bush tramways [20] emphasised the Ongarue Tramway as a nationally significant heritage site.

About 80 % of the Ongarue tramway (Fig. 7.6) survives unmodified in a natural bush environment. This gives it great authenticity and national significance as a rare historic survivor of this important facet of the timber industry. It originally had two significant engineering features: the Spiral and its tunnel, still negotiable along the tramway, and the huge Mangatukutuku viaduct (Fig. 5.12). The viaduct, built entirely of timber with a main pier standing 92 ft (28 m) above the bed of the river, was completed in March 1925 at a cost of £2900 [21: 305].

Recreational Values of the Timber Trail

The Timber Trail was opened in April 2013, and it has already won many accolades, and features prominently on tourism web sites. Those who worked so hard to create it are proud that it has qualified for a Green Flag Award, an international scheme which rates visitor experiences for their quality and maintenance.

The Trail offers panoramic views from the slopes of Pureora and along the forested and open river valleys within the Hauhungaroa Range, before it descends gently down the Spiral Creek valley to the Mangakahu Valley. It also provides access to many scenic features along the route and side trails. It is available for walkers as well as cyclists, and provides stopping points at the Piropiro campsite and in nearby private accommodation facilities. There are 15 walking/tramping tracks in the immediate vicinity of the Timber Trail, and several primary mountain biking tracks.

At the entrance of the Trail is a carving (Fig. 11.1) commemorating Te Kanawa who, during a legendary foot race across country from Tapora-roa (south of the Rangitoto Range) to Tuhua Mountain, took a route roughly equivalent to that of the present Trail [3]. Then along the whole length of the Trail, the story of the land and its people is told in some 92 innovative and informative interpretation panels (Fig. 14.14). They illustrate many aspects of the Maori and European history of forestry in the King Country as described in this book, from the volcanic geology (Chap. 1) the flora and fauna (Chap. 2), the bullock- and horse-drawn logging operations and then steam power (Chap. 5), through to the fully mechanised native and plantation forestry at Pureora (Chaps. 7 and 8), and finally the rise of

Fig. 14.14 One of many informative interpretation signs designed by Snappper Graphics for the Timber Trail, explaining the history of the famous Ongarue Tramway Spiral. See also Fig. 5.10. *Greg Martin*

the conservation movement that eventually led to the end of logging and the creation of the Pureora Forest Park in the late 1970s (Chaps. 10 and 11). The Trail amounts to an open-air, direct field experience of all the major themes of this book.

The trail opens up the Pureora Forest Park to visitors keen to see a significant area of native forest, which despite a long history of logging still contains some of the best remaining original or regenerating podocarp forest in the North Island. It makes the most of Pureora's reputation as one of only two sites on the mainland where visitors can see all the 'K birds'- kokako, kaka, kakariki, kereru and kārearea, and it has nationally significant populations of other threatened species including blue duck, *Dactylanthus taylorii* and lesser short-tailed bats (Chap. 2).

Comprehensive records including historic photographs, video footage and oral histories with former bush workers have been compiled, as well as heritage assessments of the significant remaining historic fabric or machinery which remain either on the tramway or which could be placed along it at original sites [25, 26]. There are natural records too: near the Waikoura stream, the Trail passes the site of an old logging camp whose residents used to hold rose-growing competitions. Nothing remains of the camp but a few concrete doorsteps, and the roses that still survive among the native plants reclaiming the site.

There are several other new tourism projects in the Pureora area which all complement each other. For example, the Maraeroa C Incorporation has opened its Pa Harakeke centre, which focuses on eco-tourism, arts and crafts, a carbon off-set tree planting programme, and their own Maraeroa cycleway. A canopy walk in the tree tops at Pureora (The Skywalk) is still being planned by the Hape B and Tiroa C Trusts, and this will provide an appealing new attraction right in the village. The proposal includes a visitor centre and additional accommodation.

Just as Dawson [5] predicted (Chap. 10), the recreational value of Pureora Forest Park has proved far greater than the cost (in lost revenue from timber milling) of establishing it.

References

1. Barclay, H. (2009). Mountain biking in the Pureora Forest Park: Present activity and future opportunities. Te Kuiti: Department of Conservation.
2. Clegg, S. (1986). Resurvey of Pureora north block 1981–1982. Auckland: New Zealand Forest Service.
3. Crown, P., & Te Ruki, S. (2009). Nga Tuhinga o Rereahu. Te Kuiti: Maraeroa C Incorporation.
4. Dale, R., & Fleming, J. (1978). Recreation *[Transcript of the Taupo seminar]*. Wellington: New Zealand Forest Service.
5. Dawson, J. A. (1979). Pureora Forest Park: Costs and benefits. MA thesis, Victoria University of Wellington.
6. Deuss, F. (1981). Animal survey 1980: Pureora State Forest Park. Auckland: New Zealand Forest Service.
7. Feldman, J. W. (2001). Treaty rights and pigeon poaching: Alienation of Maori rights to kereru, 1864–1960. Wellington: Waitangi Tribunal.
8. Fraser, K. W. (1996). The effect of recreational hunters on deer populations in Pureora Conservation Park. Science for Conservation, 31, 1–39.
9. Fraser, K. W. (2000). Status and conservation role of recreational hunting on conservation land. Science for Conservation, 140, 1–46.
10. Fraser, W., Cone, J. M., & Whitford, E. J. (2000). A revision of the established ranges and new populations of 11 introduced ungulate species in New Zealand. Journal of the Royal Society of New Zealand, 30, 419–437.
11. Guest, R. (1981). Observation tower–Pureora. Te Kuiti: New Zealand Forest Service.
12. Guest, R. (2013). Personal communication.
13. Halkett, J., Berg, P., & Mackrell, B. (1990). Tree people: Forest service memoirs. Wellington: New Zealand Forestry Corporation and GP Publications.
14. Harris, D. (Ed.) (2009). Timber cruising and other forestry stories. Wellington: FirstEditionPublishers.com.
15. Kaval, P. (2008). The Central North Island rail trail project: A regional economic contribution study. Hamilton: Department of Conservation.
16. Kennett, J., Kennett, S., & Kennett, P. (2011). Classic New Zealand mountain bike rides, 8th edn. Wellington: Kennett Brothers.
17. Krzystyniak, S. (1984). Pureora North Block animal survey, summer 1983. Auckland: New Zealand Forest Service.
18. Leigh, A. H., & Clegg, S. E. (1989). Maniapoto district wild animal management plan, 1989 to 1999. Hamilton: Department of Conservation.
19. Mahoney, P. (1996). The timber industry: A national heritage identification study 1840–1940 pp. Wellington: Department of Conservation, unpublished report.
20. Mahoney, P. (1998). The era of the bush tram in New Zealand. Wellington: IPL Books.

21. McMillan, L., & Walker, A. (2005). Ongarue: A place of the heart. Ongarue: Ongarue School Centenary Committee.
22. Millen, J. (1999). Through trackless bush: The story of the New Zealand National Forest Survey 1946–55. Rotorua: P. F Olsen & Co.
23. New Zealand Forest Service (1981). Pureora recreational hunting area: Wild animal control plan. Auckland: New Zealand Forest Service, Auckland Conservancy.
24. New Zealand Herald, 8 July 1983.
25. Ritchie, N. A. (2010). Historic logging machinery at Pureora: Heritage assessment and maintenance plan. Hamilton: Dept of Conservation.
26. Ritchie, N. A. (2012). Timber trail (Ongarue tramway section): Heritage assessment and maintenance plan for three elements of original historic fabric. Hamilton: Department of Conservation.
27. Robertson, H. A., Colbourne, R. M., Graham, P. J., Miller, P. J., & Pierce, R. J. (2011). Experimental management of brown kiwi *Apteryx mantelli* in central Northland, New Zealand. Bird Conservation International, 21, 207–220.
28. Scrimgeour, R. (2014). Personal communication.
29. Stafford Group (2008). A strategic and business plan for the King Country ecocultural corridor. Prepared for Maraeroa C Corporation. Sydney: Stafford Group.
30. Waikato Times, 10 March 1981.
31. Walker, A., & Cooke, E. (2003). Waimiha: People of character. Waimiha: Waimiha Reunion 2001 Committee.
32. Wilkes, O. (2005). Transcripts of oral history recordings. Hamilton: Department of Conservation.
33. Wilkes, O., Gaukrodger, J., & Ritchie, N. (2011). Chronology of Pureora. Hamilton: Department of Conservation.
34. Yonge, J. (1993). New Zealand railway and tramway atlas, 4th edition. Exeter, UK: Quail Map Company.

Conclusions 15

C.M. King, D.J. Gaukrodger and N.A. Ritchie

Why Pureora Forest?

Pureora Forest Park is one of the most significant sites of natural and cultural history interest in New Zealand. This book has explored the dramatic geological, biological and human history of this important area, drawing upon a wide range of disciplines, including volcanology, Maori and European history, ecology, sociology, forestry, and environmental economics. Some of the historical events described have been accompanied by serious environmental damage and/or social conflict, with consequences significant at national level. In many ways, the history of this one area is a microcosm of the history of New Zealand generally.

Human attitudes to the forest and its residents, especially the trees and birds, have changed over 600 years, but have always involved some mix of exploitation and reverence, matching social changes within and between Maori and European cultural traditions. The changes that took place at Pureora are not only remarkably dramatic as a human story, but also astonishingly recent and exceptionally well recorded.

Why did all this happen in Pureora in the mid twentieth century, as opposed to any other time and place?

First, the inland location (Fig. 0.1 in Preface), rugged landscape (Fig. 1.9) and difficult access meant that permanent Maori occupation of the mountain lands west of Lake Taupo was largely confined to the edges of the forests or along major rivers for most of the year, plus seasonal forays to traditional hunting areas. Then, Maori resistance to European expansion protected the forests until the completion of the North Island Main Trunk Railway in 1908.

Second, the early timber mills and their feeder networks of tramlines had to stay relatively close to the railway until the early 1940s. The development of forest roads, heavy trucks, and mechanised logging equipment put the last remote stands of native timber within reach of the new mills established at Pureora village only in the late 1940s. The peak of native timber production in that area was passed by the late 1950s. Possums and deer did not emerge as significant new pests in what forest remained until the 1960s.

Third, the environmental movement, focussed on the last few populations of the North Island kokako, an iconic endangered bird, became a force to be reckoned with only in the 1970s. They were able to force the Forest Service to stop

C.M. King (✉)
School of Science, University of Waikato, Private Bag 3105, Hamilton 3240, New Zealand
e-mail: cmking@waikato.ac.nz

D.J. Gaukrodger
3a Te Angi Street, Turua, RD4, Thames 3574, New Zealand
e-mail: beejaygauk@xtra.co.nz

N.A. Ritchie
Department of Conservation, Te Papa Atawhai, Hamilton Office, Private Bag 3072, Hamilton 3240, New Zealand
e-mail: nritchie@doc.govt.nz

C.M. King et al. (eds.), *The Drama of Conservation*,
DOI 10.1007/978-3-319-18410-4_15

logging of the last significant kokako habitat, and to commission urgent research on predator control, only in the early 1980s.

Had any of these previous events happened any sooner, it would have been too late to save the kokako, as it already was in many other, more accessible parts of the kokako's formerly very wide range. The last valuable tall native forests of the central North Island would have been clear felled and converted to farms or pine plantations, just as had those that once clothed nearly 2 million hectares of more accessible land elsewhere.

The survival of these last few forests in comparatively good condition provided the necessary theatre for drastic and widely publicised conservation action, followed immediately by internationally significant research on the biology of threatened New Zealand birds and on how we can best preserve them into the future. Looking back, we can see why Pureora was not only the *actual* place where this remarkable sequence of events took place, but the *only possible* place.

A Message for the Future

The story of Pureora Forest includes key themes in Maori and European history, ecology, sociology, forestry, conservation biology, management of human and natural resources, and environmental economics. Some of the changes that took place there were accompanied by serious social conflict, with consequences significant at national level.

The history of Pureora forest demonstrates why the task of developing ways to reconcile the needs of resource utilisation and those of biodiversity conservation requires an approach that bridges the gaps between activist passion, academic theory, industrial labour relationships, political imperatives and practical problem solving. It illustrates the way that such an approach is gradually emerging in one of the largest remaining and most iconic areas of native forest in the North Island of New Zealand.

An important dimension of its message is that national benefits can be gained for developing workable conservation solutions when conflicts of interest are managed by visionary, open-hearted people provided with well-funded, integrated and interdisciplinary support.

Surely, the spirits of all those who have loved and laboured in Pureora over the centuries must be hoping that all future managers will understand and benefit from the hard-won experience of their predecessors. It is a very special place, and it deserves very special care.

Erratum
The Drama of Conservation

C.M. King, D.J. Gaukrodger and N.A. Ritchie

DOI 10.1007/978-3-319-18410-4

In Front matter page xvii, the affiliation of the author "T. Roa" was set incorrectly earlier. The correct information is given below and also updated in the original front matter page.

School of Maori and Pacific Development, University of Waikato, Hamilton, New Zealand.

The updated original online version for this book can be found at 10.1007/978-3-319-18410-4

C.M. King (✉)
School of Science, University of Waikato, Private Bag 3105, Hamilton 3240, New Zealand
e-mail: cmking@waikato.ac.nz

D.J. Gaukrodger
3a Te Angi Street, Turua, RD4, Thames 3574, New Zealand
e-mail: beejaygauk@xtra.co.nz

N.A. Ritchie
Department of Conservation, Te Papa Atawhai, Hamilton Office, Private Bag 3072, Hamilton 3240, New Zealand
e-mail: nritchie@doc.govt.nz

C.M. King et al. (eds.), *The Drama of Conservation*,
DOI 10.1007/978-3-319-18410-4_16

Erratum
The Maori of the Central North Island Before 1860

C.M. King and T. Roa

DOI 10.1007/978-3-319-18410-4_3

In Chapter 3, the affiliation of the author "T. Roa" was set incorrectly earlier. The correct information is given below and also updated in the original chapter.

School of Maori and Pacific Development, University of Waikato, Hamilton, New Zealand.

The updated original online version for this chapter can be found at 10.1007/978-3-319-18410-4_3

C.M. King (✉)
School of Science, University of Waikato, Private Bag 3105, Hamilton 3240, New Zealand
e-mail: cmking@waikato.ac.nz

T. Roa
School of Maori and Pacific Development, University of Waikato, Private Bag 3105, Hamilton 3240, New Zealand
e-mail: tomroa@waikato.ac.nz

C.M. King et al. (eds.), *The Drama of Conservation*,
DOI 10.1007/978-3-319-18410-4_16

Erratum
The European Impact: Exploration to Conflict, 1840–1890

C.M. King and N.A. Ritchie

DOI 10.1007/978-3-319-18410-4_4

In Chapter 4 on page 87, the name "Burnand" was set incorrectly in the first reference text. The original chapter has been corrected now.

The updated original online version for this chapter can be found at 10.1007/978-3-319-18410-4_4

C.M. King (✉)
School of Science, University of Waikato, Private Bag 3105, Hamilton 3240, New Zealand
e-mail: cmking@waikato.ac.nz

N.A. Ritchie
Department of Conservation, Te Papa Atawhai, Hamilton Office, Private Bag 3072, Hamilton 3240, New Zealand
e-mail: nritchie@doc.govt.nz

C.M. King et al. (eds.), *The Drama of Conservation*,
DOI 10.1007/978-3-319-18410-4_16

Appendices

A Note About Maori Language

The Maori language was entirely oral until after 1814, when the first missionaries began to develop a written version and a Maori-English dictionary. The first missionaries spoke English, which is why Maori orthography stems from the English alphabet. It has five vowels and ten consonants: h, k, m, n, p, t, w, r, ng all pronounced as in English, plus wh (pronounced f). Maori has no equivalents to the English b, d, j, s, v, x, or z.

Maori language is unlike English in several important ways. First, Maori vowels are pronounced separately, so a word like "kiore", which a European might expect to sound as two syllables "kee-aw", is pronounced with three, as "kee-or-eh", and "poaka" as "po-ark-a" not "po-ka".

Second, Maori words do not add a suffix in the plural, so it is not correct to add an "s" to Maori nouns taken into English. The words "moa" and "kiwi" are spelled the same whether they refer to one or many individuals, just as are "deer" and "sheep" in English.

Third, the Maori of the central North Island attributed personal qualities to objects and features of great significance to them, especially to the mountains and rivers. Their sacred mountains are called simply by their names, hence "Ruapehu" or "Ruapehu maunga" (not with a title, which would be the equivalent of "Mount Ruapehu"). However, to avoid confusion for readers familiar with European-style scientific publications, we retain capitalised qualifiers which are usually regarded as part of the name, hence Waikato River. Quotations from non-Maori sources retain the normal conventions, e.g., adding an "s" to the name of a feature usually referred to in the plural, hence "Hauhungaroas", or a capital, as in "Hauhungaroa Ranges".

Because the length of vowels can sometimes distinguish words different in meaning, it is usual to mark long vowels in written Maori by means of a macron, defined as a bar inserted above the vowel. This is a very important way of ensuring that the spelling of a word accurately reflects its pronunciation and meaning.

We recognise and fully endorse the use of macrons, as an important means of ensuring respect to Maori tradition. Most New Zealand institutions, including all the major sponsors of this book, make it a firm policy always to use them in their written materials. An increasing proportion of international scientific journals use Maori macrons in texts downloadable only as PDFs, where the macrons will appear correctly even though their significance is unknown to readers in most countries.

This book is addressed to an international audience, so it is also important that it be globally accessible, especially to those searching for its content and accessing it electronically in order to read rather than to speak the text. The problem is that macrons tend to make Maori words invisible to or scrambled by standard international search engines not set up to interpret Unicode symbols. For example, Google Scholar may render Tūāpapa Pūtaiao as Tua<macron>papa Puta<macron>iao, or Māori as

C.M. King et al. (eds.), *The Drama of Conservation*,
DOI 10.1007/978-3-319-18410-4

Ma¯ori; conversely, readers not resident in New Zealand might not be able to search using the correct codes. For that reason, macrons may be omitted from the search boxes of websites containing Maori words, such as www.maoridictionary.co.nz. Macrons are also absent from authoritative general books such as Michael King's *The Penguin History of New Zealand* (23, 27).

Macrons are a potential problem for e-books, which are designed to be searchable by Google or other internet crawlers. One possible compromise would be to use macrons in the body of each chapter text but not in the headings, abstracts and keywords most likely to be searched. We anticipate that, for words designed to be read rather than spoken, such a policy risks becoming more confusing than helpful.

We have, with regret, chosen not to show the correct use of macrons in the texts, except in the glossary of Maori terms listed here. We wish to emphasise that this choice is purely pragmatic, and does not imply any disrespect for Maori protocol. We make an exception to this rule in Fig. 3.1, where addition of a macron on the "A" is crucial to distinguishing the ownership of territories belonging to two different tribes with similar names, and where a direct acknowledgement to a named Maori authority would be inappropriate without macrons (Fig. 3.5).

A Glossary of Maori Terms Mentioned in the Text

Definitions Taken from www.maoridictionary.co.nz.

Ahi kā Title to land through occupation by a group

Ake Forever

Ariki Paramount chief

Hapū Kinship group, or sub-tribe of a large tribe

Hui Assembly or meeting

Iwi Nation, people, large group descended from a common ancestor

Kāinga Home village or unfortified encampment

Kaitiaki Trustee, guardian(s)

Karakia Ritual chant or song

Kaumātua Respected tribal elders of either gender

Kuia Wise elder woman

Mana Authority, prestige, or power

Mātauranga Māori Maori traditional knowledge

Mauri Life principle, source of emotions

Maunga Mountain

Pā Fortified village

Pākehā New Zealander of European descent (originally applied to foreigners)

Pātaka Storehouse on poles

Poaka Pig, or pork

Pouihi, or pou Post or ceremonial carved pole

Rangatira Chief or noble, or (figuratively) a tree of noble lineage

Rohe Pōtae, Te King Country

Taiaha A spear of hard wood, often carved and decorated with dogs' hair

Take To originate (referring to rights of use or occupation)

Tangata whenua People of the land, indigenous people, first nations

Tangi Mourning, funeral rites

Tapu Sacred, set apart, forbidden

Tēnā koutou A greeting addressed to three or more people, thank you

Tikanga Custom, protocol, correct procedure

Tohunga Priest

Tuatara Nocturnal burrowing lizard-like reptile endemic to NZ

Tūpuna Ancestors, grandparents

Urupā Burial place

Utu Repayment, reward, satisfaction, reply, revenge

Wāhi tapu (noun) a sacred place, or (verb) to break a sacred prohibition

Wētā Large flightless insect endemic to New Zealand

Whānau Extended family, or a group of close friends

Whare House or hut

Wharepuni Sleeping house, guest house

General Glossary and Abbreviations Mentioned in the Text

AHB Animal Health Board (now TBfree New Zealand)

Basal area The area of a given stretch of forest occupied by the total cross section of tree stems at a fixed height (1.4 m in New Zealand)

Biota A general term for plants and animals

Dbh Diameter of a tree at breast height

DG Director-General

DIA Department of Internal Affairs

DOC Department of Conservation

E&B Ellis & Burnand, a prominent sawmilling company

EA Ecological Area

ECO Environment and Conservation Organisations of Aotearoa New Zealand

FBRG Forest Bird Research Group

FRI Forest Research Institute

Kya Thousands of years ago

MS&D Marton Sash & Door [Timber Co.]

Lokey A small locomotive for working narrow-gauge bush tramways

Mya Millions of years ago

NFAC Native Forests Action Council

NFRT Native Forests Restoration Trust

NIMTR North Island Main Trunk Railway

NZ New Zealand

NZDA New Zealand Deerstalkers' Association

NZFP N.Z. Forest Products

NZFS New Zealand Forest Service (disbanded 1987)

NZR New Zealand Railways

C.M. King et al. (eds.), *The Drama of Conservation*,
DOI 10.1007/978-3-319-18410-4

NZWS New Zealand Wildlife Service (disbanded 1987)

OC Officer in Charge

2iC Second in Charge

PFP Pureora Forest Park

Phenology The study of cyclic and seasonal changes in plant and animal life

PM Prime Minister

Plinian A towering eruption column of volcanic material (see Box 1.1)

PSL Pureora Sawmills Ltd

PWD Public Works Department (later, Ministry of Works)

RbM Research by Management

RFBPS Royal Forest and Bird Protection Society of New Zealand (Inc.)

RHA Recreational Hunting Area

SA Sawmill area

SF State Forest

SH State Highway

SOE State-owned Enterprise

Strainers Extra strong fence posts used as anchor points for wire fences

Swannie "Swanndri", a woollen bush shirt

Synecology The ecological study of whole plant or animal communities

TB Bovine tuberculosis, a disease of cattle carried by possums and other animals

List of Scientific Names of Biota Mentioned in the Text

Vegetation

Beech *Fucospora* and *Lophozonia* spp.

Black maire *Nestegis cunninghamii*

Bog pine *Halocarpus bidwillii*

Bracken *Pteridium esculentum*

Broadleaf *Griselinia littoralis*

Bush lawyer *Rubus* spp.

Celery pine *Phyllocladus trichomanoides*

Climbing rata *Metrosideros* spp.

Crown fern *Blechnum discolor*

Douglas fir *Pseudotsuga menziesii*

Fivefinger *Pseudopanax arboreus*

Fuchsia *Fuchsia excorticata*

Hall's totara *Podocarpus cunninghamii*

Haumakaroa *Pseudopanax simplex*

Hen and chickens fern *Asplenium bulbiferum*

Hinau *Elaeocarpus dentatus*

Horopito *Pseudowintera colorata* (and see pepperwood)

Japanese larch *Larix leptolepis*

Kahikatea *Dacrycarpus dacrydioides*

Kaikawaka/native cedar *Librocedrus bidwillii*

Kaikomako *Pennantia corymbosa*

Kamahi *Weinmannia racemosa*

Kanuka *Kunzea ericoides*

C.M. King et al. (eds.), *The Drama of Conservation*,
DOI 10.1007/978-3-319-18410-4

Kauri *Agathis australis*

Kawakawa *Macropiper excelsum*

Kohekohe *Dysoxylum spectabile*

Kohuhu *Pittosporum tenuifolium*

Lancewood *Pseudopanax crassifolius*

Lawyer see Bush lawyer

Macrocarpa *Cupressus macrocarpa*

Mahoe *Melicytus ramiflorus*

Maire *Nestegis* spp.

Mangeao *Litsea calicaris*

Manuka *Leptospermum scoparium*

Matai *Prumnopitys taxifolia*

Miro *Prumnopitys ferruginea*

Mistletoe *Peraxilla tetrapetala* (among others)

Monoao *Dracophyllum subulatum*

Mountain ash *Eucalyptus delegatensis*

Mountain toatoa *Phyllocladus alpinus*

Nikau palm *Rhopalostylis sapida*

Northern rata *Metrosideros robusta*

Pate *Schefflera digitata*

Pepperwood *Pseudowintera colorata*

Pokaka *Eleaeocarpus hookerianus*

Pukatea *Laurelia novae-zelandiae*

Putaputaweta *Carpodetus serratus*

Radiata pine *Pinus radiata*

Rangiora *Brachyglottis repanda*

Raukawa *Raukawa edgerleyi*

Red beech *Fuscospora fusca*

Rewarewa *Knightia excelsa*

Rimu *Dacrydium cupressinum*

Silver beech *Lophozonia menziesii*

Silver pine *Lagarostrobos colensoi*

Silver tussock *Poa cita*

Small-leaved coprosma *Coprosma* spp.

Square sedge *Lepidosperma australe*

Straw sedge *Carpha alpina*

Stinkwood *Coprosma foetidissima*

Supplejack *Ripogonum scandens*

Tanekaha *Phyllocladus trichomanoides*

Tangle fern *Gleichenia dicarpa*

Taraire *Beilschmiedia tarairi*

Tawa *Beilschmiedia tawa*

Tawheowheo *Quintinia serrata*

Toatoa *Phyllocladus toatoa*

Totara, thin-barked *Podocarpus cunninghamii*

Totara *Podocarpus totara*

Tree ferns *Cyathea* spp. and *Dicksonia* spp.

Turner's kohuhu *Pittosporum turneri*

Tussocks *Chionochloa pallens, Poa cita*

Upright cutty grass *Gahnia rigida*

Western red cedar *Thuja plicata*

Wheki *Dicksonia squarrosa*

Wineberry/makomako *Aristotelia serrata*

Native Birds

Australasian harrier *Circus approximans*

Bellbird *Anthornis melanura*

Blue duck *Hymenolaimus malacorhynchos*

Brown teal *Anas chlorotis*

Bush wren *Xenicus longipes*

Fantail *Rhipidura fuliginosa*

Grey duck *Anas superciliosa*

Grey warbler *Gerygone igata*

Kingfisher *Todiramphus sanctus vagans*

Laughing owl *Sceloglaux albifacies*

Long-tailed cuckoo *Eudynamys taitensis*

Morepork *Ninox novaeseelandiae*

New Zealand falcon *Falco novaeseelandiae*

New Zealand pigeon *Hemiphaga novaeseelandiae*

New Zealand pipit *Anthus novaeseelandiae*

North Island brown kiwi *Apteryx australis mantelli*

North Island fernbird *Bowdleria punctate vealeae*

North Island kaka *Nestor meridionalis septentrionalis*

North Island kokako *Callaeas wilsoni*

North Island robin *Petroica australis longipes*

North Island takahe *Porphyrio mantelli*

North Island tomtit *Petroica macrocephala toitoi*

Paradise shelduck *Tadorna variegata*

Pukeko *Porphyrio melanotus melanotus*

Red-crowned parakeet/kakariki *Cyanoramphus novaezelandiae*

Rifleman *Acanthisitta chloris*

Shining cuckoo *Chrysococcyx lucidus*

Silvereye *Zosterops lateralis*

South Island takahe *Porphyrio hochstetteri*

Spotless crake *Porzana tabuensis*

Spur–winged plover *Vanellus miles*

Titi, muttonbird *Puffinus* sp.

Tui *Prosthemadera novaeseelandiae*

Weka *Gallirallus australis*

Welcome swallow *Hirundo neoxena neoxena*

White-faced heron *Ardea novaehollandiae*

Whitehead *Mohoua albicilla*

Yellow-crowned parakeet/kakariki *Cyanoramphus auriceps*

Introduced Birds

Australian brown quail *Coturnix ypsilophora australis*

Australian magpie *Gymnorhina tibicen*

Blackbird *Turdus merula*

Californian quail *Callipepla californica*

Chaffinch *Fringilla coelebs*

Dunnock *Prunella modularis*

Eastern rosella *Platycercus eximius*

Goldfinch *Carduelis carduelis*

Greenfinch *Carduelis chloris*

House sparrow *Passer domesticus*

Myna *Acridotheres tristis*

Redpoll *Carduelis flammea*

Skylark *Alauda arvensis*

Song thrush *Turdus philomelos*

Starling *Sturnus vulgaris*

Yellowhammer *Emberiza citrinella*

Other Native Fauna

Cicada 42 species of hemipteran insects

Common skink *Cyclodina polychroma*

Copper skink *Cyclodina aeneum*

Forest gecko *Mokopirirakau granulates*

Freshwater crayfish *Paranephrops* spp.

Giant weta *Deinacrida* spp.

Green gecko *Naultinus elegans*

Hochstetter's frog *Leiopelma hochstetteri*

Mosquito 16 species (12 endemic, 4 introduced)

Puriri moth *Aenetus virescens*

Sandfly 13 species of *Austrosimulium*

Striped skink *Oligosoma striatum*

Tuatara *Sphenodon punctatus*

Vegetable caterpillar *Cordyceps robertsii*

Mammals

Australian brushtail possum *Trichosurus vulpecula*

Brown hare *Lepus europaeus occidentalis*

Dama wallaby *Macropus eugenii*

Feral cat *Felis catus*

Feral cattle *Bos taurus*

Feral ferret *Mustela furo*

Feral goat *Capra hircus*

Feral pig *Sus scrofa*

Hedgehog *Erinaceus europaeus occidentalis*

Horse *Equus caballus*

Kiore, or Polynesian rat *Rattus exulans*

Kuri, or Polynesian dog *Canis familiaris*

House mouse *Mus musculus*

Long-tailed bat *Chalinolobus tuberculatus*

Norway rat *Rattus norvegicus*

Rabbit *Oryctolagus cuniculus cuniculus*

Red deer *Cervus elaphus scoticus*

Sambar *Cervus unicolor unicolor*

Sheep *Ovis aries*

Ship rat *Rattus rattus*

Short-tailed bat, lesser *Mystacina tuberculata*

Short-tailed bat, greater *Mystacina robusta*

Sika deer *Cervus nippon*

Stoat *Mustela erminea*

Weasel *Mustela nivalis vulgaris*

Index to Persons

C.M. King et al. (eds.), *The Drama of Conservation*,
DOI 10.1007/978-3-319-18410-4

Index to Places

C.M. King et al. (eds.), *The Drama of Conservation*,
DOI 10.1007/978-3-319-18410-4

Index to Species

For Latin Names, see Appendices

C.M. King et al. (eds.), *The Drama of Conservation*,
DOI 10.1007/978-3-319-18410-4

The manufacturer's authorised representative in the EU is Springer
Nature Customer Service Centre GmbH, Europaplatz 3, 69115 Heidelberg,
Germany. If you have any concerns regarding our products, please
contact ProductSafety@springernature.com

Printed and bound by CPI Group (UK) Ltd, Croydon, CR0 4YY
06/10/2026
02228672-0001